Artificial Intelligence for Cybersecurity

Develop AI approaches to solve cybersecurity problems in your organization

Bojan Kolosnjaji

Huang Xiao

Peng Xu

Apostolis Zarras

Artificial Intelligence for Cybersecurity

Copyright © 2024 Packt Publishing

All rights reserved. No part of this book may be reproduced, stored in a retrieval system, or transmitted in any form or by any means, without the prior written permission of the publisher, except in the case of brief quotations embedded in critical articles or reviews.

The authors acknowledge the use of cutting-edge AI, in this case ChatGPT and Grammarly, with the sole aim of enhancing the language and clarity within the book, thereby ensuring a smooth reading experience for readers. It's important to note that the content itself has been crafted by the authors and edited by a professional publishing team.

Every effort has been made in the preparation of this book to ensure the accuracy of the information presented. However, the information contained in this book is sold without warranty, either express or implied. Neither the authors, nor Packt Publishing or its dealers and distributors, will be held liable for any damages caused or alleged to have been caused directly or indirectly by this book.

Packt Publishing has endeavored to provide trademark information about all of the companies and products mentioned in this book by the appropriate use of capitals. However, Packt Publishing cannot guarantee the accuracy of this information.

Associate Group Product Manager: Niranjan Naikwadi
Publishing Product Manager: Sanjana Gupta
Book Project Manager: Aparna Nair
Senior Editor: Tiksha Lad
Technical Editor: Rahul Limbachiya
Copy Editor: Safis Editing
Proofreader: Tiksha Lad
Indexer: Rekha Nair
Production Designer: Aparna Bhagat
Senior DevRel Marketing Executive: Vinishka Kalra

First published: October 2024

Production reference: 1111024

Published by Packt Publishing Ltd.
Grosvenor House
11 St Paul's Square
Birmingham
B3 1RB, UK.

ISBN 978-1-80512-496-2

www.packtpub.com

Contributors

About the authors

Bojan Kolosnjaji is a researcher working at the intersection of **artificial intelligence** (**AI**) and cybersecurity. He has obtained his master's and PhD degrees in computer science from the **Technical University of Munich** (**TUM**), where he conducted research in anomaly detection methods in constrained environments. Bojan's academic work deals with anomaly detection problems in multiple cybersecurity-relevant scenarios, and the design of AI-based solutions to these problems. Bojan is currently working as a principal engineer in cybersecurity sciences and analytics, helping various cybersecurity teams deal with large-scale data, adopt AI practices and solutions, and understand security challenges in AI systems.

Huang Xiao holds a doctorate in computer science from TUM. He is also a visiting scholar at Stanford University. His main research interests include adversarial **machine learning** (**ML**), reinforcement learning, anomaly detection, trusted AI, and AI applications in cybersecurity. Huang has published several top-tier conference and journal papers with over a thousand citations in both the ML and security domains. He led the ML research group at Fraunhofer AISEC Institute in Munich and also worked as a research scientist at Bosch Center for AI. He managed a data scientist team that designed and developed ML systems to tackle different cybersecurity problems.

Peng Xu has focused on AI for system security, **large language model** (**LLM**) security, graph neural networks, program analysis, compiler design, optimization, and cybersecurity. He completed his master's at the Chinese Academy of Science in 2013 and pursued a PhD in IT security at TUM from 2015 to 2019. He is currently awaiting his dissertation defense. Peng's research topics include malware detection, private computation, and software vulnerability mitigation using compiler-based approaches. Peng is currently working as a principal engineer in compiler optimization and programming LLMs, especially on the topics of using LLMs to generate code blocks to detect malicious code as well as bug localization.

Apostolis Zarras is a cybersecurity researcher with a rich academic background. He has served as a faculty member at both Delft University of Technology and Maastricht University. Dr. Zarras earned his PhD in IT security from Ruhr-University Bochum, where he honed his expertise in systems, networks, and web security. His research is driven by a passion for developing innovative security paradigms, architectures, and software that fortify ICT and IoT systems. Beyond his technical contributions, Dr. Zarras delves into the dark web and its underground markets, uncovering and combating malicious activities to bolster global cybersecurity. His work is dedicated to advancing IT security and protecting users and systems from emerging cyber threats.

About the reviewers

Hemanath Kumar J is a seasoned data enthusiast with extensive experience in developing and implementing ML models, GenAI models, data visualization, and analytics solutions. With a diverse background in transportation, education, finance, and healthcare, he has consistently delivered solutions with data-driven strategies, which enhanced decision-making processes with high accuracy and operational efficiency. As a technical reviewer for Packt Publications, he brings his comprehensive expertise to this book, ensuring accuracy and clarity. He would like to acknowledge his family, mentors, and friends for their unwavering support and encouragement throughout this project.

Pranav Khare is a business and technology professional with over 14 years of experience in product management, business strategy, and software engineering. Starting his journey at Infosys as a software engineer, Pranav's curiosity shifted from the "How?" to the "Why?", guiding him on a path that led from technical execution to the strategic vision of a product manager. Now, as a senior product manager at Docusign, he drives innovation in digital identity verification, employing AI/ML-based solutions to meet diverse customer, security, and compliance needs. He holds an MBA from Georgetown University and a Bachelor of Engineering in electronics and communication.

Table of Contents

Preface   xv

Part 1: Data-Driven Cybersecurity and AI

1

Big Data in Cybersecurity   3

Technical requirements	4	Advanced analytical techniques and tools	11
What is big data?	4	Resource constraints	12
Big data challenges in cybersecurity	6	Big data applications in cybersecurity	15
The velocity of data in cyberspace	6	Big data technologies for cybersecurity	16
Diverse data types in cyberspace	8	Summary	17
The veracity of data in cyberspace	8		

2

Automation in Cybersecurity   19

Tools and technologies against threats	20	Examples of automated cybersecurity tools	25
Security information and event management (SIEM)	20	Potential drawbacks and challenges of automation	26
Intrusion detection and prevention systems (IDPSs)	21	The future of automation in cybersecurity	27
Endpoint detection and response (EDR)	22	Ethical considerations	29
Security orchestration, automation, and response (SOAR)	23	Summary	30
The importance of automation in cybersecurity	24		

3

Cybersecurity Data Analytics — 31

AI in data analytics	32	Challenges of using AI	35
Types of AI used in cybersecurity data analytics	32	The role of analysts	36
		The regulatory landscape	37
Applications of AI	34	Summary	39

Part 2: AI and Where It Fits In

4

AI, Machine Learning, and Statistics - A Taxonomy — 43

Technical requirements	43	DL and its recent advances	55
A brief introduction to AI history	44	The limitation and security concern	56
The relation to statistical learning theory	46	Hallucination	56
		Privacy leakage	57
ML – classifying taxonomy	48	Intellectual property ownership	57
By learning schema	48	Bias, fairness, and their social impact	57
By learning objectives	50	Adversarial attacks	58
By model modality	52	Summary	58

5

AI Problems and Methods — 59

Supervised learning methods	59	t-SNE	75
Logistic regression	61	Semi-supervised learning methods	77
Random forest	62	Label propagation	77
Support Vector Machines (SVM)	63		
Neural networks	64	Detecting anomalies	78
Deep learning	66	Isolation Forest	79
Convolutional neural networks	68	Summary	81
Unsupervised learning methods	72	References	81
K-means	73		

6

Workflow, Tools, and Libraries in AI Projects — 83

Workflow of AI projects	83	Background of visual network traffic analysis	90
Fundamental workflow – creating an AI model from scratch	84	Tools and libraries for malware detection	102
Advanced topics – integrating an AI model into a product	86	Background of malware detection	102
Developing and creating a virtual environment	89	Summary	114
Tools and libraries for visual network traffic analysis	89	References	114

Part 3: Applications of AI in Cybersecurity

7

Malware and Network Intrusion Detection and Analysis — 119

Technical requirements	120	Exercise 1 – malware detection	123
Overcoming traditional difficulties	120	Exercise 2 – network intrusion detection	126
Proper datasets for creating an AI model	121	Moving from detection to classification	132
Malware analysis	122	Summary	133
Network intrusion detection	122		

8

User and Entity Behavior Analysis — 135

Technical requirements	136	Feature extraction	138
Shortcomings of traditional tools	136	Exercise – UEBA anomaly detection	139
Leveraging AI for UEBA	137	Other use cases	145
UEBA features	137	Summary	145

9

Fraud, Spam, and Phishing Detection — 147

Introducing fraud, phishing, and spam detection methods	147	Understanding phishing detection with a practical example	149
Fraud detection	148	Introducing the collaborative anomaly detection	151
Phishing detection	148	Summary	157
Spam detection	149	References	157

10

User Authentication and Access Control — 161

Understanding user authentication and access control	161	SELinux and access control	175
User authentication	163	Practicing user authentication and access control with Python	182
Multi-factor authentication	164	Using OAuth2.0 in mobile application authentication	182
Authentication technologies	165	Writing SELinux in Python to control the Ubuntu files	184
Access control	167	AI for user authentication – face recognition	186
Exemplifying the user authentication and access control	171	Summary	188
OAuth2.0 and user authentication	171		

11

Threat Intelligence — 189

Technical requirements	189	Data preprocessing	193
Understanding threat intelligence	190	Building the model	195
Working with AI for threat intelligence	191	Expanding on use cases of AI in threat intelligence	199
Topic modeling	192	Summary	200
Exercise – extracting CTI information from X data	192	References	200

12

Anomaly Detection in Industrial Control Systems — 203

Introducing the ICS and its components	203	Use cases and applications	212
Cyberattacks on ICSs	207	Anomaly detection for the ICS	213
Cyberattacks on the ICS	207	Ransomware detection for the ICS and its components	216
Cyberattacks on the components of ICSs	209	Challenges and future works	216
Detecting anomaly behaviors in ICSs	210	Summary	218
Classification	210	References	218

13

Large Language Models and Cybersecurity — 221

From traditional methods to LLMs	221	LLMs for threat intelligence	224
Transformers	222	LLMs for spam and phishing detection	225
Large Language Models (LLMs)	222	LLMs for a security operation center	226
Prompting	223	LLMs for offensive security	226
Retrieval augmented generation	223	The security of LLMs	227
Using LLMs for security	223	Summary	228
LLMs for vulnerability discovery	224	References	229

Part 4: Common Problems When Applying AI in Cybersecurity

14

Data Quality and Its Usage in the AI and LLM Era — 233

Data quality and its usage	233	Examples of good data quality in AI and LLMs	235
Characteristics of a high-quality dataset	234	NLP	236
Uses of high-quality datasets in real life	234	Computer vision	238
		Data quality accidents	239

Writing Python code to practice good data quality	240	Example 2 – data validation	241
		Example 3 – handling missing values	241
Example 1 – data cleansing with pandas	240	**Summary**	**242**

15

Correlation, Causation, Bias, and Variance — 243

Technical requirements	**243**	Case study 1 – correlation versus causation in phishing attacks	252
Introducing the statistical foundation	**244**	Case study 2 – managing bias and variance in IDS	252
Understanding correlation and causation	**245**	Conclusion of case studies	253
Correlation	245	**Practical applications**	**253**
Causation	246	Diagnostic tools for correlation and causation	253
Introducing bias and variance	**246**	Techniques to manage bias and variance	254
Bias	247	Advanced statistical techniques for enhanced security	254
Variance	247		
Bias and variance in polynomial curve fitting	247	Implementing responsible AI in cybersecurity	255
Managing bias and variance	251	**Summary**	**255**
Case studies and examples	**251**		

16

Evaluation, Monitoring, and Feedback Loop — 257

Technical requirements	**257**	Monitoring during testing or production	263
Evaluating models	**258**	Model monitoring tools	264
Loss functions	258	**Human in the loop**	**267**
Model metrics	259	Active learning	267
Monitoring models	**262**	**Summary**	**268**
Monitoring during training	262	**References**	**269**

17

Learning in a Changing and Adversarial Environment — 271

Technical requirements	272	Transferability of adversarial samples	284
Introduction to AML	272	**Defensive mechanisms**	**286**
The realistic learning environment	274	Defense as prevention	288
Arms race problem	274	Defense as detection	289
Learning process with data flow	276	Defense as a response	290
Adversarial threat modeling	277	Practical tools for testing on adversarial attacks	292
Attacker model	277	Summary	292
Adversarial attack taxonomy	279	References	293

18

Privacy, Accountability, Explainability, and Trust – Responsible AI — 295

Technical requirements	296	Theoretical approaches	299
Understanding the AI issues	296	Research development	300
Current challenges in AI security	296	Guidelines and standards	301
Safety concerns	297	Tools and technologies	302
Significance of AI security and safety	298	**AI risk management framework**	**302**
Impact on individual privacy	298	Main components of the AI risk management framework	303
Implications for national and global security	298	Utilizing the framework in organizations	304
Ethical considerations and public trust	298	Preparing for the future	305
Addressing AI security and safety challenges	299	Summary	305

Part 5: Final Remarks and Takeaways

19

Summary — 309

Summarizing what we've learned	309	Connecting chapters	312

Successes of AI in cybersecurity	314	Open source projects and libraries	316
Where to go from here	315	Other web links	316

Index 319

Other Books You May Enjoy 334

Preface

The cybersecurity threat landscape is evolving to include an increasing volume and variety of attacks. However, at the same time, the big data era and the proliferation of AI enable new methods and tools to tackle the challenge of detecting increasingly sophisticated malicious activity using large-scale data processing and AI for pattern recognition and enhanced data analytics. AI has already revolutionized multiple areas, some of the most well-known being healthcare, finance, and manufacturing. Currently, AI is making its way into cybersecurity as well through exciting and valuable applications.

In this book, we introduce the place of AI as a methodology in cybersecurity and how AI methods can help solve cybersecurity problems. We give a foundational knowledge of AI for those of you who are beginners in this area, starting with a theoretical underpinning. Following the theoretical chapters, we expand by looking at eight different application areas with six chapters dedicated to these applications. This way, we provide you with practical skills that can be used in concrete scenarios in real-world cybersecurity. After going through these chapters, you can immediately provide value to your organizations or advise your colleagues on how to make a difference with AI in cybersecurity.

Apart from describing the AI methods and going through various application scenarios, we include a part of the book with common pitfalls and challenges in applying AI for cybersecurity. This part helps you to be prepared for real-world applications of AI and the problems that are often overlooked by untrained practitioners, leading to inaccurate project planning and wasted time and effort. We help you recognize those problems and challenges and successfully overcome them to optimize the advantages you get with AI applications.

This book is authored by researchers with years of academic and practical experience applying AI in cybersecurity in various organizations. Publishing this book enables us to share our knowledge and experience in this area to train new experts and help improve cybersecurity overall. The combination of theoretical base and practical skills provided through this book is what has proven to be crucial in successful AI projects!

Who this book is for

This book is for cybersecurity or general IT professionals or students who are interested in AI technologies and how they can be applied in the cybersecurity context. It is useful both for readers with no knowledge about AI and experienced AI practitioners who can use it as a reference or to fill in the gaps in their skill set. It is good for both theoretically inclined and very practical readers who like hands-on exercises. This book can be used both for readers interested in solving concrete problems in their organization and for professionals who want to give advice, be a thought leader, and organize the introduction of AI as a capability in cybersecurity.

What this book covers

Chapter 1, Big Data in Cybersecurity, introduces the rising issue of handling large-scale data gathered by cybersecurity departments of various organizations and cybersecurity vendors. It describes the challenges of data processing and scale, as well as data quality, data governance, and similar.

Chapter 2, Automation in Cybersecurity, emphasizes the importance of automation as a driver for efficiency in cybersecurity. We describe tools that help achieve automation, such as SIEM, SOAR, EDR, and IDS, that help experts define workflows and automate tasks. These tools are made with data analysis problems in mind and help automation at scale.

Chapter 3, Cybersecurity Data Analytics, introduces the role of AI in advancing automation through intelligent data analytics on large-scale datasets. We describe challenges in this area that we will be solving throughout the book using AI methods and tools.

Chapter 4, AI, Machine Learning, and Statistics - A Taxonomy, helps disambiguate the terms of AI, machine learning, and statistics, which can be difficult for beginners in this area. It also helps to get the foundations and an understanding of how AI applies to various datasets, and where the important limitations and challenges are.

Chapter 5, AI Problems and Methods, builds on the basic terms of AI and helps you get more extensive knowledge and dive into concrete methods and how they work. It gives you the knowledge needed to recognize where different AI and ML methods are applicable and how to apply them.

Chapter 6, Workflow, Tools, and Libraries in AI Projects, describes the workflow of AI projects, from data collection and preprocessing to training and testing. Furthermore, it describes useful tools and libraries with examples in cybersecurity.

Chapter 7, Malware and Network Intrusion Detection and Analysis, describes the problem of malware detection and network intrusion detection and how AI is applicable to solve it. We describe how AI makes a difference to improve detection performance and provide a hands-on exercise to improve your technical skills.

Chapter 8, User and Entity Behavior Analysis, introduces the problem of finding a way to capture and analyze patterns in the behavior of users and hosts. We describe how AI methods can be used to model this behavior from raw event logs and detect anomalies that can point to cyberattacks.

Chapter 9, Fraud, Spam, and Phishing Detection, contains a description of typical methods to detect transaction fraud, as well as spam and phishing emails using anomaly-based methods. These methods heavily benefit from AI, and we clarify how AI can be applied, and what the problems and challenges are in these use cases.

Chapter 10, User Authentication and Access Control, describes the problem and solutions on how to authenticate users and how to enable them to access only the resources that we intend them to use. We also describe AI methods that are applicable to these problems.

Chapter 11, Threat Intelligence, contains an overview of cyber-threat intelligence problems and techniques to extract information from various sources important to get an understanding of cyber threats. Furthermore, we describe how AI can help solve problems in this area, and we also provide a practical exercise to practice your knowledge of AI methods.

Chapter 12, Anomaly Detection in Industrial Control Systems, shows what kind of cybersecurity-relevant anomalies happen in industrial networks and how to detect them. AI methods are useful in this scenario as well, as they help us model regular behavior and detect anomalies.

Chapter 13, Large Language Models and Cybersecurity, introduces the recently popular topic of **large language models** (**LLMs**) as generative AI methods that found applications in cybersecurity. We describe the potential of applying LLMs in cybersecurity scenarios, the challenges in making these applications successful, and how to overcome them.

Chapter 14, Data Quality and its Usage in the AI and LLM Era, is an important chapter, as contemporary AI methods are data-driven and the success of the AI application heavily depends on data being fit for purpose. We describe methods of data quality management and challenges in this area.

Chapter 15, Correlation, Causation, Bias, and Variance, covers these terms as they are important to know, and lack of understanding them often brings problems in AI applications. We give you an introduction and dive into the importance of differentiating correlation and causation, as well as describe the trade-off of bias and variance to help you avoid common pitfalls.

Chapter 16, Evaluation, Monitoring, and Feedback Loop, covers the very important parts of a machine learning workflow. We need to have proper methods for evaluation to describe performance and methods to monitor this performance. Furthermore, we often keep humans in the loop within the AI workflow to enhance our data or tune our models.

Chapter 17, Learning in a Changing and Adversarial Environment, explains how many baseline AI methods contain assumptions about a static environment, and we need new techniques that enable the handling of changes in the data that happen naturally or because of adversarial activity. We present these techniques as they are especially important in cybersecurity applications.

Chapter 18, Privacy, Accountability, Explainability, and Trust – Responsible AI, explores responsible AI – recently, a very important topic as AI applications are adopted in various areas that influence people's well-being and the development of society. We describe responsible AI and how to achieve it in general and in the cybersecurity context.

Chapter 19, Summary, contains a retrospective on what you have learned in previous chapters and helps you structure the knowledge you obtained while reading the book. Furthermore, it gives you some propositions for the next steps to enhance your knowledge and skills.

To get the most out of this book

To use this book in an optimal way, it's important to go through the introductory and theoretical chapters to get a strong basis and use this basis in the further practical chapters with hands-on exercises. This way, you can get a well-rounded knowledge that can help in a wide range of scenarios.

The technical requirements are different and described at the beginning of each chapter, but the exercises are generally done in Python 3 using various Python libraries.

Software/hardware covered in the book	Operating system requirements
Python 3	Windows, macOS, or Linux

If you are using the digital version of this book, we advise you to type the code yourself or access the code from the book's GitHub repository (a link is available in the next section). Doing so will help you avoid any potential errors related to the copying and pasting of code.

Download the example code files

You can download the example code files for this book from GitHub at https://github.com/PacktPublishing/Artificial-Intelligence-for-Cybersecurity. If there's an update to the code, it will be updated in the GitHub repository.

We also have other code bundles from our rich catalog of books and videos available at https://github.com/PacktPublishing/. Check them out!

Conventions used

There are a number of text conventions used throughout this book.

`Code in text`: Indicates code words in text, database table names, folder names, filenames, file extensions, pathnames, dummy URLs, user input, and X (formerly) Twitter handles. Here is an example: "After getting the APK object and `DalvikFormat`, more information about the Android application could be fetched up by these corresponding functions, such as `get_permission()`, `get_activities()`, and so on."

A block of code is set as follows:

```
import tensorflow as tf
import tensorflow_datasets as tfds

(pcap_data_train, pcap_data_test), pcap_ds_info = tfds.load(
    'pcap_mnist', split=['train', 'test'],
    shuffle_files=True, as_supervised=True, with_info=True,
)
```

Any command-line input or output is written as follows:

```
$ SplitCap -r example.pcap -s flow
```

> **Tips or important notes**
> Appear like this.

Get in touch

Feedback from our readers is always welcome.

General feedback: If you have questions about any aspect of this book, email us at `customercare@packtpub.com` and mention the book title in the subject of your message.

Errata: Although we have taken every care to ensure the accuracy of our content, mistakes do happen. If you have found a mistake in this book, we would be grateful if you would report this to us. Please visit `www.packtpub.com/support/errata` and fill in the form.

Piracy: If you come across any illegal copies of our works in any form on the internet, we would be grateful if you would provide us with the location address or website name. Please contact us at `copyright@packt.com` with a link to the material.

If you are interested in becoming an author: If there is a topic that you have expertise in and you are interested in either writing or contributing to a book, please visit `authors.packtpub.com`.

Share Your Thoughts

Once you've read *Artificial Intelligence for Cybersecurity*, we'd love to hear your thoughts! Scan the QR code below to go straight to the Amazon review page for this book and share your feedback.

`https://packt.link/r/180512496X`

Your review is important to us and the tech community and will help us make sure we're delivering excellent quality content.

Download a free PDF copy of this book

Thanks for purchasing this book!

Do you like to read on the go but are unable to carry your print books everywhere?

Is your eBook purchase not compatible with the device of your choice?

Don't worry, now with every Packt book you get a DRM-free PDF version of that book at no cost.

Read anywhere, any place, on any device. Search, copy, and paste code from your favorite technical books directly into your application.

The perks don't stop there, you can get exclusive access to discounts, newsletters, and great free content in your inbox daily

Follow these simple steps to get the benefits:

1. Scan the QR code or visit the link below

```
https://packt.link/free-ebook/9781805124962
```

2. Submit your proof of purchase
3. That's it! We'll send your free PDF and other benefits to your email directly

Part 1: Data-Driven Cybersecurity and AI

This part introduces how big data technology and AI are changing how we solve problems in cybersecurity. It describes the role of data and automation, as well as the opportunities that collecting large-scale data brings. Furthermore, it enumerates the cybersecurity tools and approaches where big data analytics is already making a difference.

This part has the following chapters:

- *Chapter 1, Big Data in Cybersecurity*
- *Chapter 2, Automation in Cybersecurity*
- *Chapter 3, Cybersecurity Data Analytics*

1
Big Data in Cybersecurity

In this chapter, we will explore the significance of big data in cybersecurity. More precisely, it will encompass an overview of challenges, applications, and technologies associated with big data in cybersecurity, along with considerations related to privacy and ethics. Whether you are new to the concept of big data in cybersecurity or seeking to deepen your understanding, this chapter will provide valuable insights and detailed information.

In this chapter, we're going to cover the following main topics:

- What is big data?
- Big data challenges in cybersecurity
- Big data applications in cybersecurity
- Big data technologies for cybersecurity

By the end of this chapter, you will have gained a comprehensive understanding of how big data is reshaping the landscape of cybersecurity. From grasping the fundamental concept of big data and its distinctions from conventional data processing to navigating the intricate challenges it presents in cybersecurity, you will develop a solid foundation. You'll explore diverse applications that harness the power of big data for threat detection, fraud prevention, and **incident response** (**IR**), gaining insights into the cutting-edge technologies driving these advancements. Through real-world use cases, you'll witness the tangible impact of big data in enhancing cyber resilience. Additionally, you'll be equipped to address critical ethical and privacy considerations inherent in using extensive datasets for security purposes, ensuring a well-rounded perspective on this transformative field.

Technical requirements

There are no specific technical prerequisites for delving into this chapter, apart from a basic understanding of computer science concepts. Whether you're a cybersecurity enthusiast looking to explore the broader implications of big data or a professional seeking to deepen your understanding of its applications, this chapter is designed to be accessible to a wide range of readers. It offers insights and explanations in a clear and approachable manner, making the content valuable for both technical and non-technical individuals interested in the intersection of big data and cybersecurity.

What is big data?

Before delving into the introduction of big data, it is essential to understand the concept of data. **Data** processed by a computer comprises quantities, characters, or symbols, which can be stored, transmitted, and recorded as electrical signals on magnetic, optical, or mechanical media. **Big data**, on the other hand, refers to an extensive collection of data that is massive in volume and continues to grow exponentially over time. It is characterized by its substantial size and complexity, to the extent that traditional data management tools cannot efficiently store and process it. Big data is a unique form of data that presents immense challenges and opportunities due to its sheer magnitude. Let's now explore these distinctive features, or the **four Vs** of big data, in detail:

- **Volume**: Big data refers to vast amounts of data generated, collected, and stored by various sources, including sensors, social media, transactional data, and more. The sheer volume of data is one of the defining characteristics of big data.

- **Velocity**: Big data is generated and processed at an unprecedented rate. Data can be developed in real time or near real time from various sources. The speed at which data is produced and needs to be processed is a crucial characteristic of big data. This poses challenges in capturing, storing, and processing data in real time or near real time.

- **Variety**: Big data comes in various formats and types, including structured, unstructured, and semi-structured data. Structured data contains data that can be organized in a traditional format, such as spreadsheets or databases. Unstructured data comprises data that comes with no specific format, such as text, images, audio, and video data. Semi-structured data falls in between, having some structure but not fully organized. The diverse nature of data types and formats is another characteristic of big data.

- **Veracity**: Big data can be noisy and uncertain, with varying data quality and accuracy levels. Data may be incomplete, inconsistent, or contain errors, impacting the reliability of insights and analysis derived from big data. Ensuring data veracity, including data quality, accuracy, and reliability, is a critical characteristic of big data.

Big data has become increasingly important in various domains due to its potential to unlock insights, drive innovation, and create value. In today's data-driven world, organizations across different industries leverage big data to gain deeper insights, make informed decisions, and optimize processes. From business and commerce to healthcare, finance, transportation and logistics, smart cities, social sciences, and cybersecurity, big data transforms how these domains operate and deliver value to their stakeholders. With the ability to capture, store, process, and analyze vast amounts of data, big data analytics empowers organizations to extract meaningful information, identify patterns, and make data-driven decisions, leading to improved outcomes, increased efficiency, and competitive advantage:

- **Business and commerce**: Big data transforms how businesses operate, enabling organizations to gain deeper insights into customer behavior, market trends, and operational efficiency. Through big data analytics, companies can make data-driven decisions, optimize processes, improve customer experience, and gain a competitive advantage.
- **Finance**: Big data plays a crucial role in the finance industry, where vast amounts of data are generated and analyzed for risk assessment, fraud detection, algorithmic trading, and customer profiling. Big data analytics helps financial institutions gain insights into market trends, customer behavior, and risk management, leading to improved decision-making and financial performance.
- **Healthcare**: Big data revolutionizes healthcare by enabling data-driven decision-making, personalized medicine, and predictive analytics. Analyzing large and complex healthcare datasets, including **electronic health records** (**EHRs**), medical imaging data, and genomics data, can help in disease prediction, early detection, treatment planning, and patient care optimization.
- **Transportation and logistics**: Big data alters the transportation and logistics industry by optimizing supply chain operations, improving transportation efficiency, and enhancing safety. Real-time data from sensors, telematics, and other sources can be analyzed to optimize routes, reduce fuel consumption, enhance vehicle maintenance, and improve overall operational efficiency.
- **Smart cities**: Big data is being used to create smart cities by integrating data from various sources, such as sensors, social media, and public records, to improve urban planning, transportation, energy management, and public safety. Big data analytics helps make cities more efficient, sustainable, and resilient, leading to improved quality of life for citizens.
- **Social sciences**: Big data is increasingly used in social sciences to analyze large-scale social data, such as social media data, survey data, and public records, to understand human behavior, social dynamics, and societal trends. Big data analytics in social sciences can help in political science, economics, sociology, and psychology, leading to better policy-making and decision-making.
- **Cybersecurity**: Big data plays a critical role in cybersecurity by analyzing large volumes of data from various sources, such as logs, network traffic, and user behavior, to detect and mitigate cyber threats. Advanced analytics techniques, such as **machine learning** (**ML**) and anomaly detection, applied to big data can help identify patterns, detect anomalies, and prevent cyber-attacks. We'll delve into further details about big data and cybersecurity in the remainder of this chapter.

In summary, big data has become a crucial asset in various domains, offering the potential to unlock valuable insights, drive innovation, and create value. The ability to harness and analyze large and complex datasets is transforming industries, leading to improved decision-making, enhanced operational efficiency, and better outcomes in business, healthcare, finance, transportation and logistics, smart cities, social sciences, and cybersecurity.

At this point, the concept of big data should be comprehensible to all. Let's now delve into the challenges posed by big data in the realm of cybersecurity.

Big data challenges in cybersecurity

In today's digital world, the proliferation of connected devices and the increasing digitization of information have led to a staggering volume of data generated in cyberspace. Big data presents unique challenges in the context of cybersecurity. Big data presents unique challenges in the context of cybersecurity. The volume, velocity, variety, and veracity of data generated in cyberspace can overwhelm traditional cybersecurity practices. The sheer volume of data generated by devices, networks, and applications can be massive and difficult to manage, making it challenging to detect anomalies or identify patterns indicative of cyber threats. The velocity at which data is generated and transmitted in cyberspace requires timely and efficient processing for effective cybersecurity. The variety of data types, formats, and sources, including logs, network traffic, social media, and sensor data, adds complexity to the analysis process. Moreover, the veracity or trustworthiness of data can be uncertain, as data can be incomplete, inaccurate, or deliberately manipulated by adversaries. These challenges can significantly impact cybersecurity practices, requiring organizations to adapt and evolve their approaches to effectively analyze and interpret big data for detecting, preventing, and mitigating cyber threats.

As a matter of fact, with the proliferation of connected devices and the increasing digitization of information, the volume of data generated in cyberspace is staggering. This massive volume of data poses a challenge in cybersecurity. Associations nowadays must collect, store, and process vast amounts of data to identify potential cyber threats. Traditional cybersecurity methods may struggle to handle such a large volume of data, requiring organizations to invest in robust infrastructure, storage, and processing capabilities to manage and analyze big data for cybersecurity purposes effectively. Let's start with the velocity of data.

The velocity of data in cyberspace

The velocity of data in cyberspace refers to the speed at which data is generated, transmitted, and processed digitally. With the proliferation of connected devices, the digitization of information, and the increasing reliance on real-time data processing, the velocity of data in cyberspace has reached unprecedented levels. Cyber-attacks can occur in real time or near real time. Detecting and responding to these threats requires quick and efficient data processing.

The velocity of data poses significant challenges to cybersecurity practices. Traditional cybersecurity methods that rely on batch processing or periodic analysis may struggle to keep up with the speed at which data is generated and transmitted. Real-time monitoring and analysis are essential to promptly detect and respond to cyber threats before they can cause significant damage. For instance, detecting a **distributed denial-of-service** (**DDoS**) attack or an insider threat in real time requires quickly processing and analyzing large volumes of data to identify patterns, anomalies, and malicious activities.

The velocity of data also impacts the accuracy and effectiveness of cybersecurity practices. With data being generated and transmitted rapidly, cybersecurity analysis has a higher chance of false positives and negatives. False positives refer to the incorrect identification of benign activities as potential threats. In contrast, false negatives refer to the failure to detect actual threats. The speed at which data is generated can result in a higher volume of false positives and negatives, which can overwhelm cybersecurity defenses and lead to alert fatigue, where security analysts may miss genuine threats amidst many false alarms.

Organizations need to invest in advanced technologies that enable real-time data processing and analysis to handle the velocity of data in cyberspace effectively. Automated threat detection systems that use ML algorithms can analyze data at the speed of cyber-attacks, enabling prompt detection and response to threats. Real-time monitoring tools that provide continuous visibility into networks, systems, and applications can help organizations identify potential threats as they happen. Additionally, technologies such as stream processing, **event-driven architecture** (**EDA**), and real-time data analytics platforms can enable organizations to process and analyze data in real time, mitigating challenges posed by the velocity of data in cyberspace.

Furthermore, organizations need efficient data management practices to handle the velocity of data in cyberspace. This includes data ingestion, storage, and processing capabilities that are scalable, flexible, and optimized for real-time data processing. Data pipelines and processing workflows must be designed to handle large volumes of data in real time, with appropriate data retention and archiving strategies. Data quality and integrity measures must be in place to ensure the accuracy and reliability of data being processed in real time.

In conclusion, the velocity of data in cyberspace presents significant challenges to cybersecurity practices. Traditional methods may struggle to keep up with the speed at which data is generated, transmitted, and processed, requiring organizations to invest in advanced technologies, data management practices, and skilled personnel to handle the velocity of data for cybersecurity purposes effectively. Real-time monitoring, automated threat detection, and ML algorithms are crucial in processing data at the speed of cyber-attacks. Efficient data management practices, such as scalable data ingestion, storage, and processing capabilities, are necessary to handle the volume and speed of data in cyberspace. Organizations need to continuously adapt and evolve their cybersecurity practices to effectively address the challenges posed by the velocity of data in cyberspace and ensure robust cybersecurity defenses.

Diverse data types in cyberspace

Diverse data types in cyberspace refer to the vast array of data that is generated, transmitted, and stored in the digital realm. This data can come in various formats and types and from multiple sources, making the analysis process complex and challenging. For instance, logs from different systems and applications, network traffic data, social media posts, sensor data from **Internet of Things (IoT)** devices, user-generated content, and many other data types are constantly being generated in cyberspace. Each of these data types has its unique characteristics, structures, and patterns, which can complicate the analysis process for cybersecurity purposes.

Logs, which are records of events or actions captured by systems, applications, and devices, provide valuable information for cybersecurity analysis. However, logs can vary significantly in format, structure, and content, depending on the systems or devices generating them. For example, system logs from operating systems, databases, or web servers may have different formats and fields, making it challenging to normalize and integrate them for analysis. Similarly, network traffic data, which captures the communication between devices over a network, can be complex and diverse, including different protocols, packet formats, and data payloads.

Social media data, which includes posts, comments, likes, and shares on various social media platforms, can be unstructured and vast in volume. Analyzing social media data for cybersecurity requires extracting relevant information, identifying patterns, and detecting potential threats, such as phishing attacks or social engineering attempts. Sensor data from IoT devices, such as temperature readings, motion sensor data, or location data, can also be diverse and complex, with varying formats and standards depending on the devices and manufacturers.

Furthermore, user-generated content, such as emails, documents, multimedia files, and other types of digital content, can also vary in format and structure. Analyzing user-generated content for cybersecurity may involve text mining, **natural language processing** (**NLP**), and other techniques to extract meaningful information and detect potential threats, such as malware or malicious attachments.

The diversity of data types in cyberspace presents challenges in terms of data integration, normalization, and analysis. Traditional cybersecurity methods may not be equipped to handle the complexity and heterogeneity of data types, requiring organizations to develop advanced techniques, such as data fusion, normalization, and enrichment, to effectively analyze and interpret the diverse data types for cybersecurity purposes. These advanced techniques help integrate and normalize data from various sources, making it suitable for analysis and enabling organizations to identify patterns, trends, and anomalies that may indicate cyber threats.

The veracity of data in cyberspace

The veracity of data in cyberspace refers to the accuracy, reliability, and trustworthiness of data generated, transmitted, and processed digitally. In today's interconnected world, data is constantly being generated from various sources, such as social media, online transactions, IoT devices, and other digital interactions. However, not all data in cyberspace can be trusted to be accurate, complete, or reliable. This poses significant challenges to organizations that rely on data for decision-making, analysis, and other business processes, including cybersecurity-related ones.

One of the main challenges with the veracity of data in cyberspace is the presence of misinformation, fake data, and data tampering. Malicious actors may intentionally generate and spread false information, fake news, or manipulated data to deceive, mislead, or disrupt organizations, individuals, or systems. For example, cybercriminals may alter data in a database or inject false data into a system to gain unauthorized access, steal sensitive information, or cause disruptions. Moreover, unintentional data errors, inconsistencies, or inaccuracies may also occur due to human errors, technical glitches, or data integration issues, leading to unreliable or misleading data.

Another challenge with the veracity of data in cyberspace is the difficulty in verifying the authenticity and integrity of data. With the increasing reliance on data from various sources, ensuring that data is genuine, unaltered, and trustworthy becomes crucial. However, verifying the authenticity of data can be complex, especially in cases where data is generated and transmitted across multiple systems, networks, or jurisdictions. Data may be subject to manipulation, forgery, or tampering during its life cycle, making establishing its veracity and reliability challenging.

Ensuring the veracity of data in cyberspace is critical for cybersecurity practices. Relying on inaccurate, incomplete, or tampered data can lead to false assumptions, incorrect conclusions, and flawed decisions, resulting in security breaches, financial losses, reputational damage, and other negative consequences. Therefore, organizations need to implement robust data validation, verification, and integrity checks as part of their cybersecurity strategies to mitigate risks associated with the veracity of data.

Organizations can implement various practices and technologies to address challenges related to the veracity of data in cyberspace. These may include the following:

- **Data validation and integrity checks**: Organizations can implement data validation techniques, such as checksums, digital signatures, and hash algorithms, to verify data integrity and detect any alterations or tampering attempts. Regular data validation checks can help identify discrepancies or inconsistencies in the data and ensure that it is accurate and reliable.
- **Data source authentication**: Organizations can implement authentication mechanisms to verify the authenticity of data sources and ensure that data is coming from trusted and verified sources. This may include using digital certificates, encryption, and other authentication methods to establish the credibility of data sources.
- **Data quality management**: Organizations can implement data quality management practices, such as data profiling, data cleansing, and data enrichment, to improve the accuracy and reliability of data. This may involve identifying and correcting errors, inconsistencies, or duplications in the data to ensure it is trustworthy.

- **Data lineage and auditing**: Organizations can establish data lineage and auditing practices to track data's origin, movement, and transformations across different systems and processes. This can help ensure data integrity and provide a transparent audit trail for data, making it easier to identify any potential issues or anomalies.

- **Advanced analytics and artificial intelligence (AI)**: Organizations can leverage advanced analytics and AI techniques, such as ML algorithms, anomaly detection, and pattern recognition, to identify potential discrepancies, outliers, or anomalies in data that may indicate data tampering or misinformation.

- **Collaboration and information sharing**: Organizations can collaborate with other stakeholders, such as industry partners, academia, government agencies, and cybersecurity communities, to share information and best practices related to data veracity. Collaborative efforts can help organizations stay updated with the latest threats, trends, and techniques related to data integrity and build a collective defense against misinformation and data tampering.

- **Data governance and data management**: Organizations can establish robust data governance and data management practices to ensure that data is captured, stored, processed, and shared in a secure and controlled manner. This may involve defining data ownership, access controls, data retention policies, and data handling procedures to ensure data is handled with integrity and confidentiality.

- **Employee training and awareness**: Organizations can provide regular training and awareness programs to employees to educate them about the importance of data veracity and the risks associated with misinformation and data tampering. Employees should be trained to validate data, identify data quality issues, and report any suspicions or discrepancies.

- **Encryption and data protection**: Organizations can implement encryption and data protection measures to secure data in transit and at rest. Encryption techniques, such as **Secure Sockets Layer** (**SSL**)/**Transport Layer Security** (**TLS**) for data in transit and encryption algorithms for data at rest, can help protect data from unauthorized access, tampering, or interception.

- **IR and monitoring**: Organizations should have robust IR and monitoring mechanisms to detect, respond to, and mitigate potential data veracity incidents. This may involve implementing **security information and event management** (**SIEM**) systems, **intrusion detection systems** (**IDS**), and other monitoring tools to detect and alert of any suspicious activities related to data integrity.

Ensuring the veracity of data in cyberspace is crucial for organizations to make informed decisions, maintain trust, and safeguard against potential threats. By implementing data validation, authentication, data quality management, advanced analytics, collaboration, data governance, employee training, encryption, and IR practices, organizations can enhance the accuracy, reliability, and trustworthiness of data in cyberspace, thereby strengthening their cybersecurity posture.

Advanced analytical techniques and tools

Advanced analytical techniques and tools play a pivotal role in overcoming the challenges posed by big data in cybersecurity. Traditional cybersecurity approaches may not handle the complexity, scale, and velocity of big data in cyberspace. Therefore, organizations must invest in advanced analytics to effectively analyze and interpret big data to identify patterns, trends, and anomalies that may indicate cyber threats. Here are some ways in which advanced analytical techniques and tools can address the challenges of big data in cybersecurity:

- **ML and AI**: ML and AI algorithms can be trained on large datasets to identify patterns and anomalies that may signify cyber threats. These techniques can automatically analyze vast amounts of data, such as network traffic, logs, and user behavior, to detect and respond to potential cyber threats in real time. ML algorithms can continuously learn and adapt to changing cyber threats, making them a powerful tool for cybersecurity defense.

- **Data visualization**: Data visualization techniques can help cybersecurity analysts make sense of complex and large-scale data. Analysts can easily identify patterns, trends, and anomalies by visualizing data in graphical or interactive formats. Data visualization tools enable analysts to explore and analyze data visually, helping them gain insights and quickly make informed decisions.

- **Predictive analytics**: Predictive analytics techniques can analyze historical data and identify patterns or trends that may indicate future cyber threats. By leveraging ML algorithms, predictive analytics can forecast potential cyber threats, enabling organizations to take proactive measures to prevent or mitigate attacks before they occur. Predictive analytics can also help organizations identify vulnerabilities and prioritize remediation efforts.

- **Behavioral analytics**: Behavioral analytics involves analyzing user behavior data to detect anomalies or deviations from normal behavior patterns. By analyzing user behavior data, such as login, access, and activity patterns, behavioral analytics can detect potential insider threats, unauthorized access attempts, or other suspicious activities that may indicate a cyber threat. Behavioral analytics can complement traditional rule-based approaches by detecting unknown or emerging threats based on abnormal behavior.

- **Threat intelligence (TI)**: TI involves collecting and analyzing data on known cyber threats, including malware, vulnerabilities, and attacker techniques. Advanced **TI platforms** (**TIPs**) can analyze vast amounts of data from various sources, such as threat feeds, dark web, and **open source intelligence** (**OSINT**), to identify potential cyber threats. By leveraging TI, organizations can stay updated with the latest threats, trends, and techniques cyber adversaries use and take proactive measures to defend against them.

- **Big data analytics platforms**: Big data analytics platforms provide organizations with the infrastructure, tools, and capabilities to handle the volume, velocity, variety, and veracity of big data in cybersecurity. These platforms enable organizations to ingest, store, process, and analyze massive amounts of data from various sources in real time or near real time. Big data analytics platforms can provide advanced analytics capabilities, such as ML, data visualization, and predictive analytics, to help organizations effectively analyze and interpret big data for cybersecurity purposes.

In conclusion, advanced analytical techniques and tools are essential in overcoming challenges posed by big data in cybersecurity. ML, AI, data visualization, predictive analytics, behavioral analytics, TI, and big data analytics platforms are some advanced techniques and tools organizations can leverage to effectively analyze and interpret big data for identifying cyber threats. By harnessing the power of advanced analytics, organizations can enhance their cybersecurity posture, detect threats in real time, and respond proactively to potential cyber threats.

It becomes apparent that while advanced analytical techniques hold great promise in big data cybersecurity, effectively implementing these tools often encounters resource limitations. As we explore challenges posed by resource constraints, we will delve into practical considerations organizations face when applying these advanced techniques in real-world cybersecurity scenarios. From computational resources to budgetary constraints, we will examine how organizations navigate these challenges to balance leveraging cutting-edge technologies and optimizing their available resources for robust cybersecurity practices.

Resource constraints

Resource constraints refer to limitations faced by organizations in terms of their available resources, such as budget, manpower, infrastructure, and technology, which can impact their ability to address cybersecurity challenges in the era of big data effectively. Here are some ways in which resource constraints can pose challenges in the context of big data cybersecurity:

- **Budget constraints**: Organizations may have limited budgets for cybersecurity initiatives, including investments in advanced analytical tools, technologies, and infrastructure required to handle big data. Budget constraints may limit the ability to invest in cutting-edge technologies, hire skilled cybersecurity personnel, or implement comprehensive cybersecurity measures, leaving organizations vulnerable to cyber threats associated with big data.

- **Manpower constraints**: Organizations may face limitations regarding skilled cybersecurity personnel available to handle the complexities of big data. Big data requires specialized skills, including data scientists, data engineers, and cybersecurity analysts, who can effectively analyze and interpret large-scale and complex data for identifying potential cyber threats. Manpower constraints may impact an organization's ability to handle big data in a timely and efficient manner effectively.

- **Infrastructure constraints**: Big data in cybersecurity requires robust and scalable infrastructure to store, process, and analyze massive amounts of data in real time or near real time. Organizations may face constraints regarding the availability of infrastructure, including servers, storage, and networking equipment, needed to handle big data effectively. Infrastructure constraints may limit an organization's ability to scale its cybersecurity operations and effectively manage cyberspace's volume, velocity, and variety of data.

- **Technology constraints**: Big data cybersecurity requires advanced analytical tools, technologies, and platforms to analyze and interpret large-scale and complex data effectively. Organizations may face limited access to cutting-edge technologies or tools for various reasons, such as cost, compatibility, or availability. Technology constraints may hinder an organization's ability to analyze big data and detect potential cyber threats effectively.

- **Time constraints**: Cybersecurity threats in the era of big data can evolve and propagate rapidly. Organizations need to respond on time to prevent or mitigate potential attacks. Time constraints may result in delayed or inadequate response to cyber threats, increasing the risks and impact of potential cybersecurity incidents.

Let's now turn our attention to addressing resource constraints. We will explore practical strategies and solutions that organizations can implement to effectively manage limited computational resources, budgets, and other constraints while harnessing the potential of advanced analytical techniques for bolstering their cybersecurity efforts. By understanding how to optimize available resources, organizations can strike a strategic balance between their cybersecurity objectives and the realities of resource limitations.

Addressing resource constraints

Organizations can take several steps to overcome resource constraints and effectively address cybersecurity challenges associated with big data:

- **Prioritize cybersecurity investments**: Organizations should prioritize cybersecurity investments based on risk assessment and TI. Organizations can optimize their cybersecurity investments and mitigate risks by identifying the most critical areas that require protection and allocating resources accordingly.

- **Seek cost-effective solutions**: Organizations can explore cost-effective solutions that provide value for money without compromising cybersecurity. This may include open source technologies, cloud-based services, or leveraging existing infrastructure and technologies to handle big data cybersecurity requirements within budget constraints.

- **Develop talent pool**: Organizations can invest in training and development programs to build a skilled cybersecurity workforce. This may include training existing personnel or partnering with educational institutions to foster cybersecurity skills development. Organizations can also leverage external expertise through managed security services or collaborations with cybersecurity firms to supplement their in-house resources.

- **Optimize infrastructure**: Organizations can optimize their existing infrastructure by leveraging technologies such as virtualization, containerization, or cloud computing to scale their cybersecurity operations efficiently. This can help organizations overcome infrastructure constraints and handle big data cybersecurity requirements effectively.

- **Embrace automation and AI**: Automation and AI technologies can help organizations overcome manpower constraints and improve the efficiency and effectiveness of their cybersecurity operations. Automated security tools, threat-hunting algorithms, and AI-powered security analytics can enable organizations to analyze and respond to big data cybersecurity threats in real time with limited manpower resources.

- **Collaborate with partners**: Organizations can collaborate with partners, such as other organizations, academia, or government agencies, to pool resources and expertise in addressing big data cybersecurity challenges. Collaborative efforts can lead to cost-sharing, knowledge-sharing, and resource-sharing, which can help organizations overcome resource constraints and collectively enhance their cybersecurity capabilities.

- **Implement risk-based approach**: Organizations can implement a risk-based approach to prioritize their cybersecurity efforts and allocate resources accordingly. By identifying the most critical assets, vulnerabilities, and threats, organizations can prioritize their resources on the most high-risk areas and optimize their cybersecurity measures based on the risk associated with big data.

- **Regularly assess and update cybersecurity measures**: Organizations should periodically evaluate and update their cybersecurity measures to ensure their effectiveness in addressing big data cybersecurity challenges. This may include regular vulnerability assessments, penetration testing, and security audits to identify and address potential gaps or weaknesses in the cybersecurity posture.

- **Leverage TI**: Organizations can leverage TI sources, such as cybersecurity information sharing forums, feeds, or industry reports, to stay updated on the latest cybersecurity threats and trends. This can help organizations prioritize their resources and efforts based on the most relevant and impactful threats in cyberspace.

- **Develop a comprehensive cybersecurity strategy**: Organizations should develop a comprehensive cybersecurity strategy that aligns with their business objectives, risk tolerance, and available resources. The strategy should encompass a holistic approach to big data cybersecurity, including policies, procedures, technologies, training, and **IR plans** (**IRPs**), to ensure a proactive and effective cybersecurity posture.

Resource constraints can pose challenges in big data cybersecurity. However, organizations can overcome these constraints by prioritizing investments, seeking cost-effective solutions, developing talent, optimizing infrastructure, embracing automation and AI, collaborating with partners, implementing a risk-based approach, regularly assessing and updating cybersecurity measures, leveraging TI, and developing a comprehensive cybersecurity strategy. By adopting a strategic and proactive approach, organizations can effectively address resource constraints and manage cybersecurity risks associated with big data in cyberspace.

In summary, challenges posed by big data in the context of cybersecurity are multifaceted, including the volume, velocity, variety, and veracity of data. These challenges can make it difficult for organizations to effectively manage and analyze big data in cybersecurity, requiring them to develop advanced techniques, tools, and strategies to overcome these challenges and protect their systems, networks, and data from cyber threats.

In the following section, we'll pivot from the obstacles and complexities of handling vast datasets to the practical utilization of big data solutions for enhancing cyber defenses. Having identified challenges, we'll explore how organizations leverage the power of big data analytics to proactively detect threats, respond swiftly to incidents, and strengthen their overall security posture. Through a deeper dive into real-world applications, you'll gain valuable insights into how big data is a challenge and a formidable ally in the ongoing battle against cyber threats.

Big data applications in cybersecurity

Big data has become increasingly relevant in cybersecurity because it can unlock insights and identify patterns that may indicate cyber threats. Organizations and analysts use big data to improve their cybersecurity posture by enhancing their threat detection and mitigation capabilities.

One significant application of big data in cybersecurity is **TI**. TI involves collecting and analyzing large volumes of data from various sources to identify patterns and trends in cyber-attacks. Big data techniques such as ML, data mining, and NLP are used to extract and analyze information from structured and unstructured data sources. This information is used to build threat models that help organizations and analysts identify and respond to emerging cyber threats more quickly and effectively. TI has become a critical component of cybersecurity, enabling defenders to stay ahead of cybercriminals and protect against sophisticated attacks.

Another application of big data in cybersecurity is **anomaly detection**. Anomaly detection identifies unusual or unexpected behavior in networks or systems that may indicate a security breach. Big data techniques such as ML and statistical analysis are used to identify patterns and trends in network traffic and system behavior. Anomaly detection is essential for detecting cyber threats that evade traditional security controls. With the help of big data analytics, organizations can identify suspicious activities, prioritize incidents, and take appropriate action to mitigate risks.

Behavior analysis is another critical application of big data in cybersecurity. Behavior analysis is a big data application that involves the monitoring and analysis of user behavior to detect potential threats. Cybersecurity analysts can identify deviations from normal behavior by analyzing user activity logs and detecting insider threats or other malicious activities. Behavior analysis is valuable for identifying and mitigating security risks before they can cause significant damage. It is also useful for compliance, as it can help organizations detect unauthorized access attempts and ensure that users adhere to cybersecurity policies.

Log analysis is also a popular application of big data in cybersecurity. Various systems and applications generate logs. Collecting, storing, and analyzing log data from multiple sources allows us to detect and investigate security incidents. Big data techniques such as data mining, pattern recognition, and NLP are used to identify patterns and anomalies in log data. Log analysis is a crucial component of cybersecurity as it provides organizations with insights into security incidents, enabling them to take appropriate action to mitigate risks.

In conclusion, big data is transforming the way organizations approach cybersecurity. Big data applications in cybersecurity are diverse and wide-ranging, including TI, anomaly detection, behavior analysis, log analysis, and other advanced analytics techniques. By harnessing the power of big data analytics, organizations can improve their threat detection and mitigation capabilities, enhance their overall cybersecurity posture, and stay ahead of evolving cyber threats.

In the next section, we'll investigate the technological backbone that empowers organizations to effectively harness the potential of big data in their cybersecurity endeavors. By understanding tools, platforms, and innovations that underpin data collection, processing, and analysis, you'll gain a comprehensive view of the infrastructure necessary to make informed decisions, detect vulnerabilities, and respond decisively to emerging cyber threats.

Big data technologies for cybersecurity

In recent years, big data technologies have played a significant role in advancing cybersecurity practices. With the growth of big data in cybersecurity, organizations have turned to various technologies to help manage and analyze large volumes of data to detect and respond to cyber threats. Distributed computing frameworks are a crucial technology in big data for cybersecurity. These frameworks enable processing massive amounts of data by distributing the workload across many nodes. Apache Hadoop is one of cybersecurity's most popular distributed computing frameworks. It is an **open source software** (**OSS**) framework that enables storing and processing large datasets in a distributed computing environment. **Hadoop Distributed File System** (**HDFS**) allows for the distributed storage and processing of large datasets, and its MapReduce programming model facilitates the parallel processing of data. Apache Spark is another popular distributed computing framework commonly used in cybersecurity. Spark is designed to be faster and more flexible than Hadoop, enabling real-time processing of data streams and quicker batch processing of large datasets.

Stream processing platforms are another vital technology in big data for cybersecurity. These platforms enable real-time processing of data streams, allowing organizations to detect and respond to cyber threats in near real time. Apache Kafka is a popular stream processing platform commonly used in cybersecurity. It is an OSS platform that enables real-time processing of high-throughput data streams. Kafka can handle large volumes of data streams in real time. It supports **fault tolerance** (**FT**), **high availability** (**HA**), and horizontal scalability.

Finally, data storage and management technologies have also become crucial in managing big data in cybersecurity. Traditional relational databases are unsuitable for handling massive volumes and diverse data types generated in cyberspace. NoSQL databases are a more suitable alternative, providing horizontal scalability, FT, and flexible data models. MongoDB is a popular NoSQL database that is commonly used in cybersecurity. It enables storing and retrieving large volumes of data with high performance and scalability. Other data storage and management technologies include data warehouses, lakes, and virtualization platforms.

In addition to these technologies, ML algorithms and AI tools are becoming increasingly important in cybersecurity. These technologies enable organizations to analyze large volumes of data and identify patterns and anomalies that may indicate cyber threats. By leveraging ML algorithms and AI tools, organizations can improve their ability to detect and respond to cyber threats quickly and efficiently.

Overall, big data technologies play a critical role in modern cybersecurity practices. As the volume of data generated in cyberspace grows, organizations must rely on advanced technologies to manage and analyze large volumes of data effectively. By leveraging distributed computing frameworks, stream processing platforms, data storage and management technologies, and ML algorithms, we can improve our ability to detect and respond to cyber threats, ultimately enhancing our cybersecurity posture.

Summary

Big data is critical in cybersecurity, presenting challenges and opportunities. With the proliferation of connected devices and the increasing digitization of information, modern society faces an overwhelming volume, velocity, and variety of data, making it harder to detect and mitigate cyber threats. Traditional cybersecurity methods may not be sufficient; thus, we must invest in advanced technologies and skilled cybersecurity professionals to effectively handle the complexity and scale of big data in cyberspace.

Big data applications in cybersecurity, such as TI, anomaly detection, behavior analysis, and log analysis, are used to identify and respond to cyber threats promptly. These advanced analytics techniques rely on distributed computing frameworks, stream processing platforms, and data storage and management technologies to process and analyze big data at scale.

Overall, understanding the significance of big data in cybersecurity is essential for effectively improving our cybersecurity posture and protecting against cyber threats. By leveraging advanced technologies and analytical techniques, we can harness the power of big data to enhance cybersecurity capabilities and stay ahead of cybercriminals.

In the next chapter, we will examine automation in cybersecurity, exploring how cutting-edge technologies such as AI and ML are revolutionizing the way we defend against cyber threats.

2
Automation in Cybersecurity

Cybersecurity threats are rapidly increasing in volume and complexity, making it more challenging for analysts to keep pace with the evolving threat landscape. To address this challenge, there has been a growing interest in automating cybersecurity processes to improve the speed and efficiency of threat detection, response, and mitigation. In this chapter, we aim to provide an overview of the role of automation in cybersecurity and the challenges and opportunities associated with its adoption. More precisely, we will examine the following:

- Tools and technologies against threats
- The importance of automation in cybersecurity
- Examples of automated cybersecurity tools
- Potential drawbacks and challenges of automation
- The future of automation in cybersecurity
- Ethical considerations

By the end of this chapter, you will have gained a comprehensive understanding of the pivotal role automation plays in modern cybersecurity practices. You will be equipped with insights into tools and technologies employed to combat threats and a deep appreciation for the crucial role automation plays in enhancing the speed and efficiency of threat detection and response. Additionally, we will explore real-world examples of automated cybersecurity tools, shedding light on their practical applications. Furthermore, we will examine organizations' potential drawbacks and challenges when implementing automation, offering valuable considerations for balanced cybersecurity strategies. Lastly, we will peer into the future of automation in cybersecurity, envisioning how this dynamic field is poised to evolve and shape the security landscape in the years to come.

Tools and technologies against threats

Cybersecurity automation involves using tools, processes, and technologies to automatically detect, analyze, and respond to security incidents. It can be applied to various cybersecurity tasks, including monitoring, threat detection, **incident response** (**IR**), and reporting.

One of the key benefits of automation in cybersecurity is the ability to process large volumes of data and detect threats quickly and accurately. Traditional manual monitoring and threat detection methods can be time-consuming and prone to human error. Automation tools can process and analyze data in real time, alerting security teams to potential threats and anomalies as they occur.

Automated security tools can take many forms, depending on the specific task they are designed to perform. Let's now see some common examples.

Security information and event management (SIEM)

SIEM platforms are tools to collect and analyze security event data (that is, records and information generated by various security-related activities and events) from various sources within an organization's network. This can include data from firewalls, servers, applications, and other network devices. The purpose of a SIEM platform is to provide a centralized view of an organization's security posture by correlating security event data from multiple sources and identifying potential security incidents. The platform uses advanced analytics to detect anomalous activity, identify potential threats, and generate alerts for security teams.

The process of using a SIEM platform typically involves the following four main steps:

1. **Data collection**: The SIEM platform collects event data from various sources within an organization's network. This can include logs, alerts, and other types of security event data.
2. **Data normalization**: The SIEM platform normalizes the collected data, converting it into a common format that can be analyzed and correlated.
3. **Correlation and analysis**: The SIEM platform uses advanced analytics to correlate the normalized data and identify potential security incidents. This can involve pattern recognition, anomaly detection, and other types of analysis.
4. **Alerting and reporting**: When a potential security incident is detected, the SIEM platform generates an alert for the security team. The platform can also provide reporting capabilities to help security teams understand the nature and scope of the incident.

SIEM platforms can be used for various cybersecurity tasks, including monitoring suspicious activity, detecting insider threats, and identifying compliance violations. Additionally, we can incorporate them alongside other security solutions such as **intrusion detection systems (IDSs)** and vulnerability scanners, offering a holistic perspective on an organization's security stance. Here are some examples of widely used SIEM platforms:

- **IBM QRadar**: QRadar is a leading SIEM platform that uses advanced analytics and **machine learning (ML)** to detect security threats. It provides a centralized view of an organization's security posture, with real-time visibility into potential threats and suspicious activity. QRadar also includes a range of automation features, such as automated IR and threat-hunting capabilities.

- **Splunk Enterprise Security**: Splunk is another popular SIEM platform for detecting and responding to security incidents. It collects and analyzes data from various sources, including logs, events, and network traffic, to identify potential threats. Splunk also includes a range of advanced analytics, ML capabilities, and automated IR workflows.

- **LogRhythm NextGen SIEM platform**: The LogRhythm NextGen SIEM platform provides real-time visibility into an organization's security posture, using advanced analytics and ML to detect and respond to potential threats. It includes different automation features, such as automated IR workflows and threat-hunting capabilities. The platform also consists of a **security orchestration and automation response (SOAR)** capability, which enables security teams to automate and orchestrate security tasks across multiple security tools.

- **Wazuh**: Wazuh is a free, open source SIEM platform that provides real-time security visibility and monitoring for on-premise, cloud, and hybrid environments. It is built on top of the **Elastic Stack (ELK Stack**; that is, Elasticsearch, Logstash, and Kibana) and includes a range of security capabilities, such as intrusion detection, log analysis, file integrity monitoring, and vulnerability detection. Wazuh also provides a centralized management console for security teams to view and respond to security incidents. The platform can be customized with plugins and integrations to support specific security needs.

However, no matter how capable the SIEM platforms are, at the same time, they can be complex and require significant expertise to configure and maintain. They can also generate a large volume of alerts, which can be overwhelming for security teams to manage. Proper configuration and tuning of the platforms are critical to ensure they can effectively detect and respond to security incidents.

Intrusion detection and prevention systems (IDPSs)

IDPSs are security tools engineered to swiftly identify and address potential security risks in real time. They analyze network traffic and system activity for signs of suspicious behavior, such as unauthorized access or malicious activity.

There are two main types of IDPS:

- **IDS**: IDS are designed to detect potential security threats by analyzing network traffic and system activity for signs of suspicious behavior. IDS can be network-based or host-based. **Network-based IDSs (NIDSs)** analyze network traffic to detect potential threats. In contrast, **host-based IDSs (HIDSs)** analyze system activity to detect suspicious behavior on individual devices.

- **Intrusion prevention systems (IPSs)**: IPS are designed to prevent potential security threats by analyzing network traffic and system activity and blocking traffic deemed suspicious or malicious. IPS can be network-based or host-based. **Network-based IPSs (NIPSs)** block malicious traffic before it reaches its destination, while **host-based IPSs (HIPSs)** block suspicious activity on individual devices.

The process of using IDPS typically involves three main steps:

- **Monitoring and analysis**: The IDPS continuously monitors network traffic and system activity for signs of suspicious behavior. It uses advanced analytics and ML to identify potential threats and generate alerts for security teams.

- **Alerting and reporting**: When a potential security threat is detected, the IDPS generates an alert for the security team. The platform can also provide reporting capabilities to help security teams understand the nature and scope of the incident.

- **Response and mitigation**: The security team can use the information provided by the IDPS to investigate and respond to potential security incidents. They can also use the platform's mitigation capabilities (that is, the range of actions and strategies that an IDPS can employ to actively respond to and counteract potential security threats) to block suspicious traffic and prevent potential threats from causing harm.

IDPSs can detect and respond to various security threats, including network attacks, malware infections, and insider threats. They can also be integrated with other security tools, such as SIEM platforms and vulnerability scanners, to view an organization's security posture comprehensively. However, IDPS can be complex and require significant expertise to configure and maintain. They can also generate a large volume of alerts, which can be overwhelming for security teams to manage. As such, and similar to SIEM platforms, they need proper configuration and tuning.

Endpoint detection and response (EDR)

EDR is a security technology that provides real-time visibility and monitoring of endpoint devices, such as desktops, laptops, servers, and mobile devices. EDR platforms use advanced analytics and ML to detect and respond to potential security threats on endpoints, such as malware infections, unauthorized access, and data exfiltration. EDR platforms typically include the following:

- **Endpoint monitoring**: EDR platforms continuously monitor endpoint devices for signs of suspicious behavior, such as unauthorized access or malicious activity. This includes monitoring system activity, network traffic, and user behavior.

- **Threat detection**: EDR platforms use advanced analytics and ML to detect potential security threats on endpoint devices. This includes identifying known and unknown malware and suspicious activity that may indicate an ongoing attack.

- **IR**: When a potential security threat is detected, the EDR platform generates an alert for the security team. The platform also provides tools to investigate and respond to the incident, including quarantine and remediation capabilities.

- **Forensic analysis**: EDR platforms can conduct forensic analysis after an incident has occurred on endpoint devices to understand the nature and scope of a potential security incident. This includes analyzing system activity, file changes, and network traffic.

- **Integration with other security tools**: EDR platforms can be integrated with other security tools, such as SIEM platforms and vulnerability scanners, thus offering a broader perspective on an organization's security stance.

EDR platforms are critical for detecting and responding to security threats on endpoint devices. With the increase in remote work and using personal devices in the workplace, endpoint security has become a critical component of an organization's overall security strategy.

Security orchestration, automation, and response (SOAR)

SOAR is a technology stack that helps security teams automate and streamline their IR processes. SOAR platforms provide a centralized console for managing security alerts and incidents, and they automate many manual tasks associated with IR.

SOAR platforms typically include the following capabilities:

- **Alert aggregation**: SOAR platforms aggregate alerts from multiple security tools, such as SIEMs, IDSs/IPSs, and EDR. This provides a comprehensive view of an organization's security posture and helps security teams prioritize their response efforts.

- **Incident management (IM)**: SOAR provides a centralized console for managing security incidents. SOAR platforms can create, track, and manage incidents, assign tasks, and collaborate with other security team members.

- **Workflow automation**: SOAR platforms automate many manual tasks associated with IR, such as gathering additional context and enrichment data, blocking malicious traffic, and isolating infected endpoints.

- **Threat intelligence (TI) integration**: SOAR platforms integrate with TI sources to provide additional context on potential security threats. This includes integrating with open source TI feeds and commercial intelligence providers.

- **Reporting and metrics**: SOAR platforms provide reporting and metrics capabilities to help security teams understand their IR performance. This includes tracking IR times, identifying bottlenecks, and measuring the effectiveness of response workflows.

SOAR platforms are designed to help security teams respond to security incidents faster and more efficiently. By automating many manual tasks associated with IR, SOAR platforms help free up time for security teams to focus on more strategic activities, such as threat hunting and vulnerability management.

It becomes evident that the ever-evolving threat landscape necessitates a paradigm shift in our cybersecurity strategies. In a world where threats evolve at machine speed, incorporating automation into cybersecurity practices becomes not just a preference but a critical imperative to safeguard digital assets effectively.

The importance of automation in cybersecurity

In an era of increasingly sophisticated cyber threats and a growing attack surface, the reliance on manual security practices alone is no longer sufficient. Automation is necessary to keep pace with the evolving threat landscape and the expanding digital infrastructure. By harnessing the power of automation, organizations can augment their cybersecurity capabilities, improve response times, and enhance overall resilience.

In today's interconnected world, the sheer volume and complexity of cyber threats have surpassed the capabilities of manual approaches. Automation plays a pivotal role in processing vast amounts of data, identifying patterns, and detecting threats in real time. By leveraging ML algorithms and advanced analytics, automated systems can swiftly sift through enormous datasets to pinpoint potential vulnerabilities and anomalous activities that might otherwise go unnoticed.

Furthermore, manual monitoring and IR can be time-consuming, leading to delays in identifying and mitigating cyber threats. Automation significantly reduces response times by enabling continuous monitoring, rapid threat detection, and immediate response. Automated tools can detect **indicators of compromise** (**IOCs**), abnormal behaviors, or known attack patterns, and trigger predefined reactions or alerts. This swift response not only minimizes the impact of security incidents but also prevents unauthorized access, data breaches, and system compromise.

Human error remains one of the most common causes of security breaches. Manual processes are prone to oversight, fatigue, and inconsistencies. By automating repetitive and rule-based tasks, cybersecurity professionals can reduce the risk of human error and improve the accuracy and consistency of security operations. Automated systems follow predefined protocols, adhere to best practices, and execute tasks precisely, significantly reducing the chances of misconfigurations or oversight.

At the same time, maintaining a robust security posture becomes increasingly challenging as organizations grow and their digital infrastructures expand. Automation allows for scalability by efficiently managing security operations across large and complex environments. Automated tools can handle a higher volume of security events and incidents, thereby optimizing the use of limited resources. By automating routine tasks, security teams can allocate their time and expertise to strategic initiatives and higher-value activities, such as threat hunting, vulnerability management, and proactive risk assessment.

Finally, regulatory frameworks and compliance requirements place an additional burden on organizations to demonstrate effective security practices and IR capabilities. Automation simplifies compliance by automating data collection, analysis, and reporting processes. By streamlining compliance workflows, organizations can ensure consistency, accuracy, and auditability in meeting regulatory obligations. Automated reports can provide comprehensive insights into security incidents, trends, and remediation efforts, facilitating compliance assessments and regulatory reporting.

It should be clear by now that automation is a critical enabler in addressing the evolving challenges of cybersecurity. By leveraging advanced technologies and streamlining security operations, automation enhances threat detection, response times, and overall security posture. The benefits of automation include improved efficiency, reduced human error, scalability, and better compliance. As cyber threats evolve, embracing automation becomes increasingly vital to protect digital assets and defend against emerging risks effectively.

Let's now see some real-world examples.

Examples of automated cybersecurity tools

Automation has revolutionized the shift in cybersecurity, empowering organizations to keep pace with the evolving threat landscape and the increasing complexity of digital environments. By leveraging automated tools and technologies, organizations can achieve greater efficiency, accuracy, and agility in their cybersecurity operations. These tools streamline routine tasks, allowing security professionals to focus on more strategic and complex initiatives. More precisely, automation enables organizations to detect and respond to threats in real time, reducing the time between detection and remediation. It also helps organizations scale their cybersecurity capabilities and protect their systems and data as they grow. Organizations can also reduce human error and ensure consistency in their security practices by automating repetitive and rule-based processes. Let's now see some real-world automated cybersecurity tools:

- **Splunk Enterprise Security**: Splunk is a popular SIEM platform that provides real-time monitoring, TI integration, and IR capabilities. It automates collecting and correlating security event logs from various sources, enabling organizations to detect and respond to security incidents effectively.
- **Snort**: Snort is an open source IDPS tool known for its intrusion detection and prevention capabilities. It analyzes network traffic in real time, detects suspicious patterns or signatures, and generates alerts or takes proactive actions to block potential threats.
- **CrowdStrike Falcon**: Falcon is an EDR solution that provides advanced endpoint protection and TI. It uses behavior-based analytics, ML, and real-time monitoring to detect and respond to malicious activities on endpoints, helping organizations mitigate risks and prevent breaches.

- **Demisto (now part of Palo Alto Networks)**: Demisto is a leading SOAR platform that automates and orchestrates security operations and IR processes. It integrates with various security tools, collects and correlates data, and automates response actions, enabling security teams to handle incidents more efficiently.
- **Nessus**: Nessus is a widely used vulnerability management tool that scans systems and applications for vulnerabilities, misconfigurations, and compliance issues. It provides prioritized vulnerability reports, remediation recommendations, and integration with patch management systems to help organizations address security weaknesses.
- **Cisco Umbrella**: Cisco Umbrella is a cloud-delivered security platform that provides **Domain Name System (DNS)** and web filtering, TI, and automated threat response. It helps organizations protect against malware, phishing, and **command-and-control (C2)** callbacks by blocking malicious domains and IP addresses.

These are just a few examples of the many automated cybersecurity tools available in the market. Organizations must evaluate their specific needs, consider scalability, integration capabilities, and vendor support, and choose tools that align with their cybersecurity objectives and infrastructure.

It's crucial to recognize that while automated cybersecurity tools offer significant advantages, they are not without complexities. We will now explore nuances and potential pitfalls associated with the widespread adoption of automation in cybersecurity practices. By understanding these challenges, organizations can make informed decisions about navigating them and optimizing their automated defenses while mitigating associated risks.

Potential drawbacks and challenges of automation

While automation in cybersecurity brings significant advantages, it is crucial to approach its implementation with a clear understanding of potential drawbacks and challenges. Organizations must be aware of the risk of false positives and false negatives, as automated systems may generate incorrect alerts or miss genuine threats. Complex implementation and integration processes can pose challenges, requiring careful planning and expertise. Moreover, analysts' contextual understanding and decision-making capabilities cannot be fully replaced by automation. The rapidly evolving nature of cyber threats also necessitates continuous updates and enhancements to automated systems. Finally, overreliance on automation without maintaining human expertise can lead to losing critical thinking and vigilance. Let's now see in more detail all the aforementioned limitations:

- **False positives and false negatives**: One of the primary challenges of automation in cybersecurity is the risk of false positives and false negatives. Automated systems heavily rely on algorithms and predefined rules to identify and classify security incidents. However, these systems may generate false positives, flagging benign activities as malicious, leading to unnecessary investigations and resource wastage. Conversely, false negatives occur when automated tools fail to detect genuine threats, allowing malicious activities to go unnoticed. Organizations must fine-tune automated systems to minimize false alerts and continuously update them to adapt to emerging threats.

- **Complex implementation and integration**: Implementing automation in cybersecurity can be complicated and challenging. Organizations must carefully select and integrate appropriate tools and technologies into their security infrastructure. Integration issues may arise when connecting different automated systems, leading to interoperability problems and gaps in security coverage. Additionally, automation implementation often requires expertise and specialized skills, which may not be readily available within the organization. Adequate planning, testing, and collaboration with vendors or experts can help mitigate these challenges.

- **Lack of contextual understanding**: Automated systems rely on algorithms and predefined rules to analyze security events. However, they may lack human analysts' contextual understanding and situational awareness. Cyber threats often require nuanced analysis, a knowledge of business operations, and the ability to make judgment calls based on a broader context. While automation can speed up processes and flag suspicious activities, human expertise is still crucial for accurate threat analysis, decision-making, and taking appropriate actions.

- **Advanced and evolving threats**: Cyber threats evolve rapidly, with attackers employing sophisticated techniques to evade detection. Automated systems may struggle to keep pace with the ever-changing threat landscape. Threat actors constantly adapt their **tactics, techniques, and procedures** (**TTPs**) to bypass automated defenses. Consequently, organizations must regularly update and enhance their automated security systems to effectively combat new and emerging threats. Continuous monitoring, TI sharing, and a proactive approach to system updates and patching are essential to stay ahead of evolving threats.

- **Overreliance and loss of human expertise**: Excessive reliance on automated systems can lead to a loss of human expertise in cybersecurity. While automation can handle routine tasks and augment human capabilities, it cannot replace skilled security professionals' critical thinking, creativity, and intuition. Overreliance on automation may result in complacency, decreased vigilance, and a potential blind spot to novel attack vectors that automated systems have not yet learned to detect. Organizations must balance automation and human involvement, ensuring that human analysts remain integral to cybersecurity decision-making.

While automation significantly benefits cybersecurity, organizations must carefully consider and address potential drawbacks and challenges. False positives and false negatives, complex implementation and integration, lack of contextual understanding, advanced and evolving threats, and overreliance on automation are among the key challenges to be addressed. Organizations can optimize their cybersecurity practices, effectively mitigate risks, and maintain a robust defense against evolving cyber threats by understanding these challenges and adopting a holistic approach that combines automation with human expertise.

The future of automation in cybersecurity

With the ever-increasing complexity and volume of cyber threats, automation is becoming a necessity rather than an option for organizations. We envision that the future of automation in cybersecurity will witness the integration of advanced technologies such as **artificial intelligence** (**AI**) and ML, enabling more intelligent and adaptive security systems. IR processes will be streamlined and expedited through automated workflows, minimizing the time between threat detection and containment. SOAR platforms will mature, offering enhanced capabilities for threat hunting, IM, and security operations. The rise of **Internet of Things (IoT)** devices and cloud computing will drive the need for automation in securing these environments, ensuring their protection at scale. Ultimately, automation will empower cybersecurity professionals by augmenting their capabilities, enabling them to focus on strategic initiatives and proactive threat hunting rather than mundane and repetitive tasks.

In more detail, AI and ML technologies are expected to drive the future of automation in cybersecurity. These technologies enable automated systems to learn from vast amounts of data, identify patterns, and make intelligent real-time decisions. AI-powered solutions can enhance threat detection by continuously analyzing evolving attack vectors and adapting to new threat patterns. ML algorithms can identify anomalies, detect previously unseen threats, and provide proactive defense measures. As AI and ML continue to advance, automation will become more intelligent, enabling cybersecurity systems to anticipate and respond to emerging threats effectively.

Automation will also increasingly streamline IR processes, enabling faster and more efficient remediation of security incidents. Automated **IR platforms (IRPs)** can detect and classify threats, trigger predefined response actions, and facilitate coordinated incident handling. The future of automation in IR will involve intelligent automation, where systems can autonomously investigate and contain threats, reducing the need for manual intervention. Additionally, automation will enable seamless collaboration and information sharing across security teams and stakeholders, improving IR capabilities and reducing time to contain and mitigate the impact of security breaches.

Furthermore, SOAR platforms will continue to gain prominence in the future of cybersecurity automation. These platforms integrate security tools, automate workflows, and enable centralized security operations management. The future of SOAR will involve tighter integration with AI and ML capabilities, allowing for more intelligent and automated decision-making. SOAR platforms will become the backbone of cybersecurity operations, facilitating seamless coordination between people, processes, and technology to enhance IR, threat hunting, and overall security management.

As the proliferation of IoT devices continues, automation will become crucial in securing these interconnected devices and networks. The future of automation in IoT security will involve intelligent systems that can automatically discover and onboard IoT devices, continuously monitor their behavior, and detect and respond to security threats immediately. Automation will play a vital role in managing the security posture of IoT devices at scale, mitigating risks associated with vulnerabilities and unauthorized access. This includes automating device patching, access control, and security policy enforcement, ultimately enhancing the overall security of IoT ecosystems.

Finally, cloud computing presents unique security, scalability, and complexity challenges. Automation will address these challenges by enabling automated security controls, continuous monitoring, and compliance enforcement in cloud environments. The future of automation in cloud security will involve the integration of security controls within the cloud infrastructure, leveraging APIs and cloud-native automation tools. This will enable organizations to achieve greater visibility, control, and agility in securing their cloud-based assets, workloads, and data.

The future of automation in cybersecurity is promising, driven by advancements in AI, ML, and the increasing need for efficient and effective security practices. AI and ML will empower automated systems to detect and respond to threats with higher accuracy and speed. IR will become more streamlined and automated, improving the efficiency of incident handling. SOAR platforms will continue to evolve, enabling organizations to centralize and automate security operations. Automation will also play a vital role in securing IoT devices and managing cloud security. As organizations adapt to the evolving threat landscape, the future of automation in cybersecurity will be instrumental in maintaining robust defenses, enabling proactive threat detection, and safeguarding digital assets and sensitive information.

Ethical considerations

Automation in cybersecurity introduces a plethora of ethical considerations that necessitate thorough examination. First and foremost, automated cybersecurity systems often rely on vast amounts of data to detect and respond to threats effectively. However, collecting, processing, and storing this data raises significant privacy concerns. Ethical practices dictate that individuals should be aware of the data being collected about them and give explicit consent for its use in cybersecurity measures. Moreover, ensuring that automated systems anonymize or pseudonymize data wherever possible helps protect individual privacy rights.

Next, algorithms powering automated cybersecurity tools may inherit biases in the training data. This can lead to discriminatory outcomes, such as unfairly targeting certain demographics or disproportionately flagging specific groups as potential threats. Ethical considerations demand ongoing scrutiny of these algorithms to identify and mitigate biases and transparency in how decisions are made to ensure fairness and accountability.

While automation can expedite threat detection and response, humans must retain oversight to intervene in complex or ambiguous situations. Ethical cybersecurity practices require clear delineation of responsibilities between automated systems and human operators, along with mechanisms for human intervention when necessary. Moreover, establishing accountability frameworks is essential to address errors or failures in automated systems and ensure that appropriate corrective actions are taken.

Finally, ethical cybersecurity practices dictate that automated systems should be transparent in their operations and outcomes. This includes clearly explaining how automated decisions are reached and disclosing any limitations or uncertainties associated with these decisions. Building trust in automated cybersecurity tools requires transparency, accountability, and a commitment to continuously improving the reliability and effectiveness of these systems.

Summary

Automation has emerged as a transformative force within the dynamic and ever-changing cybersecurity landscape, revolutionizing the very foundations of security practices and redefining the boundaries of defense capabilities. With cyber threats growing increasingly sophisticated and pervasive, traditional manual approaches to cybersecurity have become inadequate in effectively countering these evolving risks. This is where automation enters the scene as a powerful protagonist, empowering organizations to protect their digital assets with unparalleled efficiency and precision proactively.

By harnessing the power of automation, organizations can unleash a new era of defense, where intelligent and adaptive technologies shield digital systems and data. This radical shift from manual to automated processes enhances the speed and accuracy of threat detection and bolsters the effectiveness of IR and mitigation. Automation acts as a formidable guardian, tirelessly monitoring vast amounts of data, swiftly identifying anomalies, and responding quickly to mitigate potential risks.

The impact of automation extends far beyond IR alone. It permeates every facet of cybersecurity, from proactive threat hunting to compliance management. With automation as an ally, cybersecurity professionals have an arsenal of tools and technologies that amplify their abilities, enabling them to focus on strategic initiatives, proactive defense, and developing robust security strategies.

Furthermore, automation stands as a sentinel against human error, minimizing risks associated with oversight, fatigue, or judgment lapses inherent to manual processes. By automating routine and repetitive tasks, organizations can eliminate the potential for human fallibility and ensure consistent and reliable execution of critical security operations.

This chapter has comprehensively explored the transformative force of automation within cybersecurity. We learned that automation is pivotal for modern cybersecurity practices, offering diverse tools and technologies that empower organizations to protect their digital assets efficiently. Real-world examples illustrated how automation enhances threat detection and IR, emphasizing its vital role as a guardian against evolving risks. Yet, we also delved into potential drawbacks and challenges, underscoring the importance of a balanced approach. Looking forward, we glimpsed the future of automation in cybersecurity, foreseeing its continued evolution and integration as organizations strive to maintain robust defense strategies in an ever-changing threat landscape, a landscape where cybersecurity data analytics will play a complementary role, harnessing the power of data to enhance protection and resilience further.

In the next chapter, we will investigate the critical role of cybersecurity data analytics and its essential importance in protecting against evolving cyber threats.

3
Cybersecurity Data Analytics

In this chapter, we delve deeply into the highly significant and rapidly evolving domain of cybersecurity data analytics, uncovering its pivotal role in bolstering defense against an incessantly morphing array of cyber risks that threaten the very fabric of digital security. As we embark on this exploration, we will traverse a comprehensive landscape encompassing various facets of this dynamic field. Our journey will examine the complex interaction between cybersecurity and data analytics, shedding light on the symbiotic relationship underpinning their efficacy in safeguarding against multifaceted challenges of the cyber realm.

In this chapter, we're going to cover the following main topics:

- AI in data analytics
- Types of AI used in cybersecurity data analytics
- Applications of AI
- Challenges of using AI
- The role of analysts
- The regulatory landscape

By the end of this chapter, you will have gained an understanding of the intricate interplay between data analytics and cybersecurity, recognizing how they combine forces to shield our digital world from an ever-evolving cyber threat landscape. We will have explored the pivotal role of AI in fortifying our defenses and dissected its various types and applications, all while acknowledging the complex challenges it presents. Moreover, you will appreciate the irreplaceable role played by analysts, whose expertise complements AI technologies, making our defenses more robust and adaptable. Lastly, we will navigate the regulatory landscape, emphasizing the need for a harmonious coexistence of technological innovation and ethical responsibility. Armed with this knowledge, you'll be better equipped to comprehend the nuanced realm of cybersecurity data analytics and contribute to ongoing efforts to safeguard the digital realm for future generations.

AI in data analytics

AI can meticulously scrutinize vast volumes of cybersecurity data, discerning intricate patterns and deviations that could signify potential security breaches. This data pool comprises network traffic, system logs, and user behavioral information. In this realm, AI is an invaluable asset to security teams, expediting identifying and mitigating threats. This expeditiousness is attributed to AI's unparalleled aptitude for swiftly parsing through expansive datasets—a feat beyond human capacity. Consequently, security teams are empowered to expedite threat detection and response, a dual-purpose achievement.

Furthermore, AI is an astute sentinel, adeptly uncovering threats that might otherwise evade human notice. It reveals hidden correlations and irregularities within data, a skill that often eludes human perception. By doing so, security teams are equipped with the means to unveil latent threats that might otherwise escape detection—an invaluable advantage.

AI's role in minimizing false positives is of equal significance—a bane that often burdens security teams. AI's deployment can significantly curtail the influx of erroneous alerts, allowing security analysts to concentrate their efforts on authenticating and addressing genuine threats. This strategic resource allocation optimizes security teams' investigative and responsive capacities.

Among the array of AI applications in cybersecurity data analytics, several stand out prominently. Foremost, AI proves its mettle in malware detection, pinpointing malicious elements within network traffic and system logs. Additionally, its prowess extends to phishing detection, effectively identifying fraudulent emails and deceptive websites. Equally noteworthy is AI's role in intrusion detection, capably discerning unauthorized access to critical networks and systems. In parallel, AI facilitates risk assessment, efficiently gauging the vulnerability to potential breaches. Finally, AI's automation capabilities shine in **incident response** (**IR**), streamlining tasks such as prioritizing alerts and delving into incident investigations.

In essence, the utilization of AI in cybersecurity data analytics epitomizes an era of heightened efficacy and accuracy. Its discerning eye, swift analysis, and automated capabilities collectively augment the prowess of security teams, ushering in a new frontier of safeguarding digital landscapes.

As we wrap up our discussion on AI in data analytics, it is clear that AI's pivotal role in cybersecurity data analysis is transforming the digital security landscape. The upcoming section will delve deeper into specific AI branches that power the defense against evolving cyber threats, exploring their unique applications and contributions to cybersecurity.

Types of AI used in cybersecurity data analytics

It is evident at this juncture that the integration of AI is swiftly cementing its status as an indispensable instrument within cybersecurity. AI's remarkable utility lies in its capacity to sift through voluminous datasets, discern patterns, and unveil threats that would prove arduous, if not impossible, for human agents to uncover. In the sphere of cybersecurity data analytics, three prominent branches of AI take the spotlight: **machine learning** (**ML**), **natural language processing** (**NLP**), and **deep learning** (**DL**).

ML, a subset of AI, endows computers with the ability to learn autonomously, unburdened by explicit programming. Having been trained on data, ML algorithms can subsequently leverage this knowledge to make predictions and informed decisions. In the domain of cybersecurity, ML plays a pivotal role in a myriad of tasks, encompassing the detection of malware, identification of phishing attempts, intrusion monitoring, and analysis of user behavior.

NLP, an integral facet of AI, focuses on the interplay between computers and human languages. Within cybersecurity, NLP is a stalwart ally in multifarious endeavors, including unearthing malicious code within text, discerning threats concealed within social media posts, and parsing through **threat intelligence** (**TI**) reports.

Finally, DL, a variant of ML, harnesses **artificial neural networks** (**ANNs**) to glean insights from data. Armed with this capability, DL algorithms are primed to unravel intricate data patterns, often tackling tasks that prove formidable for other AI modalities. Within cybersecurity, DL finds applications in diverse domains, such as image recognition, voice authentication, and malware detection.

The potency of AI as an augmentation to cybersecurity is undeniable. Organizations stand to expedite threat detection, bolster incident responsiveness, and fortify data protection through judicious employment. A repertoire of concrete illustrations highlights the manifold roles AI assumes in the arena of cybersecurity data analytics:

- **Malware detection**: AI-imbued malware detection tools meticulously scrutinize extensive data repositories to unveil telltale patterns synonymous with malicious code. This proactive approach empowers organizations to intercept and neutralize malware before its malevolent impact.

- **Phishing detection**: AI-infused phishing detection mechanisms dissect emails and websites with acumen, exposing phishing endeavors in their nascent stages. This proactive stance safeguards users, thwarting their susceptibility to phishing ploys.

- **Intrusion detection**: AI-powered **intrusion detection systems** (**IDS**) vigilantly oversee network traffic, promptly flagging anomalous activity. This vigilant surveillance equips organizations to forestall intrusions, averting potential havoc.

- **User behavior analysis**: AI-driven user behavior analysis tools meticulously chart user actions, discerning patterns that hint at nefarious intent. This vigilance extends to the identification and mitigation of insider threats.

In summary, AI emerges as a formidable arsenal poised to elevate cybersecurity to unprecedented heights. Its prowess lies in its capacity to harness data for discernment, enabling organizations to anticipate, counteract, and safeguard against a multifaceted array of threats. Having explored fundamental types of AI employed in cybersecurity data analytics, we now transition into a comprehensive examination of diverse applications of AI in this dynamic field.

Applications of AI

AI has emerged as a versatile asset, wielding its capabilities across a spectrum of cybersecurity data analytics tasks. AI's impact reverberates through the digital security landscape, from malware detection and phishing identification to intrusion monitoring, TI, and risk assessment. AI's prowess extends to task automation, heightening security awareness, and fortifying system integrity. As AI evolves, its potential to foster novel applications within cybersecurity only grows more pronounced.

Foremost among these tasks is malware detection—a cornerstone of cybersecurity. AI leverages code analysis, behavioral insights, and communication patterns to unveil the presence of malware, even anticipating emerging threats hitherto unseen. AI-driven malware detection employs ML to decode intricate patterns within malicious code, deftly spotting indicators of nefarious intent. Notably, this approach extends to discerning obfuscated or modified malware, evading traditional detection methods.

Equally pivotal is phishing detection, where AI lends its acumen. Phishing emails, engineered to trap users, fall under AI's vigilant gaze. By dissecting content, sender information, and recipient details, AI-powered solutions employ NLP to identify phishing cues such as specific keywords or phrases. Moreover, through ML, patterns within sender-recipient dynamics indicative of phishing tactics are discerned.

In the realm of intrusion detection, AI once again demonstrates its prowess. Unauthorized system or network access is promptly unveiled through AI's analysis of network traffic and system logs. Employing ML to identify patterns heralding intrusion attempts, AI also employs anomaly detection to pinpoint irregular activity signaling a potential breach.

Data collection and analysis of TI to safeguard organizations is another crucial application. AI's data-crunching capabilities enable it to assimilate threat data from diverse sources, uncovering nascent threats and trends. Utilizing ML, AI discovers patterns in TI data indicative of emergent perils, while NLP extracts pivotal insights from TI reports.

Furthermore, AI informs risk assessment by evaluating the likelihood and repercussions of a cybersecurity assault. Armed with ML, AI pinpoints markers signaling high-risk security incidents, enabling organizations to prioritize their defensive strategies. Meanwhile, NLP distills pivotal insights from security risk reports, facilitating well-informed risk mitigation decisions.

Beyond these spheres, AI is indispensable in various other cybersecurity domains:

- **Botnet detection**: AI scrutinizes network traffic, system logs, and user behavior to unmask networks of compromised computers—botnets—under the sway of malicious controllers. Employing ML, AI identifies botnet activity through network traffic patterns and deploys anomaly detection to pinpoint suspicious behavior.
- **Data breach detection**: AI scrutinizes data logs, system events, and user behavior to unearth unauthorized exposure of sensitive information. Using ML, AI unveils patterns in data logs indicative of a breach and detects anomalous activities suggesting potential breaches.

- **Vulnerability management**: AI automates vulnerability assessment by scanning, prioritizing, and suggesting remedies. Employing ML, AI unearths vulnerabilities in code, while NLP identifies vulnerabilities detailed in security reports.

- **Identity and access management (IAM)**: AI streamlines IAM by automating user account management, enforcing password protocols, and identifying unauthorized access. AI-driven IAM solutions use ML to spot suspicious login patterns and employ NLP to identify IAM-related keywords within security reports.

- **Security orchestration, automation, and response (SOAR)**: AI enhances SOAR, automating IR, threat hunting, and compliance tasks. By deploying ML, AI detects attack patterns within security events and harnesses NLP to extract pivotal insights from security reports.

As AI's evolution unfolds, the prospect of innovative applications in cybersecurity intensifies. AI stands poised to reshape our approach to safeguarding systems and data from the ever-evolving landscape of cyber threats. We will now navigate through the intricate landscape of these obstacles, ranging from data quality and quantity to adversarial attacks and continuous adaptation, shedding light on nuanced issues that organizations and cybersecurity practitioners encounter in harnessing the full potential of AI for defense against cyber threats.

Challenges of using AI

Implementing AI for security data analytics can be incredibly beneficial as it can assist organizations in detecting and responding to threats more effectively. However, it comes with various challenges that organizations need to address. Some of these challenges include the following:

- **Data quality and quantity**: AI models, especially DL ones, require large amounts of high-quality data to perform effectively. Obtaining sufficient data that accurately represents various cyber threats can be challenging in cybersecurity. Additionally, the data needs to be well labeled and up to date to ensure the AI model's accuracy and reliability.

- **Imbalanced data**: Cybersecurity datasets often suffer from class imbalance, where malicious activities are much less frequent than normal activities. This can lead to biased models that perform well on normal data but struggle to detect rare threats. Addressing class imbalance is crucial to avoid false negatives.

- **Model bias**: Bias can emerge in AI models, leading to unfair or inaccurate predictions. In cybersecurity, biased models might disproportionately flag certain user groups or ignore specific threat vectors. This bias can result from biased training data or the algorithm's design, and it must be actively mitigated to ensure equitable and effective threat detection.

- **Adversarial attacks**: Cyber attackers can potentially manipulate AI models by feeding them data designed to exploit vulnerabilities. Adversarial attacks can lead to AI systems misclassifying malicious inputs as benign, undermining security measures.

- **Expertise gap**: Developing and deploying AI models for cybersecurity requires a specialized skill set combining domain expertise in cybersecurity with ML and AI techniques. Finding and retaining professionals with this hybrid skill set can be a challenge.
- **Interpretability and explainability**: Many AI models, especially **deep neural networks** (**DNNs**), are often considered black boxes, making it difficult to understand why they make specific decisions. In cybersecurity, it is essential to have transparent and interpretable models so that security analysts can trust and act on the model's findings.
- **False positives and false negatives**: AI models can produce false positives (flagging benign activities as threats) and false negatives (failing to detect actual threats). Achieving the right balance between these two metrics is critical to avoid overwhelming security teams with false alarms while ensuring that genuine threats are not overlooked.
- **Scalability and performance**: Cybersecurity systems operate in real time, requiring AI models to analyze vast amounts of data quickly. Ensuring that the AI system can scale to handle the volume of data and still maintain high performance is a significant technical challenge.
- **Regulatory and compliance concerns**: Depending on the jurisdiction and industry, the use of AI for cybersecurity might be subject to various regulatory frameworks. Organizations must ensure that their AI solutions meet these compliance requirements while enhancing security.
- **Continuous adaptation**: Cyber threats constantly evolve, requiring AI models to adapt and learn from new data. Implementing a system that can effectively learn and evolve is essential to staying ahead of emerging threats.

In conclusion, while AI has the potential to revolutionize cybersecurity data analytics, organizations must be aware of these challenges and develop strategies to address them effectively. This often involves a combination of robust data management practices, algorithmic innovations, and collaboration between cybersecurity experts and AI practitioners. We will now examine the importance of human analysts in this ecosystem.

The role of analysts

The indispensable role of human analysts within AI-powered cybersecurity data analytics cannot be overstated. While AI models excel at detecting patterns and anomalies, their comprehension of contextual nuances remains limited. Human analysts bridge this gap, interpreting AI results and making informed decisions regarding threat responses.

In tandem, human analysts are instrumental in instilling trust and confidence in AI-driven solutions. AI models can be very complex, making it difficult for people to understand how they work. Human analysts can help build trust and confidence in AI-powered solutions by explaining how the models work and by demonstrating the value of the models.

Moreover, human analysts are guardians of ethical and responsible AI applications. AI models can be used maliciously, so ensuring they are used ethically and responsibly is crucial. Human analysts can help ensure that AI-powered solutions are used ethically and responsibly by setting policies and procedures and monitoring the use of models.

In addition, human analysts need to define the exact problem that AI will solve. This includes understanding the organization's security goals, the types of threats that the organization faces, and the available data. Furthermore, human analysts steer AI model training, curating training data and monitoring model performance. Effective communication of AI results to stakeholders—security engineers, incident responders, and executives—also falls within their purview.

As AI advances, the role of human analysts will ascend in significance. While AI models mature, human analysts will remain vital, offering context, judgment, trust building, ethical stewardship, creativity, and adaptability. This multifaceted role encompasses the following:

- **Creative and innovative insights**: Human analysts infuse creativity and innovation, a trait AI models often lack. They conceptualize novel ways to leverage AI for enhancing cybersecurity, devising ingenious solutions beyond mere rule-based instructions.
- **Adaptability and resilience**: In an ever-evolving cybersecurity landscape, AI models' agility relies on human analysts' adaptability. Their capacity to navigate dynamic threats ensures AI solutions remain relevant and responsive.

In conclusion, human analysts are the cornerstone of AI-powered cybersecurity data analytics. Their contributions span the realms of context enrichment, trust establishment, ethical supervision, creative thinking, adaptability, and resilience. As AI evolves, its role remains pivotal, securing a future where human-machine collaboration fortifies our digital defenses.

Let's now take a look at the regulatory landscape.

The regulatory landscape

The regulatory landscape for AI in cybersecurity data analytics is still evolving. As AI-powered solutions become more sophisticated, regulators must develop new regulations to address challenges associated with this technology.

Data privacy is one of the most critical regulatory challenges associated with AI in cybersecurity. AI-powered solutions often collect and analyze large amounts of data, which raises concerns about privacy and data protection. For example, AI-powered malware detection solutions may collect data about users' browsing behavior, which could be used to track users' online activity.

To address this challenge, regulators are developing **new regulations** requiring organizations to obtain user consent before collecting their data. They also require organizations to implement appropriate technical and organizational measures to protect the privacy of users' data.

Another important regulatory challenge is **transparency**. AI-powered solutions often make decisions that are opaque to humans, which raises concerns about transparency and accountability. For example, an AI-powered IDS may decide whether to block a network connection. Still, it may not be clear why the decision was made.

To address this challenge, regulators require organizations to be transparent about how they use AI. They also require organizations to explain the rationale behind their decisions so that users can understand why they were made.

Bias is also a crucial regulatory challenge. AI-powered solutions can be biased, which means that they may make decisions that are unfair or discriminatory. For example, AI-powered malware detection solutions may more likely flag files associated with certain minority groups.

To address this challenge, regulators require organizations to mitigate bias in their AI systems. They also require organizations to monitor their AI systems for signs of bias so that they can take corrective action if necessary.

Finally, **regulatory compliance** is another important regulatory challenge associated with AI in cybersecurity. AI-powered solutions must comply with various regulations, such as the **General Data Protection Regulation (GDPR)** and the **California Consumer Privacy Act (CCPA)**. Failure to comply with these regulations can result in significant fines and penalties.

To address this challenge, organizations must ensure that their AI systems comply with the relevant regulations. They also need to have a process in place for monitoring their AI systems for compliance issues so that they can take corrective action if necessary.

The regulatory landscape for AI in cybersecurity data analytics is still evolving. As AI-powered solutions become more sophisticated, regulators need to develop new regulations to address the challenges associated with this technology. Organizations using AI in cybersecurity need to stay up to date on the latest regulatory developments to ensure compliance with the relevant regulations and use AI responsibly.

In addition to the aforementioned regulatory challenges, a number of other ethical considerations need to be taken into account when using AI in cybersecurity. For example, organizations need to be careful not to use AI in a way that violates human rights or discriminates against certain groups of people. They also need to be transparent about how they utilize AI so that users can understand how their data is used.

The use of AI in cybersecurity is a powerful tool that can help protect organizations from cyber-attacks. However, it is essential to operate AI responsibly and ethically. By understanding the regulatory landscape and the ethical considerations involved, organizations can use AI to improve their cybersecurity posture while protecting their users' rights.

Summary

This chapter underscored the pivotal role of cybersecurity data analytics and its paramount importance in defending against evolving cyber threats. It revealed the symbiotic partnership between cybersecurity and data analytics, powered by integrating AI technologies. AI's transformative influence is showcased, enabling proactive cyber risk detection, mitigation, and prevention through advanced algorithms, ML, and DL.

The chapter explored AI's multifaceted applications in cybersecurity, encompassing anomaly detection, behavior analysis, threat prediction, and IR. It acknowledged challenges, such as AI model biases and integration complexities, emphasizing the need for collaborative efforts between AI and human analysts, combining algorithmic precision with human judgment.

Furthermore, it addressed the complex regulatory landscape governing data privacy and compliance in cybersecurity data analytics, emphasizing ethical considerations organizations must navigate.

In essence, the convergence of cybersecurity and data analytics, bolstered by AI, offers hope in the battle against cyber adversaries. This chapter not only unveiled their synergistic potential but also stressed the importance of continuous evolution, collaboration, and vigilance in the face of ever-changing cyber threats. Embracing opportunities while addressing challenges allows organizations to steer toward a more resilient and secure digital future.

Next, we'll dive deeper into the fascinating world of AI.

Part 2: AI and Where It Fits In

This part provides you with the required AI background to be able to recognize opportunities to apply AI in cybersecurity. Furthermore, this part will enable you to understand the AI methods and know where different methods are applicable. Lastly, we describe a typical AI workflow and give some examples of where AI workflows are applied in cybersecurity scenarios.

This part has the following chapters:

- *Chapter 4, AI, Machine Learning, and Statistics – A Taxonomy*
- *Chapter 5, AI Problems and Methods*
- *Chapter 6, Workflow, Tools, and Libraries in AI Projects*

4

AI, Machine Learning, and Statistics - A Taxonomy

In the current literature and in media articles, there is often a misrepresentation of the meanings of the terms **artificial intelligence** (**AI**) and **machine learning** (**ML**). These terms are often used interchangeably and as synonyms. However, in this chapter, we differentiate between AI and ML. We define AI as a capability, and ML is defined as a set of techniques that can enable us to achieve some form of AI. For many readers and practitioners, it might also be interesting to reiterate the difference between ML and statistics, more specifically, statistical machine learning. Although it is not required to master the technical details of all these definitions, we believe you could benefit a lot from understanding the key aspects of the disciplines, which will certainly help you with the rest of the book.

In this chapter, we will cover the following topics:

- A brief introduction to AI history
- The relation to statistics (statistical machine learning)
- ML — Taxonomy
- Deep learning and its recent advances
- The limitations of ML and its security concerns

Technical requirements

Although there are no hard requirements for any technical skills, it is generally beneficial for you to have a basic understanding of modern engineering concepts. Previous chapters should prepare you with the knowledge required to follow this chapter.

A brief introduction to AI history

In classic AI literature, there are broadly eight definitions of AI, and we can categorize them into four categories based on the named properties of the AI systems:

- Thinking humanly
- Acting humanly
- Thinking rationally
- Acting rationally

AI systems are expected to perform functions that humans can perform, but also to act and think rationally, thereby exhibiting intelligent behavior. On the other hand, ML is a set of techniques that enables AI. The techniques of ML are based on designing methods and systems that use data to learn how to perform tasks that exhibit AI. Apart from ML, many other methods contribute to AI, such as expert systems and fuzzy logic. We show a very simple illustration of their relationship in *Figure 4.1*. By no means is this comprehensive, but we want to ensure you're aware that AI is a top-level concept that contains many other domains. This book focuses on ML techniques under AI, and how they are applied in cybersecurity. Let's start by introducing the history of AI briefly, and then give you a taxonomy system to navigate the subject of ML.

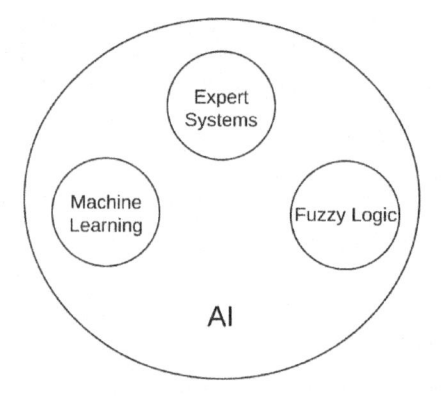

Figure 4.1 – A simple illustration of AI taxonomy

The modern history of AI can be traced back to the Dartmouth conference in 1956, where a group of leading researchers in this field coined the term "artificial intelligence". Prior to the workshop, Alan Turing wrote a thesis in 1950 that laid the groundwork for the birth of AI. Turing's thesis explored the idea of whether machines could be made to "think" like humans, and he proposed a test (now known as the Turing test) to determine whether a machine could exhibit intelligent behavior indistinguishable from that of a human. This idea was a major inspiration for the researchers who attended the Dartmouth conference, and it set the stage for the development of AI as a field of study. Over the years, the field of

AI has grown exponentially, and today, it is a rapidly evolving and exciting field that holds tremendous promise for the future. While the formal definition of AI can be somewhat contentious, it generally involves the development of algorithms and approaches that simulate human thought processes and decision-making in order to create intelligent machines. The goal of modern AI is to create software or hardware that can think, learn, and reason like humans, and that can perform a wide range of tasks in a variety of domains.

During the early days of AI research, **symbolic processing** and **logic programming** were the focus of attention. Symbolic processing represents knowledge in terms of symbols and rules and uses logical reasoning to manipulate those symbols in order to solve problems. Logic programming, on the other hand, involves defining a set of logical rules and using them to derive new knowledge from existing knowledge. While symbolic processing and logic programming were successful in solving certain types of problems, they had limitations. One of the main limitations of symbolic processing was its inability to handle uncertainty and incomplete information. This led to the development of **probabilistic reasoning**, which is the basis of many modern ML algorithms. Another limitation of symbolic processing was its inability to learn from experience, which led to the development of connectionism and neural networks, which later formed the essence of **deep learning** (**DL**). Nevertheless, symbolic processing and logic programming remain important areas in early research of AI, and they continue to be used in various applications.

Despite the early successes, the first AI winter occurred in the late 1970s and early 1980s. This was a period of reduced funding and interest in AI research, due to a lack of significant progress in the field. However, with the introduction of connectionism and expert systems, AI research experienced a resurgence in the 1980s, leading to the second AI boom. Connectionism involves the use of **artificial neural networks** (**ANNs**) to simulate the functioning of the human brain, while expert systems are computer programs that are designed to mimic the decision-making ability of a human expert in a particular field. With the dramatic development of personal computing devices, the second AI winter occurred in the late 1980s and early 1990s. At that time, PC and Mac machines were overtaking the world, which made expert systems and other commercial AI systems less practical. After late 1990, there was a shift towards neural networks and probabilistic reasoning, which was introduced by Jude Pearl. These developments paved the way for the rise of ML and DL, which are the primary focus of this chapter. ML is a method of teaching computers to learn from data, without being explicitly programmed. On the other hand, DL is a subset of ML that involves training deep layered neural networks to solve more complex tasks. As ML and DL have advanced, more and more mathematical tools have been formally introduced into this area. These tools include probability theory, linear algebra, calculus, optimization, and information theory. Learning algorithms use these mathematical tools to identify patterns and relationships within data and create models that can be used for prediction and decision-making.

The following figure summarizes the timeline of the development of the field of AI.

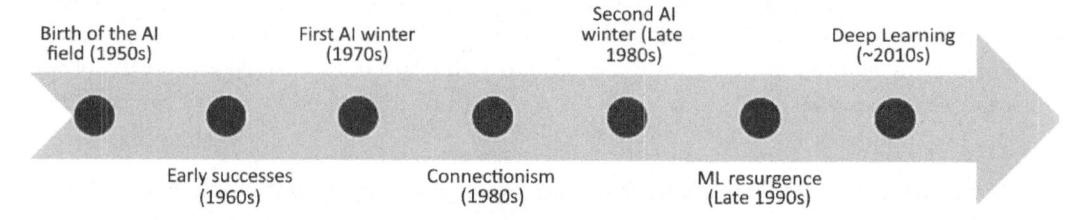

Figure 4.2 – Timeline of the development of AI

In recent years, ML and DL have revolutionized the field of AI, leading to significant advances in computer vision, speech recognition, natural language processing, and other areas. For instance, ML algorithms have been used to develop self-driving cars, predict consumer behavior, and improve healthcare outcomes. In finance, ML is being used to detect fraud, predict market trends, and optimize investment portfolios. Furthermore, in the field of cybersecurity, AI is being used to combat adversarial threats by simulating various defensive techniques, typically conducted by security experts. However, there are still challenges that need to be addressed, including ethical and societal implications.

As AI becomes more advanced and ubiquitous, it raises important ethical questions related to privacy, security, bias, and accountability. It is important for individuals, organizations, and governments to consider these issues and work together to ensure that AI is developed and used in a responsible and ethical manner. Note that AI itself is also vulnerable to certain adversarial attacks that change the behavior of AI systems to produce harmful outcomes. It is therefore important to consider these vulnerabilities and develop strategies to mitigate them. This includes developing robust and secure AI systems, as well as developing methods for detecting and responding to adversarial attacks.

In conclusion, the history of AI has been characterized by a series of breakthroughs and advances in various technologies and approaches. The recent advancements in AI have been closely associated with the increase in computing power and the availability of big data technologies. These have enabled AI algorithms to process and analyze vast amounts of data quickly and efficiently, leading to significant improvements in the performance of AI systems. Furthermore, as computing power continues to increase and big data technologies become more sophisticated, we can expect even more exciting developments in the field of AI in the years to come. While there have been challenges and setbacks along the way, the field of AI is making rapid progress and holds great promise for the future. With continued research, innovation, and collaboration, we can create intelligent machines that can help us solve some of the world's most challenging problems and improve our lives tremendously.

The relation to statistical learning theory

We have seen a brief history of how AI has developed and the modern rise of ML as the most promising approach to many applications. Apparently, the technology shift isn't an isolated event; it is being realized along with the advance of computing technologies, such as high-performance computing and big data technology. The mathematical foundation of ML algorithms utilizes a heavy implementation of modern statistics. For a long time, statisticians thought computer scientists were reinventing the wheel, while computer scientists were seeking more practical realization of statistics-driven methods. The truth is that statistics is studying the properties of data generation processes. However, ML is a relatively new paradigm that is quite the opposite: that is, it studies the data generation process given some observations. In other words, they are closely related and complementary to each other. In *Figure 4.2*, we relate statistics and ML together to give readers a general idea of them.

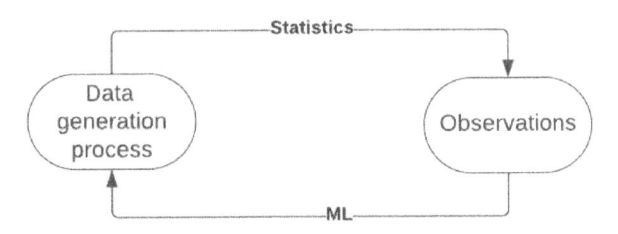

Figure 4.3 – Connection between statistics and ML

Let's elaborate a bit more on the key differences. In the real world, there are numerous data generation processes that generate samples to form certain populations. Statisticians are often interested in establishing a mathematical language to describe the properties of the population. For instance, expectation, variance, and correlation are examples of descriptive statistics that describe properties of the data generation process. By formal verification of these properties, we can test different hypotheses and derive outcomes that are of interest. On the other hand, ML concentrates on learning or estimating the data generation process from observations, and utilizing it for various task completion. A common interest of both statistics and ML is inference or statistical inference; that is, given a data generation process, we draw conclusions from it with a certain level of reasoning. The key to inference is understanding why we get certain outcomes, rather than estimating the outcome values. This led to the prominence of the area called statistical learning theory or computational learning theory in the modern age of computing. The whole idea of it is that how closely we can approximate the ground truth, which translates into the aforementioned concept of data generation process. This is how accurately we can build a model or function that gives promising inference. In the 1990s and 2000s, statistical learning became the core paradigm of ML. You will learn about the algorithms that fall under the category in the next chapters, such as logistic regression and **support vector machines** (**SVMs**). Deep learning has begun to overtake statistical learning in almost every line of AI research; nevertheless, statistical learning still finds wide applications in the industry thanks to its theoretical soundness, which is yet to be established in the DL community.

It can be challenging for readers to fully grasp the difference between them; however, it is useful to pull them together to clarify some common confusions in the AI and ML communities. Next, we will look at a systematic description of AI and ML that will help you understand the chapters in the rest of the book.

ML – classifying taxonomy

In this section, we provide a taxonomy of ML and the basic framework of an ML system. There are various ways to define the taxonomy of ML. Broadly speaking, we can define the taxonomy as follows:

- How is the ML model trained?
- What is the learning objective?
- What to expect during model inference
- Model modality

This is a general classification of ML that is widely accepted in the community. It should be noted that there may be other taxonomies available as well. However, this should be sufficient for you to navigate all different ML algorithms.

By learning schema

Depending on how we train a model, we can place it in one of four main categories: supervised learning, unsupervised learning, semi-supervised learning, and reinforcement learning, each with its own unique set of algorithms and applications.

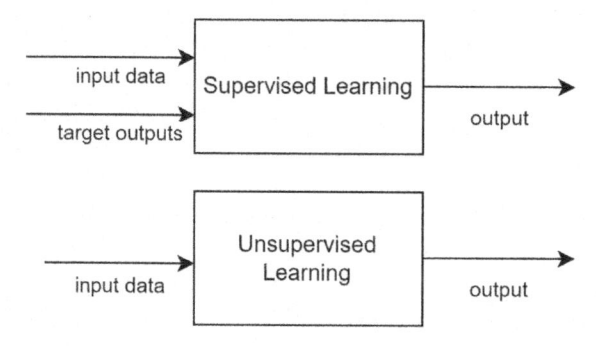

Figure 4.4 – Difference between supervised learning and unsupervised learning

Supervised learning

Supervised learning algorithms are trained on labeled data and are typically used for classification and regression tasks. Labels are considered the ground truth that guides the model to learn a meaningful pattern between many input and output pairs. Examples of supervised learning algorithms are decision trees and SVMs. A decision tree is a model that uses a tree-like structure to represent decisions and their possible consequences. Its variants, such as random forest and XGBoost, are popular algorithms for various applications, especially when the data is in a tabular format. On the other hand, SVMs are a family of learning algorithms that can be used for classification, regression, and even anomaly detection. The main idea behind SVMs is to find a **hyperplane** that best separates the data into different classes. They are also commonly used in many applications.

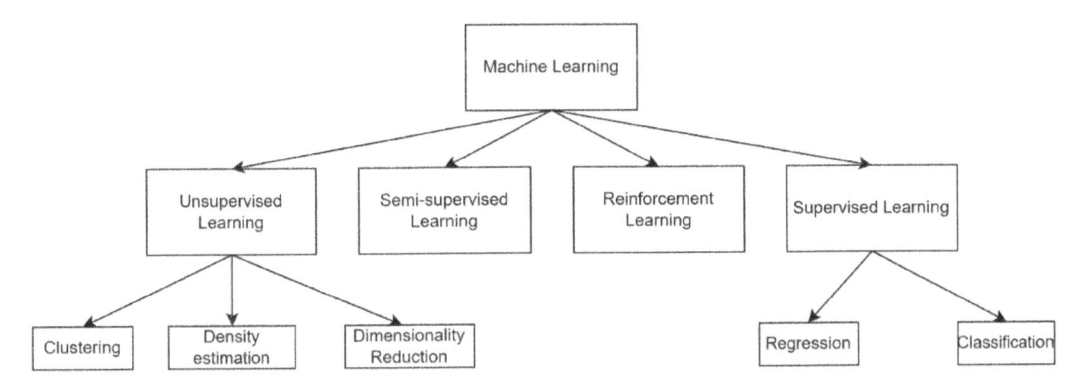

Figure 4.5 – Taxonomy of ML by learning schema

Unsupervised learning

Unsupervised learning algorithms are trained on unlabeled data and are typically used for clustering and dimensionality reduction tasks. This means the data does not provide a ground truth, and there is no explicit guidance on how to train the model. Most unsupervised learning algorithms involve a definition of criteria that can be used as a learning objective, for instance, the probability density of input or the geometric distance of pairs of data points.

Examples of unsupervised learning algorithms are K-means clustering, **Principal Component Analysis** (**PCA**), and autoencoders. K-means clustering is an algorithm that partitions data into K clusters based on their geometric similarity. PCA is a popular algorithm that reduces the dimensionality of data while preserving its variance. It is also commonly used in data visualization and feature extraction, or as an initialization step for other complex models. **Generative adversarial networks** (**GANs**) consist of a generator and discriminator, which work against each other to learn how to generate data that cannot be distinguished from the ground truth. It is notable that the key difference between supervised and unsupervised learning is solely whether we have the ground truth in the training data.

Semi-supervised learning

Semi-supervised learning algorithms are trained on both labeled and unlabeled data and are typically used when labeled data is scarce or expensive to obtain. In practice, usually we do not have enough labeled data, therefore semi-supervised learning has practical benefits. Examples of semi-supervised learning algorithms are self-training and co-training, where self-training is a simple and effective algorithm that involves training a model on the labeled data and using it to label the unlabeled data. The model is then retrained on the combined labeled and pseudo-labeled data. Co-training is an algorithm that involves training two models on different subsets of the data and using each model to label the other's unlabeled data. This approach is commonly used in text mining tasks.

Reinforcement learning

The last one, reinforcement learning, has gained popularity in recent years. It has got attention from both research and industry, as it learns toward intelligent agents that interact with a given or known environment where agents are trained to make decisions based on rewards and punishments. Many real-life scenarios involve dynamic environments, such as autonomous driving, robotics, and gaming. A typical reinforcement learning setting can be defined by a tuple of agent, reward function, and state-action space. Depending on the use case, there may also be a transition matrix advising the actions of the agent. Reinforcement learning algorithms can be used to optimize complex systems with a large number of possible actions and states, and can learn to make decisions based on long-term rewards rather than just short-term gains. Furthermore, reinforcement learning algorithms can be used to learn from experience and adapt to new situations, making them ideal for applications where the environment is constantly changing.

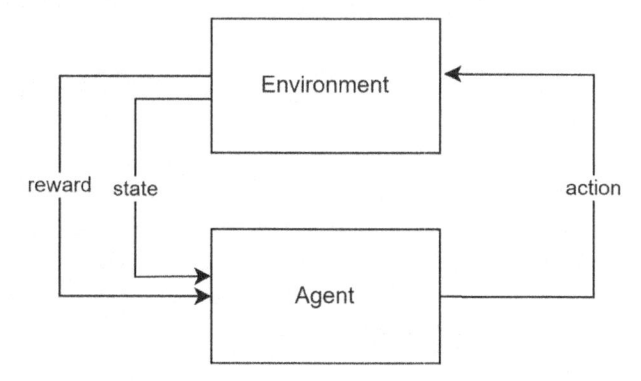

Figure 4.6 – A representation of reinforcement learning

By learning objectives

Besides the learning schema, ML algorithms can also be categorized based on the learning objectives. In the following sections, we list the taxonomy by the widely recognized learning objectives that will cover most of the learning tasks in practice.

Classification

The goal of classification is to assign input data points to one of several predefined classes. Here are some example use cases:

- Recognition of handwritten digits (classification of images into one of the 10 digits)
- Classification of malware into malware families
- Attribution of attacks to different threat actors
- Spam detection – classification of e-mails into spam/not spam

Regression

When the goal is to predict a continuous output variable based on one or more input variables, we say the learning task is regression. Here are some example use cases:

- Prediction of house prices based on location, size, and market trends
- Estimation of the vulnerability score of software based on code properties
- Prediction of the number of cyberattacks based on the security posture of the organization

Clustering

The goal of clustering is to group similar data points together based on their similarity or distance. Here are some example use cases:

- Clustering Netflix users based on their watching habits
- Malware clustering to discover new malware groupings
- Clustering phishing messages to determine groups of similar phishing attempts

Anomaly detection

For many cybersecurity use cases, anomaly detection plays a central role. As the name suggests, anomaly detection seeks to identify data points that are significantly different from the rest of the data, although the minority is not necessarily malicious. However, the consequence of misclassification might be very different for positive and negative samples. Many clustering or classification algorithms can be used in anomaly detection as well, such as one-class SVM, DBSCAN, and Local Outlier Factor.

Broadly speaking, anomaly detection can also be supervised, unsupervised, or semi-supervised. In supervised anomaly detection, we have target labels from majority and minority classes, based on which we can use ML methods for classification with some additional tweaks to account for class imbalance. In the unsupervised methods, we use either clustering or some other dedicated methods to discover regular patterns in the data and find anomalies, which are observations that don't conform to that pattern. Semi-supervised learning methods are employed if only part of the data is labeled, and the methods applied in this case are a combination of supervised and unsupervised learning approaches. Here are some examples of anomaly detection use cases:

- Detecting faults in industrial control systems
- Determination of user account takeover in web applications
- Detecting lateral movement or command and control techniques employed by attackers in enterprise networks

Generative models

Generative AI has gained significant attention. It differs from the previously most common learning objectives, where learning models only generate atomic output. Generative models are used to learn the underlying distribution of the data and generate content that conforms to the original data. With the popularity of transformer networks, generative approaches are progressing at an unprecedented pace. Some examples of generative models are **Variational Autoencoders** (**VAEs**), **GANs**, and the most recent **large language models** (**LLMs**), such as Llama and GPT-4. Here are some example use cases:

- Generation of conversational text as answers to questions by chatbots
- Image generation based on textual description
- Malware code generation based on desired characteristics

These are the most common learning objectives you need to learn in practice. As you should have noticed, the learning objectives focus on the type of model outputs. It is usually straightforward to formulate the learning task in any one of these categories.

By model modality

Model modality refers to the type of input data that an ML algorithm uses to make predictions or decisions. The input data can be of various modalities, such as text, image, audio, video, or a combination of these modalities. Recently, multimodal models have gained significant attention. They use multiple modalities to improve performance on a given task. We'd like to introduce the taxonomy of ML methods based on the model modality. Before we dive into that, let's first introduce various data formats that relate to ML models, since ML methods are often designed for specific types of data so that the models can exploit the inherent data structure and capture the prior knowledge about how the data is generated. This enables easier and more effective recognition of patterns during the training of machine learning models. Furthermore, it enables the transfer of empirical knowledge in application areas that share common data types, such as vision and natural language.

Data types

In this section, we will list some data types.

Sequential data

Sequential data is a type of data where there is a sequential dependency between data points, that is, the probability of a value of one data point (x_i depends on the values of previously generated data points (x_{i-k}, $0 < k < i$). Examples of this data are DNA sequences, sequences of characters in text, and time series data.

Time series data

Time series data is a special type of sequential data where the sequence of data points has a time dependency. Each data point is associated with an absolute or relative timestamp, and the machine learning methods used on this type of data need to take account of the time context applied to this data. The patterns in this data that the model should learn are time-dependent. Some examples of this type of data are EEG scans, video data, meteorological measurements, and vehicle sensors.

Graph data

ML is often applied to learn structures in graphical data, such as social networks, computer networks, and website structure graphs. This data is represented with nodes and edges that describe the relationships between nodes. The goal of the ML methods is to observe the relationship between nodes and find patterns in their structure. ML methods can be used to classify graphs, classify nodes in graphs, find clusters of nodes, or find anomalies in graph data. The most famous example is finding structure in social networks, which helps us discover groups of users with similar interests or connections.

Textual data

Textual data assumes that the data is structured in a hierarchy, where words are structured in sentences according to a certain grammar, and sentences are structured in documents following the writing style of the text author. Some ML methods also work at the granularity level of characters, which are combined into words. With these methods, we often take this structure into account either by defining features based on discovering the grammatical facts associated with words or sentences (e.g., finding nouns, verbs, etc.), by preprocessing the data to remove the words or characters that don't contribute towards the accuracy of the model (e.g., word stemming), or by adding constraints on ML models to explicitly encode grammar (e.g., syntax trees). This enables us to have more accurate models for textual data, for example, for text classification, author recognition, and e-mail spam detection.

Image data

There are multiple applications of ML in processing image data, such as image classification, image generation, and object recognition. These applications exploit the structure of image data, such as the hierarchy of objects in an image. For instance, **convolutional neural networks** (**CNNs**) are inspired by convolutional filters for image processing that were traditionally used to detect object edges in images. The ML methods for video data combine characteristics of ML methods for images with ML methods for sequence data.

Now, let's learn the taxonomy based on model modality.

Single-modal models

Single-modal models are those that use a single modality as input data. For instance, a **natural language processing** (**NLP**) model that takes in only text data is a single-modal model. Other examples are image classification models that take in only image data and audio classification models that take in only audio data. Most ML models are single-modal models that only accept one type of input, that is, tabular data, images, text, or audio, and they only work in one particular domain. It is only recently that researchers have progressed with multi-modal models that show some early evidence of artificial general intelligence.

Multi-modal models

Multi-modal models, on the other hand, are models that use multiple modalities as input data. For instance, a model that takes in both text and image data to classify an image is a multi-modal model. Another example is a model that takes in both text and audio data to perform speech recognition. Multi-modal models have shown remarkable improvements in performance on various tasks, such as image captioning, visual question answering, and speech recognition. Some examples of multi-modal models are vision-language models, such as ViLBERT and LXMERT, and text-image-audio models, such as MMT.

Overall, the taxonomy of ML algorithms based on model modality is a relatively new one, but it has shown great potential in improving the performance of ML models. It is reasonable to believe that as more data becomes available across different modalities, there will be significant improvement in the field.

The following table disambiguates the classical ML methods:

Method	Category	Data	Purpose
Clustering	Unsupervised learning	Unlabeled	Grouping data in a set of clusters
Density estimation	Unsupervised learning	Unlabeled	Estimating probability density based on data
Dimensionality reduction	Unsupervised learning	Unlabeled	Reducing the dimensionality of data (compression)
Classification	Supervised learning	Labeled with discrete labels	Classifying data based on the model
Regression	Supervised learning	Labeled with continuous labels	Making numerical predictions
Anomaly detection	Supervised or unsupervised learning	Can be labeled or unlabeled	Detecting outliers in data

Table 4.1 – Classical ML methods

So far, we have introduced different ways of categorizing ML algorithms. In the next section, we will shift our focus to DL. Note that DL is a subset of ML; the taxonomy is applicable to DL as well. However, DL is much more powerful than traditional ML models in terms of predictive capability and its flexibility in modeling complex tasks. It has found its applications in some of the most challenging scenarios.

DL and its recent advances

DL is a subset of ML that has gained significant attention in recent years due to its state-of-the-art performance on many complex tasks, such as image recognition, machine translation, and speech recognition. While traditional ML algorithms are limited by their inability to learn from vast amounts of data, DL algorithms can identify complex patterns and relationships within data by using deep neural networks, which consist of multiple layers of interconnected nodes that can learn to recognize increasingly abstract features.

The key difference between DL and traditional ML algorithms is the ability of deep neural networks to extract high-level features from raw input data. For example, in image recognition tasks, deep neural networks can learn to recognize edges, shapes, and textures in an image by processing the raw pixel data through a series of convolutional layers. Similarly, in NLP tasks, deep neural networks can learn to generate meaningful representations of words and phrases by processing the raw text data through layers of **recurrent neural networks** (**RNNs**).

The core training process of DL algorithms is via the **backpropagation algorithm**, which is used to adjust the weights and biases of the neural network during training. Backpropagation works by computing the gradient of the cost function with respect to each weight and bias in the network and using this gradient to update the weights and biases in the direction that minimizes the cost function. By iteratively adjusting the weights and biases in this way, the neural network can learn to recognize complex patterns and relationships within the data. In addition to backpropagation, the development of software libraries such as Autograd has also greatly accelerated the development of DL. Autograd is a powerful tool that automatically computes the gradients of numerical functions, which is extremely useful in DL because it enables the automatic differentiation of complex functions. This allows developers to focus on designing and implementing models, rather than manually computing gradients for each individual component of the model. As a result, DL has become more accessible and easier to develop with the help of software libraries such as Autograd. Besides, the success of DL is also largely due to the availability of computing power and big data. With the rise of cloud computing and the availability of powerful **graphics processing units** (**GPUs**), DL can now be trained on massive amounts of data in a relatively short amount of time. Furthermore, big data technologies such as Apache Spark and Hadoop have made it easier to store, process, and analyze large-scale datasets, which is essential for training DL algorithms.

Nowadays, DL algorithms are used to achieve state-of-the-art performance on many complex learning tasks. For example, in computer vision, CNNs have been used to achieve human-level performance on tasks such as image classification, object detection, and facial recognition. In natural language processing, RNNs have been used to achieve state-of-the-art performance on tasks such as language modeling, machine translation, and text classification. In speech recognition, deep neural networks have been used to achieve significant improvements in accuracy and speed, making it possible to develop real-time speech recognition systems. One recent development in DL is the transformer architecture, which has revolutionized the field of NLP. The transformer architecture was introduced in the seminal paper *Attention Is All You Need* by Vaswani et al. in 2017, and it has since become the standard architecture for many NLP tasks. The transformer architecture is based on the concept of self-attention, which allows the model to weigh the importance of different input tokens when generating output tokens. This attention mechanism has been shown to be highly effective for tasks such as language modeling, machine translation, and text summarization. DL algorithms have also been used in cybersecurity to detect and prevent cyberattacks. For example, deep neural networks have been used to detect malware and phishing emails, while deep reinforcement learning algorithms have been used to develop adaptive security policies that can respond to changing threats. Furthermore, DL algorithms have been used to analyze network traffic and detect anomalies that may indicate a security breach.

Overall, DL is a rapidly evolving field that holds tremendous promise for the future of AI. With the availability of computing power and big data, DL algorithms have achieved state-of-the-art performance on many complex tasks, and they are likely to play an increasingly important role in many areas of our lives, from healthcare to finance to transportation. While there are still challenges to be addressed, such as ethical and societal implications, the future of DL is bright, and continued research and innovation will be essential for unlocking its full potential.

In the rest of the book, you will learn more about DL, such as some widely used networks and how they can be implemented and applied in the cybersecurity domain. You will learn about their limitations and some of their security concerns in the rest of this chapter.

The limitation and security concern

As AI becomes more advanced and ubiquitous, there are growing concerns about its security and safety. These concerns include issues related to hallucination, privacy leakage, intellectual property ownership, ethics, bias, fairness, social impact, and adversarial attacks, among others. In this section, you will learn the basics of those issues and concerns; we will cover them in more depth in the rest of this book.

Hallucination

One of the main concerns about AI is the potential for hallucination. This refers to situations where AI systems generate output that is not based on reality or does not reflect the true nature of the data they are processing. For example, an AI system trained on images of dogs may generate images of dogs with six legs or blue fur, which are not reflective of reality. This can be particularly problematic in applications where AI systems are used to make critical decisions, such as in healthcare or finance. For example, an AI system could generate the wrong treatment recommendation that a user might take as credible advice. Furthermore, it could give advice about investing money to a user that would cause financial losses. It heavily relies on users who already have the domain knowledge to determine whether the generated content is legitimate, or otherwise it could lead to harmful consequences.

Privacy leakage

Another concern related to AI is privacy leakage. As AI systems are trained on large amounts of data, there is a risk that sensitive information may be leaked or exposed. For example, an AI system trained on medical records may inadvertently reveal personal information about patients, such as their medical history or genetic information. This can have serious implications for privacy and can lead to ethical concerns. There is also research that provides evidence that a generative model is able to recover social security numbers or passwords, given a few hints.

Intellectual property ownership

Foundation models have become increasingly prevalent in the field of AI; however, the use of foundation models raises important questions about intellectual property ownership. As models become more advanced, they may generate output that is considered to be original work, such as music or art. This raises questions about who owns the rights to these works and how they should be protected. Additionally, there is a concern that foundation models may infringe on the intellectual property rights of others, such as by using copyrighted material to train the models. This can lead to legal challenges and disputes over ownership and licensing rights. Therefore, it is important for researchers and organizations to consider these issues and work together to ensure that foundation models are developed and used in a responsible and ethical manner.

Bias, fairness, and their social impact

AI bias and fairness are important considerations in the development of AI systems. These systems can amplify or perpetuate bias in the data they are trained on, leading to discrimination and unfairness. To address these issues, it is crucial to ensure that AI systems are developed and utilized in a responsible and ethical manner. This requires using diverse and regularly updated data to train these systems. Additionally, it is important to develop methods for detecting and mitigating bias in real time. The decision-making process of AI systems should also be transparent and explainable. These steps are necessary to ensure that AI benefits society as a whole and does not perpetuate or amplify bias. The social impact of AI is another concern. As AI systems become more advanced, they may significantly affect employment, the economy, and society. It is essential to consider the potential impact of AI on these areas and strive towards a future where AI enhances human capabilities rather than replaces them.

Adversarial attacks

Adversarial attacks are a growing concern in AI. These attacks involve manipulating inputs to an AI system to cause unexpected and potentially harmful behavior. For example, an attacker could manipulate an image to cause a facial recognition system to misidentify a person. They could add small perturbations to an image of a stop sign to cause an autonomous vehicle to misinterpret it as a yield sign. Adversarial attacks can also be used to manipulate the output of NLP systems. An attacker could craft a message that appears benign to an email filter but could be a phishing attempt against a human.

There are several types of adversarial attacks, including evasion, poisoning, and model stealing attacks. Evasion attacks involve modifying inputs to an AI system in order to evade detection or classification. Poisoning attacks involve modifying the training data used to train an ML algorithm, in order to bias the algorithm towards a specific outcome. Model stealing attacks involve reverse-engineering an ML algorithm in order to obtain information about its architecture and parameters.

To mitigate the impact of these attacks, researchers are developing new methods and techniques, such as adversarial training and ensembles of ML models. It is important for individuals, organizations, and governments to work together to ensure that AI is developed and used in a responsible and ethical manner to minimize the risks associated with these attacks.

You have had an overview of the security and safety concerns around modern AI systems. From a cybersecurity perspective, every emerging technology always comes with emergent security risks. We should analyze the risks and adopt the technology carefully, as this is the only way to be responsible for society and the future while maximizing the benefits of new technology.

Summary

Let's recap what we have learned in this chapter. We began our discussion with a brief introduction to the history of AI. You learned how it has evolved into ML and then DL. Then, we looked at different ways of defining the ML taxonomy, which gave you a systematic way of describing ML methods. In the taxonomy, you learned various types of ML algorithms and their applications, for instance, supervised learning, unsupervised learning, semi-supervised learning, and reinforcement learning. More details about those algorithms will be provided in the following chapters. In addition, you also learned about various data types that are common in ML and the definition of model modality. Finally, we shifted our focus to the recent advances in DL and reiterated its limitations and the security and safety concerns for AI.

The next chapter will further explore ML methods, including where they are applicable. You will extend your knowledge of this chapter and learn more about the formal definitions of different AI and ML methods.

5
AI Problems and Methods

In the previous chapters, you learned some basic definitions and taxonomies regarding artificial intelligence. This chapter will help you enhance this foundation and learn about concrete AI methods that are applied in cybersecurity. The knowledge from this chapter will help you get an idea of how different methods work and in what type of problems they are applicable. Furthermore, you will learn the advantages and disadvantages of applying these methods in particular conditions.

We'll start by looking at the main categories of machine learning methods and describe example methods relevant to this book:

- Supervised learning methods
- Unsupervised learning methods
- Semi-supervised learning methods
- Anomaly detection

We'll show some examples of various methods, including those from the supervised machine learning (random forest and neural networks), unsupervised learning (K-means, DBScan, and t-SNE) and semi-supervised learning categories. We'll also describe anomaly detection and introduce you to convolutional neural networks, deep learning, and **generative adversarial networks** (**GANs**), all of which are popular methods at the time of writing. We'll cover anomaly detection as a separate topic as it can be both supervised and unsupervised, and it often contains distinct methods that are not used in other types of scenarios.

Supervised learning methods

This section describes the concrete supervised learning methods that can be applied in cybersecurity scenarios. We'll describe where they can be applied, how they work, and what their advantages and disadvantages are. Supervised learning methods are used to solve problems where we have input data (X) and target outputs (Y), which means the training and validation data needs to be labeled with its associated output. In cybersecurity, an example would be data that is labeled as an attack or benign network traffic. The goal is to train the machine learning model so that it can find the pattern that produces the target output.

In other words, we are looking for a mathematical function (f) with a parameter set (W), where Y=f(X, W). For instance, a simple example would be a linear function, Y=WX. Depending on our assumptions, we can use different types of functions, such as linear, polynomial, sine, and so on. During training, the parameters of the function are optimized to produce outputs that are close to the target training outputs.

The following figure illustrates a linear model that's been fitted to the training inputs and target outputs:

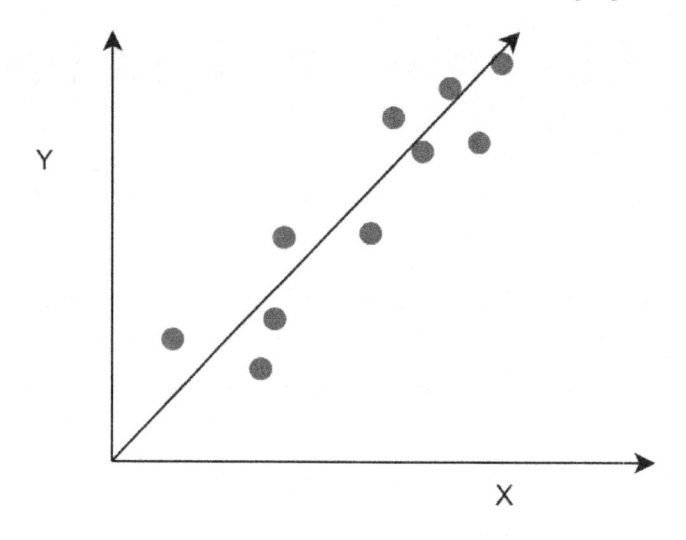

Figure 5.1 – A linear regression model trained based on 2D inputs (blue points)

In the preceding figure, we're looking for a model that can find the line that describes the pattern created by the data points.

Multiple types of models can address a supervised learning problem. Depending on the complexity of the patterns, we can apply more complex methods that are more powerful but more computationally demanding and difficult to optimize, or simpler methods that are easier to use but only apply to simple supervised learning problems.

We'll start by looking at a simple method called logistic regression, which is used for classification problems. These are problems where we have input data to which we want to assign a label from a predefined list. In terms of cybersecurity, this could involve assigning labels for malicious or benign behavior or classifying input data to one of the users for authentication in biometrics.

Logistic regression

One of the methods of supervised learning that is often used for classification is logistic regression. In classification problems, we assign discrete labels to data as associated output. For example, when classifying network traffic, we can assign attack data as 1 and benign data as 0. It's similar to linear regression, which was illustrated in the previous example, except that we assign discrete labels rather than continuous ones.

To adapt the method for classification, we add an additional part to the function that converts the result of the linear function into a value between 0 and 1. This allows us to use it for classification at a later time by thresholding those values and discretizing them to 0 or 1.

We can use the following equation (go to *References* section and go to [1]) to define the logistic function:

$$f(x) = \frac{1}{1 + e^{-(x-\mu)/s}}$$

In our case, the parameters, μ and s, are the parameters of the model, and x is the input.

We can also rewrite the expression in the following way:

$$f(x) = \frac{1}{1 + (e^{w_0 + w_1 x})}$$

Here, it's more obvious that we have a linear function under exponentiation. Given that this function returns a value between 0 and 1 as a result, it's a good candidate to be used for a binary classification model.

To have an accurate model, we need to find the parameters, w_0 and w_1 such that $f(x)$ provides a value as close as possible to the target labels for every input, x, from the training set. In practice, we create a *loss function* that's optimized using a gradient descent procedure. A common loss function that's used for logistic regression is cross-entropy [1]:

$$y(x) = y\log(f(x_i)) + (1 - y)\log(1 - f(x_i))$$

The closer the function values are to the target outputs, the lower the loss function value is. The gradient of this function helps us find the direction of the fastest change of the function toward the optimal value – in our case, the minimum. By iteratively changing the parameters, w_0 and w_1, in the direction of the negative gradient, we find the parameters that get the minimal cross-entropy value. This way, we find the function, $f(x)$), that represents the optimal classifier based on our training set.

Logistic regression can be a very good solution for simple classification problems, but as it's a simple model, it can struggle with more complex patterns. In the following subsection, we'll describe random forest, a method that is more robust and can give better results for more difficult classification problems.

Random forest

Another approach to classification is to base it on decision trees. Decision trees are often used in real life to make decisions by iteratively considering different factors. For instance, when we are deciding whether to go on vacation, we consider our financial status, then available vacation days, weather forecasts, and so on. We can translate this way of thinking into machine learning-based classification, where the classification process is guided by thresholding different values of features. However, the question is how to set the thresholds for feature values for each decision.

Furthermore, we need to consider in which order we consider the feature values when making iterative decisions. Algorithms such as ID3 and CART can help us optimize decision trees so that they can classify data.

The following figure shows an example of a decision tree with two variables, x_1 and x_2. The path in the tree and the resulting decision depends on the value of the two variables:

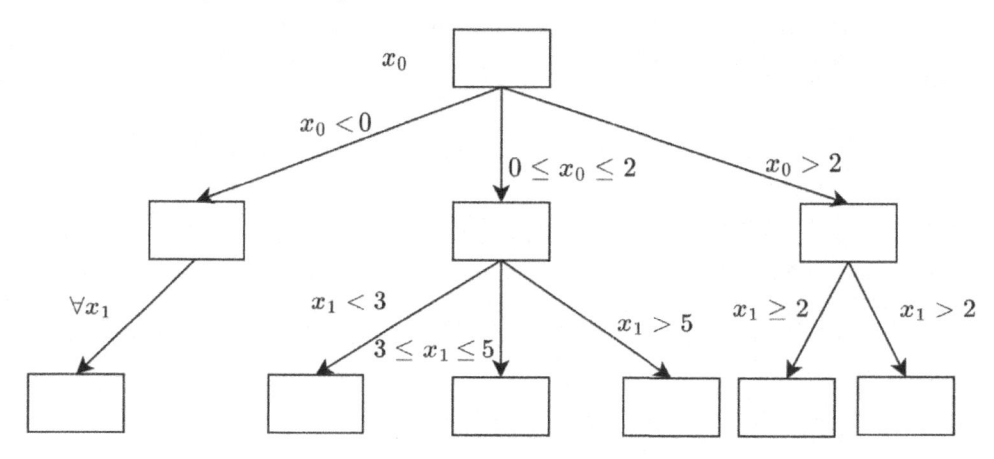

Figure 5.2 – Example of a decision tree with three attributes

In practice, one decision tree often doesn't get us close to the globally optimal solution for classification due to the size of the search space. Therefore, in most cases, in practice, we use a large number of decision trees and combine their results. These decision trees are constructed using a subset of data and a random subset of features. We combine their decisions by averaging in the case of regression and majority voting in the case of classification. Majority voting means that, for the same input, we take the result that most decision trees produce. This approach of combining multiple decision trees is called random forest. The approach of combining different singular classifiers into one is generally called **ensemble learning** [15] and is often used with other classifiers as well. Ensemble learning often produces superior solutions compared to training one classifier because it makes the model more robust. However, it causes an additional computational burden as we need computational power and time to train a large set of classifiers.

The following figure shows an example of a random forest, where the outputs of multiple decision trees are combined:

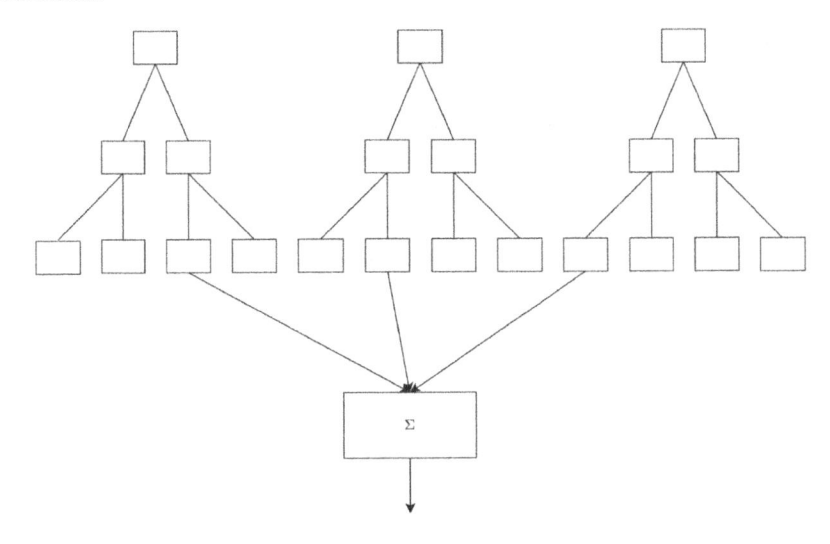

Figure 5.3 – Visualization of a random forest – an ensemble of decision trees

Next, we'll cover SVMs.

Support Vector Machines (SVM)

Another traditionally popular algorithm for classification is SVM. This approach relies on identifying an ideal separation boundary within the dataset, with the boundary strategically positioned to maximize the distance between points of distinct classes and the boundary itself. This enables good generalization on the test set. We can imagine the classification boundary as a line in the 2D space, and a hyperplane in the n-dimensional space.

The following figure shows the classification boundary that's defined by training an SVM, together with two classes separated by that boundary. The boundary is set to create the maximum distance between the boundary and the dotted line, which represents the closest points of different classes:

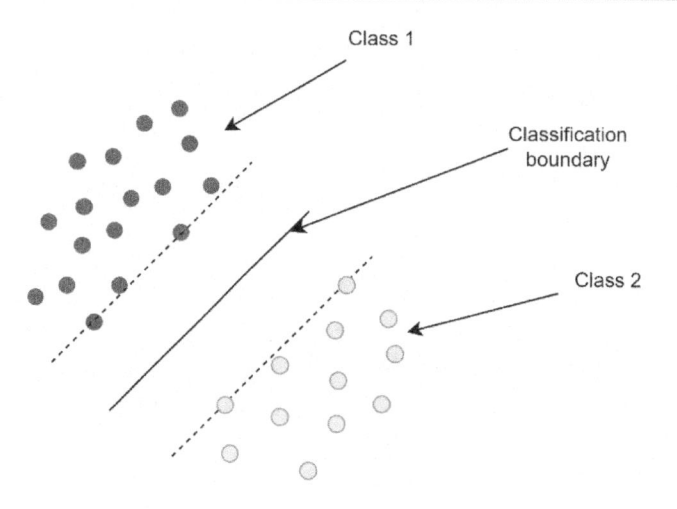

Figure 5.4 – The maximum margin boundary found by an SVM

The points closest to the boundary are called **support vectors** and they determine the position of the boundary.

Mathematically, the training process can be defined like so [1]:

$$\min_{w,b,\zeta} w^2 \; C \sum_{i=1}^{n} \zeta_i$$

$$y_i\left(w^T x_i - b\right) \geq 1 - \zeta_i$$

In the first expression, the vector, w, represents the margin – that is, the smaller the value of w; this is the larger margin between the classes. The second expression represents the separation between the classes – that is, the target class, y_i, should be the same as the class retrieved by the model.

If the data in different classes is not linearly separable, there is a way to transform it into a different feature space, where they can be separated. This is called a **kernel trick**. The most used transformation is the Gaussian kernel, where we use the Gaussian radial basis function:

$$k\left(x_i, x_j\right) = \exp\left(-\gamma \left\|x_i - x_j\right\|^2\right)$$

Neural networks

Neural networks can be used in both supervised and unsupervised learning settings, but they are most famous for their use in supervised learning, in particular in classification. This method is inspired by the way biological neural networks behave, where you have multiple constructs (units), and the input is transformed through multiple connected units (neurons) to produce the final result.

The following figure shows a typical neural network architecture. This model architecture is called a **multilayer perceptron (MLP)**:

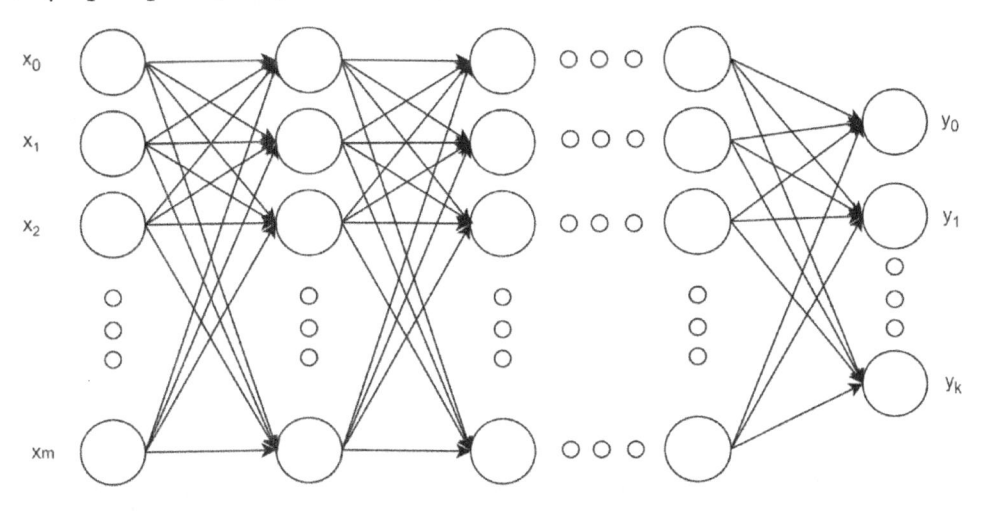

Figure 5.5 – An MLP network

In an MLP, the inputs are forwarded through multiple layers of parameterized transformation, which in the end produces an output that should comply with the pattern based on the training set. Each connection represented by arrow connectors in the preceding figure is parameterized by weight and a bias parameter. In a typical MLP architecture, the input into a unit (represented by a circle) is multiplied by a weight and processed with an activation function, and a bias is added to this product. This makes it a linear transformation. Furthermore, to account for non-linear dependencies between input and output, we typically use a non-linear activation function.

A common example of an activation function is the sigmoid function, which is used to model nonlinear patterns:

$$f(x) = \frac{1}{1 + e^{-x}}$$

The following plot shows the shape of the sigmoid activation function:

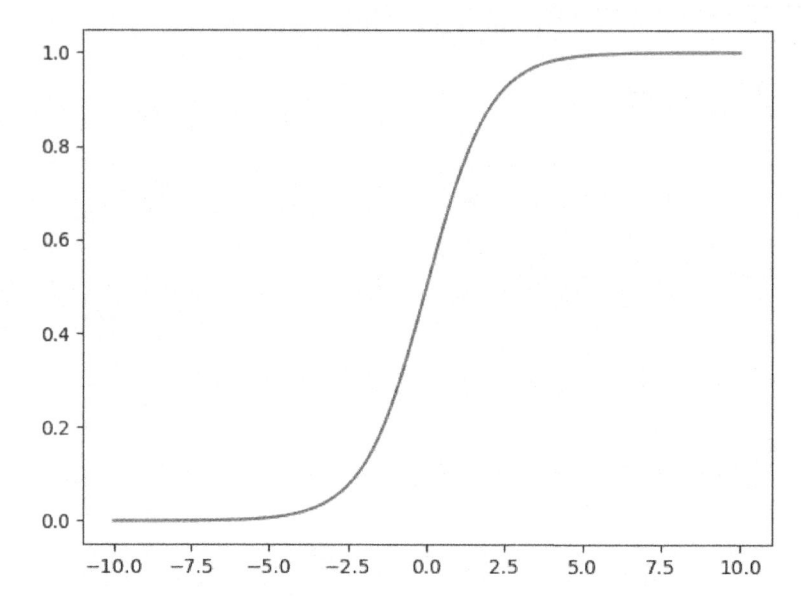

Figure 5.6 – Example plot of a sigmoid function

From the preceding plot, you can see that lower values of x produce an f(x) close to 0, whereas higher input values produce an output closer to 1. This makes the sigmoid function control the values of the output to be between 0 and 1, and at a value of 0.5 in case the input is 0. If we use this function in the last layer of transformation, the sigmoid function is perfect for binary classification as it can play a role in the decision function. For values over 0.5, the data example is classified as class 1, whereas for output values under 0.5, the data example is classified as belonging to class 0.

Deep learning

Deep learning is a machine learning paradigm that has enjoyed great popularity in the last decade. It is based on the idea that creating a neural network with a large number of layers enables hierarchical learning of a representation of the input data, where each subsequent layer learns increasingly more sophisticated features, such as going from raw pixels to being able to detect objects in an image. This concept is inspired by neuroscience research regarding how the human brain works. However, many deep learning methods are based on empirical knowledge from testing various deep neural network configurations on large-scale datasets. Similar to what we described in the Neural networks section, deep neural networks consist of multiple layers of transformation. In deep learning, the idea is to increase the number of layers and use various types of units and activation functions to allow the layers to be used optimally to produce the right performance.

For instance, in addition to using fully connected layers like in MLP, there are other popular layers such as convolutional layers, which enable convolutional filters to be used in each layer to capture localized patterns. This is based on convolution filters in image processing or speech recognition. Another option is to vary the activation functions. For instance, one popular activation function is the **rectifier linear unit** (**ReLU**), which enables better propagation of the gradient during training.

The following figure shows a plot of the ReLU activation function. Notice the constant part for inputs lower than zero and the linear increase for inputs higher than zero:

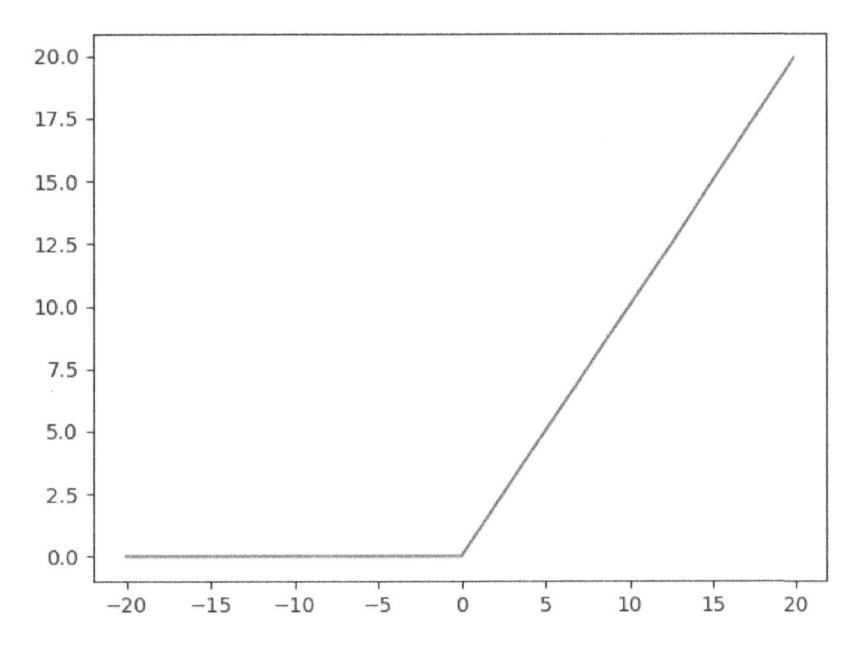

Figure 5.7 – Example plot of a ReLU activation function

It enables us to tackle the well-known problem of **vanishing gradients**. This is the problem characteristic of deep learning architecture, where the gradients sometimes become too small while being propagated through the neural network and the information for updating the neural network weights is not transmitted properly. This is because the number of updates depends on the size of the gradient. Apart from activation change, we can also create variations in training techniques to improve the solutions we obtain while training the deep neural networks. One such technique is **dropout** [6]. The idea of dropout is to use only a random subset of units in the network – that is, drop out some of the units – to introduce more variation and reduce the chance of overfitting. This is implemented by activating the units in each training iteration with a certain probability, where a lower probability means that more units will be dropped out.

The following figure shows an example of connections with dropout during MLP training:

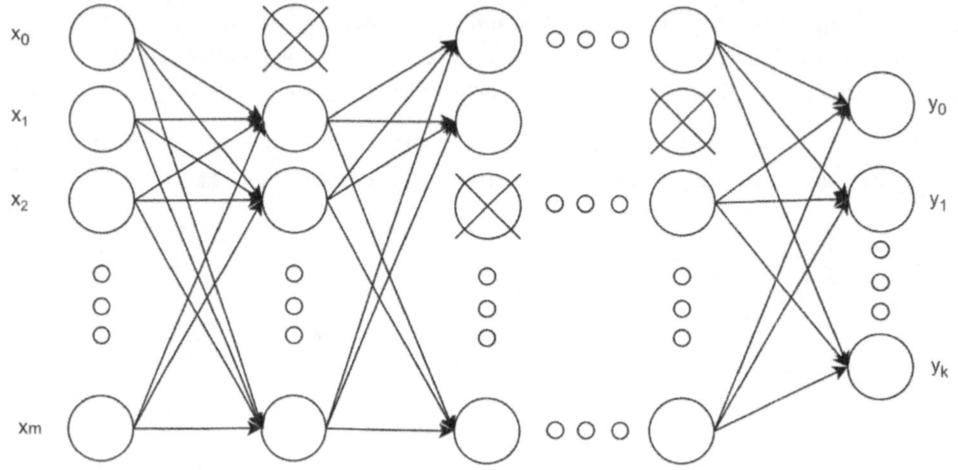

Figure 5.8 – The effect of dropout on the neural forward pass during training

After training, in the test period, all the units are activated and we use the weights that were determined during training.

Various types of deep neural networks fit into the deep learning paradigm. One such type is convolutional neural networks.

Convolutional neural networks

Convolutional networks are commonly used in contemporary deep learning architectures. They are typically based on alternated use of *convolution* and *pooling* layers and have both a biological basis and a basis in previous research – in areas of signal processing that deal with pattern recognition in natural signals such as images, speech, video, and similar.

Convolution layers are based on using a filter matrix that's applied to different portions of an image using matrix multiplication to detect patterns and transfer the knowledge about them to the subsequent neural network layers. As a difference from feedforward neural networks (such as MLP), in convolutional networks, the layers are not fully connected – that is, not all the units in neighboring layers are connected. Here, we use a transformation matrix (kernel) that has smaller dimensions than the layer size, and we apply this matrix by repeatedly using matrix multiplication in different parts of the network layer. This means that the same parameters are shared in different parts of the network (*sparse interaction*). This improves the efficiency of the neural network because there are a smaller number of parameters that we need to store and fewer operations to execute during training. Furthermore, the transformations in the convolution layers have **translation invariance**. This is achieved because the same transformation is applied in different parts of the network. So, independently of where an input pattern appears (for example, an object on an image), this pattern will be detected by the convolution operation:

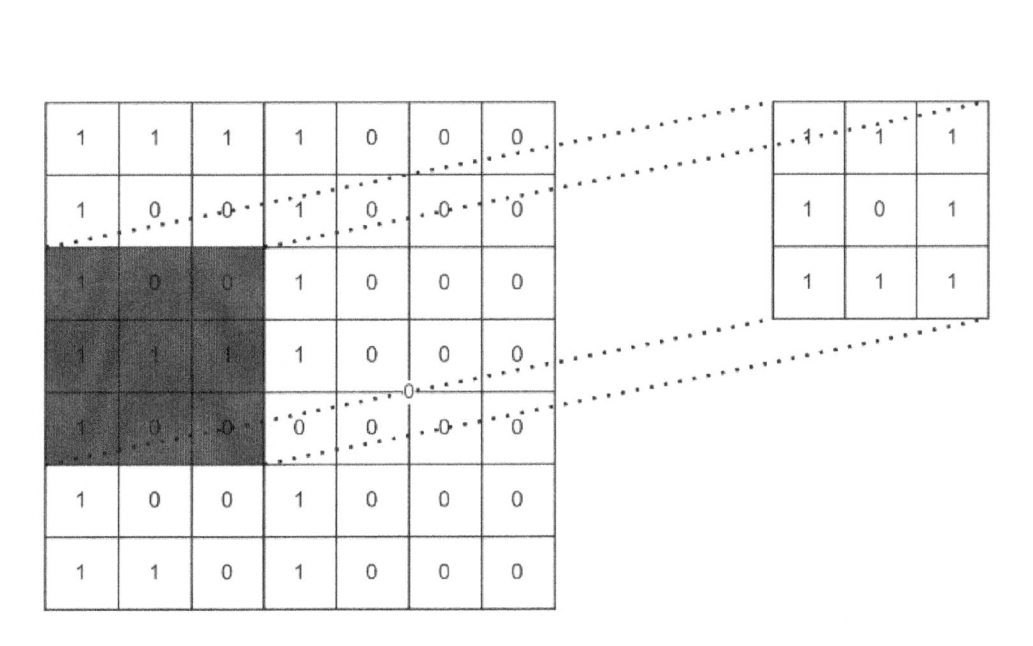

$$X[:,:,0] \qquad W[:,:,0]$$

1	1	1	1	0	0	0
1	0	0	1	0	0	0
1	0	0	1	0	0	0
1	1	1	1	0	0	0
1	0	0	0	0	0	0
1	0	0	1	0	0	0
1	1	0	1	0	0	0

1	1	1
1	0	1
1	1	1

Figure 5.9 – The convolution filter that's used in a convolutional neural network

In a typical series of layers in the convolutional network, the results of the convolution layer are forwarded to an activation function and then to a pooling layer. The pooling layer reduces the size of the input by applying an aggregation, such as maximum (max-pooling), average (average pooling), or similar. Pooling allows the amount of features to be reduced and helps make the result invariant to small translations in the input.

The following figure shows the architecture of AlexNet, a famous neural network architecture that was used in the *ImageNet Large Scale Visual Recognition Challenge* competition and produced one of the top results:

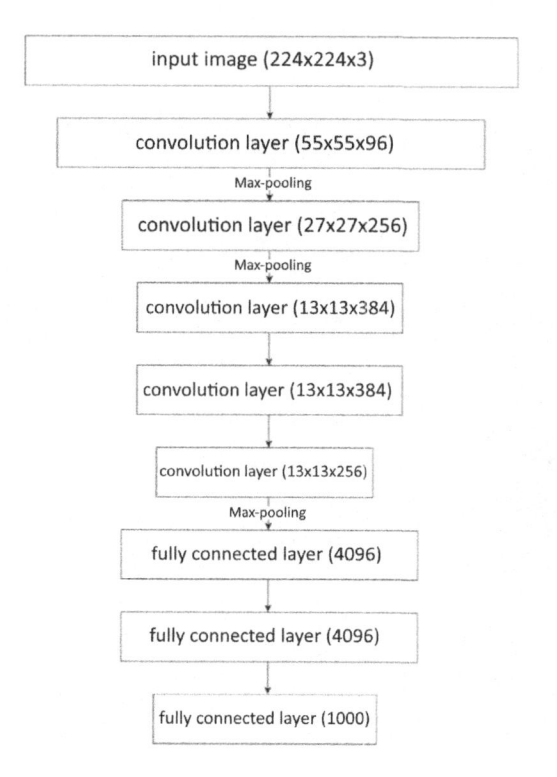

```
input image (224x224x3)
          ↓
convolution layer (55x55x96)
          ↓ Max-pooling
convolution layer (27x27x256)
          ↓ Max-pooling
convolution layer (13x13x384)
          ↓
convolution layer (13x13x384)
          ↓
convolution layer (13x13x256)
          ↓ Max-pooling
fully connected layer (4096)
          ↓
fully connected layer (4096)
          ↓
fully connected layer (1000)
```

Figure 5.10 – Neural network architecture of AlexNet (inspired by https://papers.nips.cc/paper_files/paper/2012/file/c399862d3b9d6b76c8436e924a68c45b-Paper.pdf)

After AlexNet, there were multiple research efforts to improve the architecture and gain better neural network performance by varying the number of layers and the types of units and connections used in the neural network. The following table explains some of the most famous neural network architectures:

Neural Network	Number of Parameters	Year of Publication
AlexNet [7]	60 million	2012
GoogLeNet [8]	4 million	2014
VGGNet [9]	138 million	2014
ResNet [10]	60.3 million	2015
Inception v3 [11]	6.4 million	2019

Table 5.1 – Some of the most famous neural network architectures

Convolutional networks are typically used in discriminative learning scenarios such as classification. However, increasingly important are the generative models, in particular GANs, where instead of output labels, new data points are generated based on the detected patterns.

A **GAN** [12] is an example of a neural network that's used for generative models. It differs from typical MLP and convolutional neural networks, which are discriminative models. Even though generative models are typically unsupervised, due to the configuration of the network, a GAN works more like a supervised model.

Furthermore, a GAN is special in that it consists of two networks: a generator and a discriminator. The idea is that the generator and discriminator are optimized with different criteria, and are conditioned to be mutually adversarial. The goal of the generator is to produce an output that closely resembles the distribution of data from the training set. On the other hand, the goal of the discriminator is to predict whether a data sample comes from the training set or has been generated by the generator part of the network. At the beginning of training, the generator only generates random noise and observes the discriminator's output. The parameters of the generator and discriminator are gradually adjusted using a gradient descent procedure in an alternating manner to improve their goals. The idea is that the generator needs to improve to fool the gradually better discriminator to generate data that cannot be distinguished by a sophisticated discriminator network. This means that the generated data is very close to the distribution of the training set (p(x)).

This is depicted in the following figure:

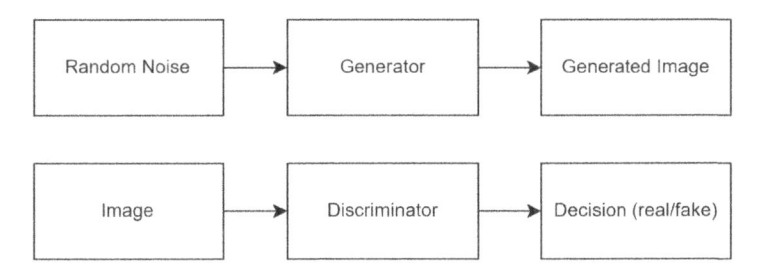

Figure 5.11 – High-level GAN architecture

The goal of training the GAN is to learn the distribution, $p_g(x)$. We start by defining the prior noise distribution, $p_z(z)$, and a network architecture that maps the noise to the image, $G(z, \Theta_g) G(z, \theta_g)$. This function depends on the parameter set, Θ_g, that gets optimized during training. The discriminator network, $D(x, \theta_x)$, has a distinct architecture, with parameters, Θ_x, and outputs a probability of an image belonging to the set of real (or fake) images.

Mathematically, the optimization procedure is defined in the following way [12]:

$$\min_G \max_D V(D, G) = \mathbb{E}_{x \sim p_{data}(x)}[\log D(x)] + \mathbb{E}_{z \sim p_z(z)}[\log(1 - D(G(z)))]$$

Here, D(x) is the objective function of the discriminator and G(z) is the objective function of the generator. The first part of the function, V, specifies the discriminator's attempt to successfully differentiate the generated images from the real ones. The second part encodes the objective of the generator to generate images that the discriminator has difficulty classifying. In essence, it is a zero-sum game where two neural networks compete to optimize their objective function.

The equation for training the generator and discriminator is as follows [12]:

$$\nabla_{\theta_d} \frac{1}{m} \sum_{i=1}^{m} \left[\log D(x^{(i)}) + \log\left(1 - D(G(z^{(i)}))\right)\right]$$

$$\nabla \theta_d \frac{1}{m} \sum_{i=1}^{m} \left[\log D(x^{(i)}) + \log\left(1 - D(G(z^{(i)}))\right)\right]$$

The preceding expression contains the gradient update for the discriminator. The parameters of the discriminator are the weights of the discriminator neural network [12]:

$$\nabla_{\theta_g} \frac{1}{m} \sum_{i=1}^{m} \log\left(1 - D(G(z^{(i)}))\right)$$

In the preceding expression, we define the part of the loss that is included in the update of the generator neural network.

In the next section, we'll take a deep dive into unsupervised learning methods and describe how to uncover patterns in unlabeled input data.

Unsupervised learning methods

In unsupervised learning, the task is to train a model that represents the input data, without having the target data available. This section will help you understand the most prominent unsupervised learning methods and give you an idea of when they are useful in cybersecurity scenarios.

Typically, the learned model in unsupervised learning provides a compressed representation of the input data that abstracts away the noise and uncovers the latent structure in the dataset. For instance, in clustering, the input data is represented by the clusters that have been uncovered by the clustering model. The parameters of the model govern the mapping between the input data and the cluster IDs.

The following figure illustrates the underlying structure of input data. In this case, the data can be naturally grouped into three clusters:

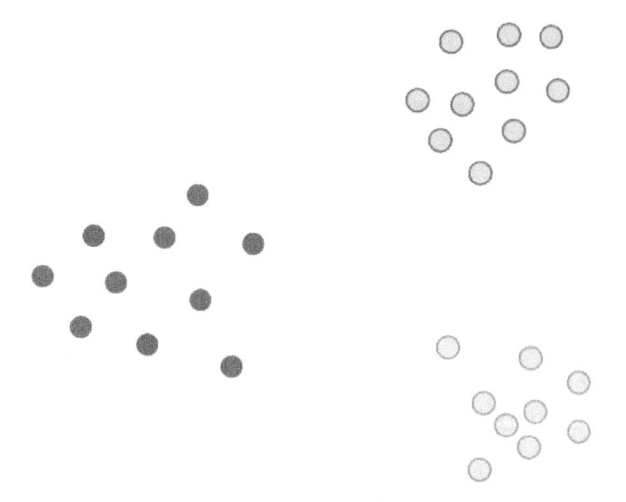

Figure 5.12 – Example of clusters in data

In the following subsection, we'll describe K-means, one of the most well-known unsupervised learning methods.

K-means

One of the most commonly used algorithms for clustering is the K-means method. When using this method, the data is partitioned into K clusters, where each cluster is characterized by its central point (mean). Each data point is classified to its nearest mean. The goal is to find a partition where the distance between the points in the same cluster is minimal – that is, where the distance between the points and the mean of their assigned cluster is the smallest possible.

The simplest model training procedure is built on alternating between two steps:

- **Cluster assignment to training points**: Each point is assigned to its nearest mean
- **Recalculating the cluster centers**: The position of the center for each cluster is computed as the mean for all points in that cluster

The result of applying the K-means method highly depends on the preassigned number of clusters (K). Here, K needs to be known in advance, or it can be optimized by trying out different values of K and observing the result. This can be a disadvantage if it's difficult to estimate the number of clusters in advance and it's too time-consuming to run the clustering with multiple values of K. In this case, sometimes, it's better to use alternative clustering methods. One of these methods is DBSCAN.

DBSCAN is a density-based clustering method. This means that the clusters are detected as areas where there is high density in the data. The advantage is that for this method, you don't need to provide the number of clusters in advance. However, two hyperparameter values need to be known and that the user needs to enter before training:

- `minPts`: The minimum number of points that constitute a region of high density
- `eps`: The maximum distance between points in the same region of high density (neighborhood)

Here, we create a neighborhood around each point, recursively expand them, and merge them to create clusters. More specifically, the algorithm consists of the following steps:

1. Find the core points as a basis for the neighborhoods. The core points are the points where at least `minPts` points are within `eps` distance of them.
2. Expand the neighborhoods by adding points that are less than `eps` distance from the points already in the neighborhoods. Furthermore, points in the same neighborhoods are in the same cluster.
3. Points that are not assigned to any neighborhoods (not within `eps` distance from any point) are considered outliers.

The additional feature of this algorithm is that it can be used for anomaly detection as well since it's possible to discover outliers in the data that don't belong to the defined clusters.

The following plot shows an example of the results of DBSCAN on a sample dataset. Here, we can see that the DBSCAN detected the existence of three clusters:

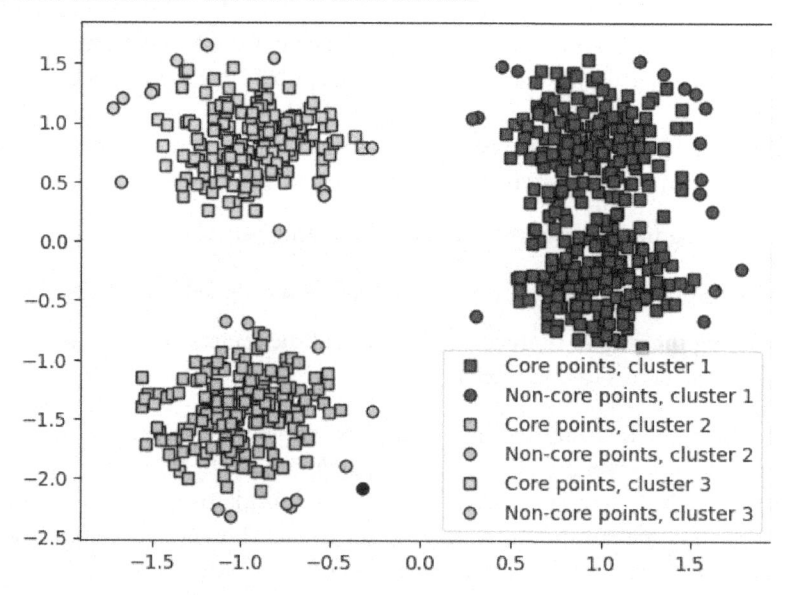

Figure 5.13 – Example of data clusters detected by the DBSCAN method

t-SNE

t-SNE [13] is an unsupervised algorithm for compressing data that's often used for visualizing high-dimensional datasets. It's based on finding a compressed (lower-dimensional) representation of data, where relative distances between data points are preserved. This means that the data points that are similar – that is, with lower pairwise distance in the original data – should also be similar in the compressed representation. To do this, we must use a conditional probability given by the following equation [13]:

$$p_j = \frac{\exp\left(-\frac{|x_i - x_j|^2}{2\sigma_i^2}\right)}{\sum_{k \neq i} \exp\left(-\frac{|x_i - x_k|^2}{2\sigma_i^2}\right)}$$

$$p_{j|i} = \frac{\exp\left(-\frac{\|x_i - x_j\|^2}{2\sigma_i^2}\right)}{\sum_{k \neq i} \exp\left(-\frac{\|x_i - x_k\|^2}{2\sigma_i^2}\right)}$$

We define a similar similarity measure for the lower dimensional representation [13]:

$$C = \sum_{i=1}^{m} K(P_i \| Q_i) = \sum_{i=1}^{m} \sum_{j=1}^{m} (P_{j|i}) \log \frac{p_{j|i}}{q_{j|i}}$$

The assumption is that if the t-SNE mapping is correct, the probabilities, p_{ij} and q_{ij}, will be the same (or similar). Therefore, in the t-SNE training algorithm, we gradually adjust the mapping to increase the similarity between the two distributions.

P_i and Q_i are the distributions over data points, x_i, and mapped points, y_i, respectively.

The training is done using gradient descent, where we adjust the compressed representation, y_i, to gradually minimize the KL divergence.

Given that we can compress data while keeping the underlying similarity structure, this method can be used to visualize clusters in the data and provide a first impression of what high-density areas exist. However, since the method is stochastic, when the algorithm is run multiple times, the results can look different each time.

The following two figures show three separately generated groups of three-dimensional points initialized with random Gaussians centered at 3, 5, and 10, respectively, with a standard deviation of 1, and their 2D compressed representation generated using t-SNE:

76　　AI Problems and Methods

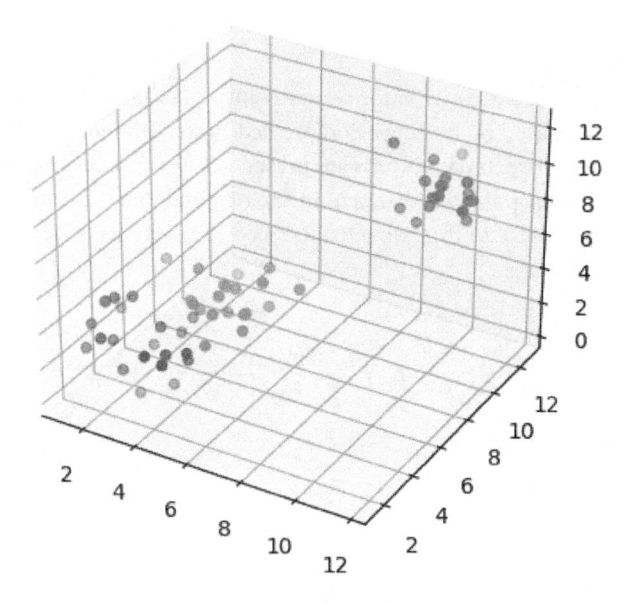

Figure 5.14 – Example of 3D data with three clusters

The following figure shows a 2D compression of the data with three clusters:

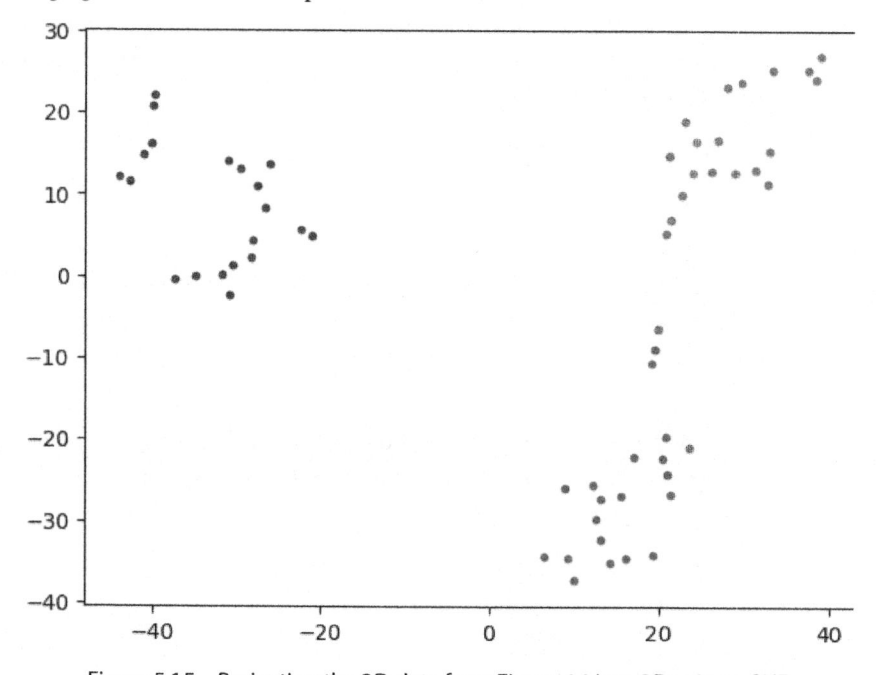

Figure 5.15 – Projecting the 3D data from Figure 14 into 2D using t-SNE

Apart from supervised and unsupervised methods, we have semi-supervised problems. Since they often appear in many areas, including cybersecurity, we'll take a closer look at them in the following section.

Semi-supervised learning methods

Semi-supervised learning [5] is applied when a dataset consists of both labeled and unlabeled data. This is often because labeling data is expensive or difficult to obtain. Therefore, part of the data is unlabeled. However, we still want to take this data into account by using a method that generates outputs according to the target labels, but also conforms to the latent structure patterns that are discovered via unlabeled data.

Semi-supervised learning methods are usually a combination of supervised and unsupervised learning strategies and enable a more accurate model than the ones that are obtained using only labeled or unlabeled data. For this combination to provide added value compared to using only labeled data and purely supervised training, some assumptions (to an extent equivalent) about data need to hold:

- **Smoothness assumption**: If points x_1 and x_2 have similar values, their corresponding target values, y_1 and y_2, need to be similar
- **Low-density separation**: The decision boundary should lie in a low-density region
- **Manifold assumption**: The data lies in a low-dimensional manifold

These assumptions show that the data needs to have an underlying structure that can be exploited by performing unsupervised learning on unlabeled data, to cause the semi-supervised model to provide more accurate results.

In the next subsection, we'll describe label propagation, one of the most well-known instances of semi-supervised learning methods.

Label propagation

A subgroup example of semi-supervised learning methods that is often used and available in popular machine learning libraries is label propagation. The idea of label propagation methods is to propagate labels from labeled examples to unlabeled ones using similarity to infer labels for the unlabeled data. This way, we can classify the unlabeled points with the labels that were previously associated with similar points. This is based on the previously named smoothness assumption. For this, we need a matrix (W) that contains a similarity measure between each pair of data points (w_{ij}).

The following algorithm describes the training process for label propagation:

1. Compute the affinity matrix, W.
2. Compute the diagonal matrix, D, where we have the following [5]:

$$D_{ii} = \sum_j W_{ij}$$

3. Initialize $Y^{(0)} = (y_1, y_2, \ldots, y_l, 0, 0, \ldots 0)$.

4. Iterate until convergence:
   - $Y^{(t+1)} = D^{-1} W Y^{(t)}$
   - $Y_l^{(t+1)} = Y_l$

5. Label point x_i with the sign of y_i.

This means that we iteratively assign labels to the unlabeled examples until they satisfy our smoothness assumption optimally. There are multiple more sophisticated versions of this algorithm, but they all have a similar basis.

In the next section, we'll cover various anomaly detection methods. These methods are useful for finding outlier points or patterns in data.

Detecting anomalies

As described in the previous chapter, anomaly detection methods are used to detect outliers, novelty, and anomalies in datasets that don't conform to the dominant patterns in the data. To do this, in general, we need to detect the dominant pattern during training. This can be done both supervised and unsupervised, depending on whether we have labeled data. Therefore, anomaly detection methods are often extensions of conventional supervised or unsupervised learning methods. In the case of supervised learning, multiple variants of anomaly detection methods are based on supervised classification models:

- **Methods where anomalies belong to the minority class**: Here, we can use conventional supervised learning methods, with additional methods being used to tackle the problem with unbalanced data, for example by undersampling the majority class or oversampling the minority class.

- **Methods where anomalies don't conform to existing classes present in the training labels**: Here, we can detect anomalies by observing the output of the classifier, where the class probabilities don't show a high enough value for available classes. Another option is to use a loss function that allows points that are outside the pattern of current classes in the training set to be detected.

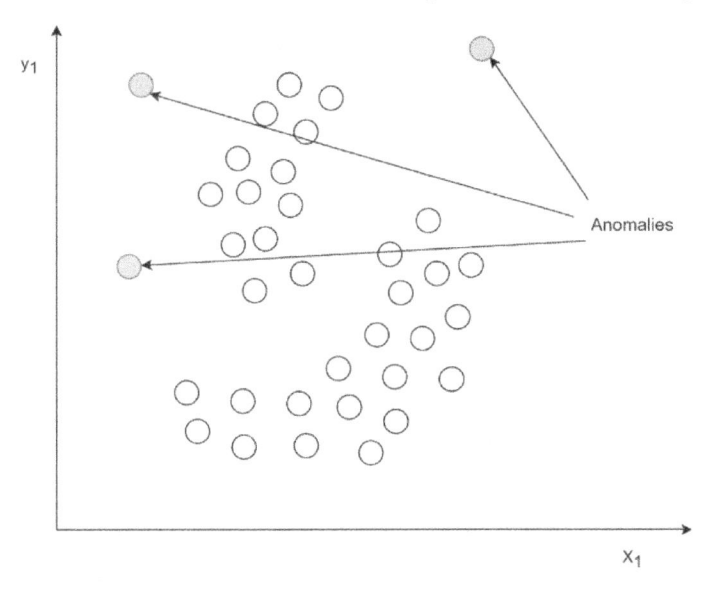

The following figure illustrates a dataset with anomalies compared to the dominant high-density regions:

Figure 5.16 – Anomalies in data that have been detected using anomaly detection

In the following subsection, we'll describe isolation forest, one of the most used instances of anomaly detection methods.

Isolation Forest

An example of an unsupervised anomaly detection method is isolation forest [17]. This method can be understood as a variant of the previously described random forest method, although random forest is a supervised method.

The idea is to split the data points using a binary tree iteratively by one-by-one attribute. In each iteration, the split is done by an attribute, q, and a random split value, p, and they are split into two subtrees, $q<p$ and $q \geq p$. We do this procedure on random subsets of data and create multiple such trees. In the test phase, we run each point through the constructed tree. The expectation is that, for anomalous points, it takes fewer splits to isolate a point on average compared to *normal* points. The anomaly score in isolation forest is based on this number of splits:

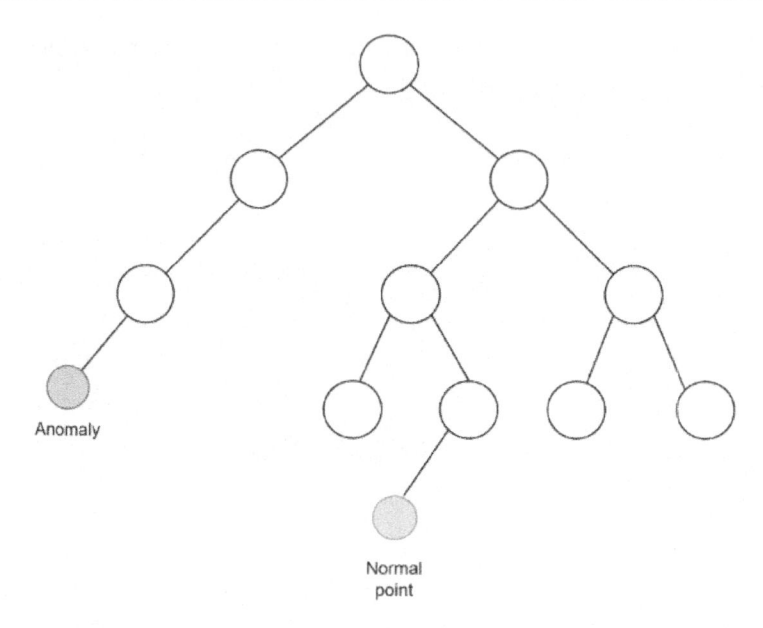

Figure 5.17 – Running a normal point and an anomaly through a tree via isolation forest

In the next subsection, we'll cover another anomaly detection method known as one-class SVM. This method is a variant of SVM that's used to encircle the points that belong to the majority pattern and detect outliers outside of that pattern.

One-class SVM

One-class SVM can be understood as a variant of SVM that's used for anomaly detection scenarios. In this case, we only have one class of normal data, and we are trying to separate this normal data from outliers by a large margin. The normal data is limited by a hypersphere with a specific center (c) and radius (r). During training, the optimization is directed to finding the smallest hypersphere that contains most of the data points. Therefore, it should contain all the normal points, and only anomalies will stay outside of the hypersphere.

This can be encoded in the following mathematical expressions for constrained optimization [14]:

$$\min_{r,c,\zeta} r^2 + \frac{1}{vn} \sum_{i=1}^{n} \zeta_i$$
$$|| \Phi(x_i) - c ||^2 \leq r^2 + \zeta_i, \forall i = 1,2,\ldots n$$

In the first expression, the radius (r) is minimized, whereas the second expression contains the constraint that most of the data needs to satisfy to be within the hypersphere of that radius. The value, v, determines the proportion of outliers that are accepted in the training set, whereas the constant, ζ_i, is there to allow outliers to be outside of the hypersphere.

The following plot contains an example of the hypersphere defined by training a one-class SVM, with the regular observations within the sphere and outliers outside of it:

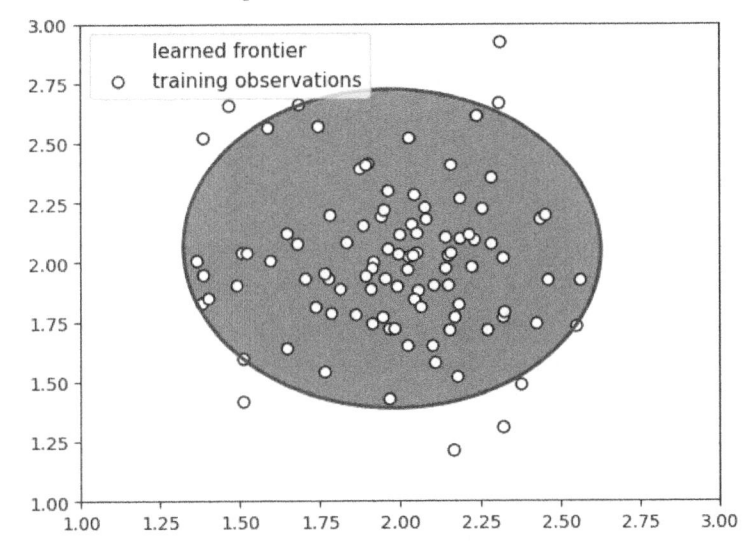

Figure 5.18 – A dataset with a trained anomaly detection boundary when using one-class SVM

Now that we've covered the various methods that will be specified in this book, let's summarize this chapter.

Summary

In this chapter, we described various unsupervised learning, supervised learning, semi-supervised learning, and anomaly detection problems and presented concrete methods that are used for them. You need to understand these concrete methods and the problems that they address to be able to understand when and how to apply them. Even though we rarely implement these methods from scratch, knowing how these methods work will enable you to save time and cost when applying the wrong tools to the job. Furthermore, it will enable you to gain a better understanding of the results you'll see when applying these methods and help you explain them to your project stakeholders.

In the next chapter, you will learn where these methods fit into an artificial intelligence project, how the workflow works, and how to get optimal results by using the available tools.

References

- [1] Murphy, Kevin P. *Machine learning: a probabilistic perspective.* MIT press, 2012.
- [2] James, Gareth, Daniela Witten, Trevor Hastie, and Robert Tibshirani. *An introduction to statistical learning.* Vol. 112. New York: springer, 2013.

- [3] Chandola, Varun, Arindam Banerjee, and Vipin Kumar. "Anomaly detection: A survey." *ACM computing surveys (CSUR)* 41, no. 3 (2009): 1-58.

- [4] Aggarwal, Charu C., and Charu C. Aggarwal. *An introduction to outlier analysis*. Springer International Publishing, 2017.

- [5] Learning, Semi-Supervised. "Semi-supervised learning." *CSZ2006. html 5* (2006): 2.

- [6] Bengio, Yoshua, Ian Goodfellow, and Aaron Courville. *Deep learning*. Vol. 1. Cambridge, MA, USA: MIT press, 2017.

- [7] Krizhevsky, Alex, Ilya Sutskever, and Geoffrey E. Hinton. "Imagenet classification with deep convolutional neural networks." *Advances in neural information processing systems* 25 (2012).

- [8] Szegedy, Christian, Wei Liu, Yangqing Jia, Pierre Sermanet, Scott Reed, Dragomir Anguelov, Dumitru Erhan, Vincent Vanhoucke, and Andrew Rabinovich. "Going deeper with convolutions." In *Proceedings of the IEEE conference on computer vision and pattern recognition*, pp. 1-9. 2015.

- [9] Simonyan, Karen, and Andrew Zisserman. "*Very deep convolutional networks for large-scale image recognition*." arXiv preprint arXiv:1409.1556 (2014).

- [10] He, Kaiming, Xiangyu Zhang, Shaoqing Ren, and Jian Sun. "*Deep residual learning for image recognition*." In Proceedings of the IEEE conference on computer vision and pattern recognition, pp. 770-778. 2016.

- [11] Szegedy, Christian, Vincent Vanhoucke, Sergey Ioffe, Jon Shlens, and Zbigniew Wojna. "*Rethinking the inception architecture for computer vision*." In Proceedings of the IEEE conference on computer vision and pattern recognition, pp. 2818-2826. 2016.

- [12] Goodfellow, Ian, Jean Pouget-Abadie, Mehdi Mirza, Bing Xu, David Warde-Farley, Sherjil Ozair, Aaron Courville, and Yoshua Bengio. "*Generative adversarial networks*." Communications of the ACM 63, no. 11 (2020): 139-144.

- [13] Van der Maaten, Laurens, and Geoffrey Hinton. "*Visualizing data using t-SNE*." Journal of machine learning research 9, no. 11 (2008).

- [14] Schölkopf, Bernhard, Robert C. Williamson, Alex Smola, John Shawe-Taylor, and John Platt. "*Support vector method for novelty detection*." Advances in neural information processing systems 12 (1999).

- [15] Kuncheva, Ludmila I. Combining pattern classifiers: methods and algorithms. John Wiley & Sons, 2014. Scikit-learn

- [16] Pedregosa, Fabian, Gaël Varoquaux, Alexandre Gramfort, Vincent Michel, Bertrand Thirion, Olivier Grisel, Mathieu Blondel et al. "*Scikit-learn: Machine learning in Python*." the Journal of machine Learning research 12 (2011): 2825-2830.

- [17] Liu, Fei Tony, Kai Ming Ting, and Zhi-Hua Zhou. "*Isolation forest*." In 2008 Eighth IEEE international conference on data mining, pp. 413-422. IEEE, 2008.

6
Workflow, Tools, and Libraries in AI Projects

So far, we have discussed the background of big data, AI, AI problems, and methods. This chapter will discuss the workflow, tools, and libraries in AI projects, especially topics associated with security such as anomaly detection, spam detection, and so on. Technically speaking, nearly all AI projects have the same workflow, which includes the pre-processing (data collection, data cleaning, data annotation, etc.), the model training and testing/prediction, and fine-tuning (especially with the development of a transformer-based large model). However, there are various tools and libraries for different tasks. Therefore, we will discuss the tools and libraries within specific tasks.

All in all, in this chapter, the following topics are discussed:

- Workflow of AI projects
- Tools and libraries for visual network traffic analysis
- Tools and libraries for malware detection

Workflow of AI projects

Starting in November 2022, the workflow of AI projects shifted and extended to a more extensive scope because of OpenAI's ChatGPT [1]. After OpenAI released ChatGPT, AI-powered projects attracted technical persons and people without technical backgrounds. That means the workflow of AI projects not only covers its technical content but also extends to how to integrate them into products. Traditionally, when we discuss the workflow of AI projects, we mainly define this process as combining data preparation, AI modeling, simulation and evaluation, and deployment. That is most often defined as workflow in AI projects. Meanwhile, currently, especially with the development of **large language** (pre-trained) **models** (**LLMs**) and how to integrate them into AI products, the definition of "workflow" in AI projects shifts to AI-powered projects, which also includes how to incorporate the AI projects into a production environment.

In this section, we have structured our discussion into two distinct parts. The first part delves into the fundamental workflow, focusing on the technical aspects of AI projects. The second part explores the extended workflow, which details the process of integrating AI projects into products. This clear division allows for a comprehensive understanding of the entire AI project workflow.

Fundamental workflow – creating an AI model from scratch

In this subsection, we concentrate on the workflow of traditional **machine learning** (**ML**) as the first step and then move on to the fundamental workflow of the pre-trained language model with additional steps.

Workflow for the traditional machine learning model

To obtain an available machine learning model, normally, there are three steps: pre-processing, model training, and model testing, which are illustrated in *Figure 6.1*. It shows the workflow of traditional machine learning:

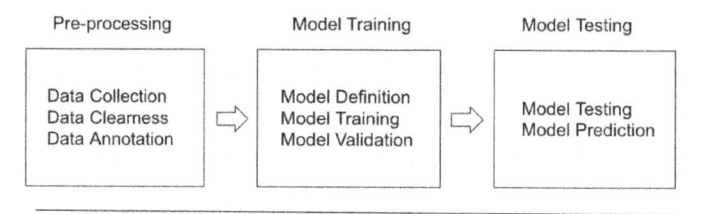

Figure 6.1 – Fundamental workflow for the traditional machine/deep learning system

It includes three primary steps, as we mentioned. The pre-processing step includes data collection, data clearness, and data annotation. The stages of data collection and data clearness are different due to the different tasks and most of them are associated with the original data collection, task-specific data formatting, and so on. The data annotation stage is essential for traditional machine learning, especially for supervised and semi-supervised machine learning because of the annotated data. Finally, dataset splitting is also an important step in the pre-processing process that is not shown in *Figure 6.1*. The primary purpose of dataset splitting is to prepare the dataset for training, validation, and testing.

The model training step includes the model definition, the model training, and the model validation. The process of model definition is the procedure of model design. For example, for image processing tasks, we normally define the **convolutional neural network** (**CNN**) structure as the fundamental model; meanwhile, for **natural language processing** (**NLP**) tasks, we mostly define the **recurrent neural network** (**RNN**) as the basic structure. Using the split training dataset to train the defined model, this process is called model training, and one term, **Epoch**, is used here to express the number of the training round. One of the three split datasets mentioned at the pre-processing stage is used to train the model and the second one is used to validate the trained model. Model validation refers to the process of confirming that the model obtains the designed purpose.

Finally, testing/prediction of the trained model with the last split dataset is the last step. The purpose of model testing is to verify the performance of a fully trained model.

We will move on to the workflow for the pre-trained model in the next section.

Workflow for the pre-trained AI model

Pre-trained LLMs, especially after ChatGPT, Claude, and Bard were released, increasingly attract attention not only from researchers, engineers, and students in the fields of computer science, engineering, and data analysis but also from AI model users without a technical background because of the lower bar for AI-model-based development. Although those LLMs belong to deep learning models, there are also several differences. From the perspective of the workflow, the significant differences are the pre-training and fine-tuning steps.

In *Figure 6.2*, the basic workflow for a pre-trained AI system (e.g., an LLM) is presented. Similar to the workflow of traditional machine learning or deep learning models, the fundamental idea of training and testing the AI model is the same, and that is the reason *Figure 6.2* also includes pre-processing and model testing modules. However, contrary to the model training in *Figure 6.1*, the fundamental workflow in *Figure 6.2* for the pre-trained AI system has pre-training and fine-tuning steps. Let us discuss these steps one by one:

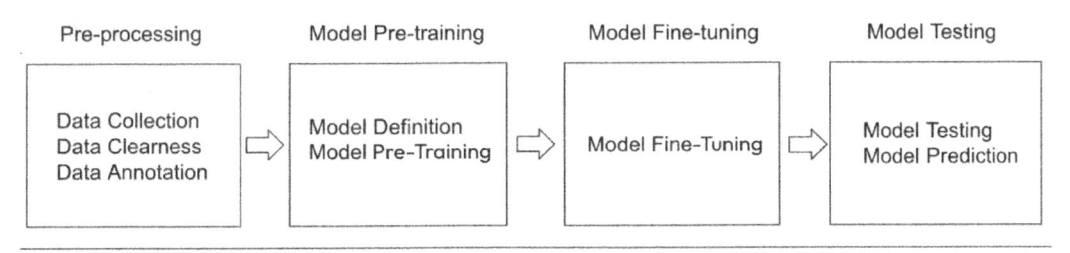

Figure 6.2 – Workflow for the Pre-Trained AI system

The pre-processing step of the pre-trained AI system is similar to that of the traditional machine/deep learning systems, which include data collection, data clearness, and data annotation. The differences are reflected in the original data amount at data collection, and the annotated dataset. Generally, the requirement of data pre-processing of pre-trained systems is looser than that of traditional machine/deep learning systems.

For the model pre-training step in *Figure 6.2* and the training step in traditional machine/deep learning in *Figure 6.1*, both situations have the same model definition modules, although the models' architectures are different. The traditional machine/deep learning systems and the models' architecture vary due to the difference in tasks. For instance, for image processing, CNN models are normally used, whilst RNN models are utilized for language processing tasks. However, for the pre-trained AI systems, the models' architecture is very similar and nearly all of them are based on the transformer structure, no matter whether it is for the image processing or NLP tasks. The architecture difference of pre-training models is presented by the downstream tasks. The generative models, such as GPT-2 [2], GPT-3 [3], Pangu-alpha [4], OPT [5], Llama [6], and so on, are based on the decoder-only architecture. Meanwhile, the encoder-only architecture is primarily used for understanding tasks, such as BERT [7], Code-BERT [8], RoBERTa [9], and so on. Besides them, there are also many models that have both encoder and decoder components and comprise the encoder-decoder architecture, such as the T5 [10] and BART models [11].

The pre-training step in this model is referred to as transformer-based model learning. Normally, we can define the pre-trained model in this phase as the fundamental one. The pre-training tasks normally include the masked language model, the next token prediction, the next sentence prediction, the random token substitution, and so on. After defining the model architecture and the pre-training tasks, this phase uses large amounts of unlabeled data to build out the model parameters to learn the data pattern.

Here, we take RoBERTa [9] as our example to illustrate this step. Building on BERT's language masking strategy, where the system learns to predict hidden sections of text within annotated language examples, RoBERTa uses five English-language (BookCorpus, CC-News, OpenWebText, Stories, and Common Crawl) corpora of varying sizes and domains, totaling over 160 GB of uncompressed text. Similarly, we can also use RoBERTa as the fundamental model and re-train the new fundamental with specific pre-training tasks and datasets, such as CodeBERT. Fine-tuning the CodeBERT [8] model for the programming domain on the fundamental of RoBERTa introduces new pre-training tasks (such as **masked language modeling** (**MLM**) and **replaced token detection** (**RTD**)) and new datasets with 2.1 million bimodal data points and 6.4 million unimodal codes across six programming languages (Python, Java, JavaScript, PHP, Ruby, and Go).

The fine-tuning phase is the process that narrows the fundamental model down to specific domains. For example, we can utilize CodeBERT as the fundamental model and fine-tune it with the natural language code search dataset to get the code search capability on CodeBERT.

The last step is model testing or prediction. Following all this, we can use the CodeBERT model for auto-complete code programming, bug detection, and so on.

Advanced topics– integrating an AI model into a product

In this section, we assume that the process of obtaining available trained models is done, and we discuss how to integrate them into the product environment as the primary components. Our discussion here is based on the workflow of the pre-trained LLM.

Here, LangChain [12] and Semantic Kernel can be used to integrate the trained model into the product environment.

LangChain is a framework for developing applications powered by language models. It is designed to help users create powerful applications for various purposes that are not only calling out to an LLM via an **application programming interface (API)** but also being data-aware and agentic. Data-aware means connecting a language model to other sources of data, and Agents (one of LangChain's modules) allows a language model to interact with its environment. Generally, LangChain provides out-of-the-box support to build NLP applications. It is released as a Python library that uses LLMs such as GPT-3 [12].

Figure 6.3 shows an example of how to integrate the aforementioned CodeBERT pre-trained model with LangChain and, finally, construct a code recommendation:

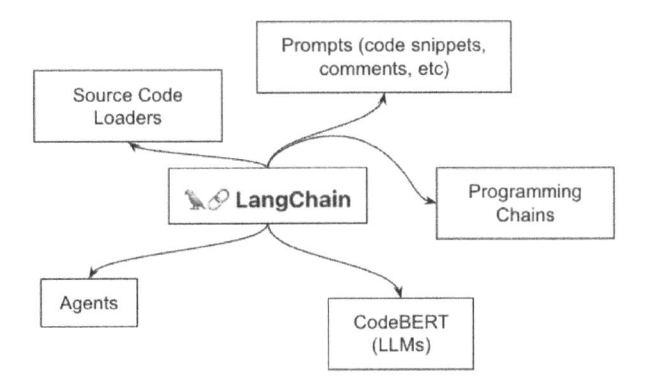

Figure 6.3 – LangChain for a code recommendation with CodeBERT as LLMs

As shown in *Figure 6.3*, LangChain connects to the adhesion agent, which binds all relevant core components to LLMs. For the code recommendation system, LLMs are the primary components that are used to respond to users' prompts. The prompts component is normally represented by code snippets, users' questions, comments, and so on. The programming chains are mostly connected with prompts, and with chain structure, the semantics of users will be maintained step by step. Other components such as source code loaders, agents, and so on are also controlled by LangChain providing the corresponding functionalities.

LangChain has several significant features, which are summarized as follows:

- **Prompt engineering**: LangChain helps to manage, optimize, and chain prompts for various tasks and applications
- **Chains**: LangChain allows users to interface with LLMs or services based on LLMs sequentially and forms the entire semantic expression
- **Data-augmented generation**: LangChain enables the AI system to fetch data from external sources and use it to enhance the downstream tasks

- **Agents**: LangChain creates agents that can interact with the working environment, make decisions, and take action with the help of the corresponding LLMs
- **Memory**: LangChain supports persisting the state among calls of various components of LangChain, such as the calls between a chain and an agent, using different memory implementation
- **Evalution**: LangChain provides some prompts and chains for evaluating the results of tasks using LLMs themselves

Semantic Kernel has been engineered to allow LLM developers and users to flexibly integrate AI services into their existing applications, which is similar with LangChain and is supported by Microsoft. It is an open source SDK that is used to help users integrate AI services such as OpenAI, Azure OpenAI, and Hugging Face with conventional programming languages such as C# and Python.

As *Figure 6.4* shows, Semantic Kernel is located between the users' prompts and fundamental models and provides a set of connectors that makes it easy to add memories and models. In other words, Semantic Kernel is able to add a simulated "brain" to your existing application. Additionally, Semantic Kernel lowers the bar for developers to integrate AI capabilities through AI plugins with the real world. These plugins are composed of prompts and native functions that can respond to perform actions:

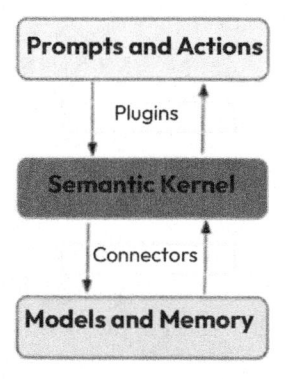

Figure 6.4 – Semantic Kernel

By using multiple AI models, plugins, and memory all together within Semantic Kernel, developers can create sophisticated pipelines that allow AI to automate complex tasks for users.

> **Important note**
> Now that we are done with the introduction of the general workflow of AI projects, we will illustrate the preceding steps in the specific use cases later in this chapter. But first, we will describe the steps to build up the working environment.

Developing and creating a virtual environment

Before explaining all of our text and examples in this book with suitable code samples, we need to build a suitable developing environment. In this book, we will select Conda as our fundamental framework and write all of the code samples with the Python programming language, since that is more popular than others and has an easy automatic package management mechanism.

We will write and test all the samples in Conda version 4.11.0. For the various tasks, we'll use different libraries and introduce them in the relevant sections. In this part, we will see how to install the Conda environment for use in the upcoming sections.

Detailed information can be found at the following link: https://docs.conda.io/projects/conda/en/latest/user-guide/getting-started.html

There are several steps to creating a virtual environment:

1. Install Conda.
2. Create a new environment named ai_security_books (or any name you want) that contains Python 3.9:

    ```
    $ conda create -n ai_security_books python=3.9
    ```

3. Activate the new environment:

    ```
    $ conda activate ai_security_books
    ```

4. Install the following libraries:

    ```
    $ pip install langchain==0.0.263
    $ pip install openai==0.27.8
    $ pip install transformers==4.31.0
    $ pip install tensorflow==2.13.0
    $ pip install tensorflow_datasets==4.9.2
    $ pip install scikit-learn==1.3.0
    $ pip install numpy==1.25.2
    $ pip install matplotlib==3.7.2
    ```

So far, we have prepared the environment and installed basic libraries for use in the upcoming sections.

Tools and libraries for visual network traffic analysis

Starting from this section, we'll present our tools and libraries for AI projects for the analysis of Android applications and concentrate on anomaly network traffic analysis and malware detection. Since typical AI projects mostly focus on computer vision and NLP, we will select topics relevant to visual network traffic analysis and instruction-embedding-based malware detection as our AI projects. In those projects, we'll use methods inspired by computer vision and NLP, but we'll use them in the cybersecurity context.

Background of visual network traffic analysis

In the global landscape, Android stands as the predominant mobile operating system, a position that consequently attracts a myriad of cyber adversaries. Such malicious actors employ pernicious applications to compromise mobile devices, subsequently facilitating a spectrum of illicit activities. These encompass the dissemination of unsolicited emails, propagation of further malware, capitalization of online advertisements through click-fraud tactics, and even duping individuals into divulging sensitive personal information.

Within the realm of Android malware research, both commercial sectors and scholarly communities actively engage in endeavors encompassing malware detection and classification, aiming to counteract the aforementioned malicious trends. Numerous advanced methodologies in this domain emphasize the contextual analysis of Android applications, predominantly focusing on the code within Android APKs. Alternatively, some techniques advocate for a static behavioral examination of Android malware, encompassing an assessment of its components, permissions, and sensitive API interactions. Further contributions have proffered classifiers predicated on the **Factorization Machine** (**FM**) architecture, extracting a plethora of heuristics from both manifest documents and the intrinsic source code. Notably, these techniques predominantly engage in a static evaluation of the Android application, eschewing the execution of the program in their analysis. Concurrently, other studies gravitate toward harnessing the dynamic attributes of Android, generated during the execution of applications within a controlled environment or Sandbox, such as CopperDroid [13] and DroidBox [14].

In the scholarly landscape, there exists a subset of research that illustrates methodologies to transmute executable files into two-dimensional images, subsequently facilitating malware detection via 2D image classification. Additionally, capturing network traffic emerges as an alternative modus operandi to discern between benign and malevolent behaviors. The former method, involving monitoring applications at varying call depths—such as system-level, function-level, and so forth—is resource-intensive, entailing numerous low-tier operations throughout the application's runtime. Conversely, the network traffic analysis avenue is more cost-efficient. Nonetheless, a significant portion of extant research [15, 16] in this domain leans heavily on manually defined rules, constructing classifiers rooted in traditional machine learning. These classifiers typically encompass parameters such as network port utilization, deep packet inspection, statistical markers, and behavioral indicators. An emergent quandary within this paradigm, however, is the discernment of optimal feature selection.

Within this section, we'll elucidate a framework centered on network traffic-pattern-oriented malware detection and classification. Initially, we'll undertake the transformation of network packets into two-dimensional grayscale images. Subsequently, CNNs are employed to pre-train the classification network, specifically tailored to the network traffic attributes. Following this, a bi-directional **Long Short-Term Memory** (**LSTM**) network is harnessed to address the continuous stream of network traffic, executing malware classification in a manner analogous to the classification task of 2D image sequences.

Tools and libraries for visual network traffic analysis

All in all, we summarize the preceding context as our purpose, which is introducing the AI tools and libraries relevant to computer vision by an example of visual network traffic analysis. We'll present these tools and libraries in detail as follows.

Tools of visual network traffic analysis

In a fundamental workflow for traditional machine/deep learning systems, there are three stages: pre-processing to prepare the dataset, model training to train the model, and model testing to evaluate the trained model. We divide our tools of visual network traffic analysis with the same policy. *Figure 6.5* shows one of the workflows to a convert network traffic package to an image and then do the network flow classification to get the benign and malicious network flow:

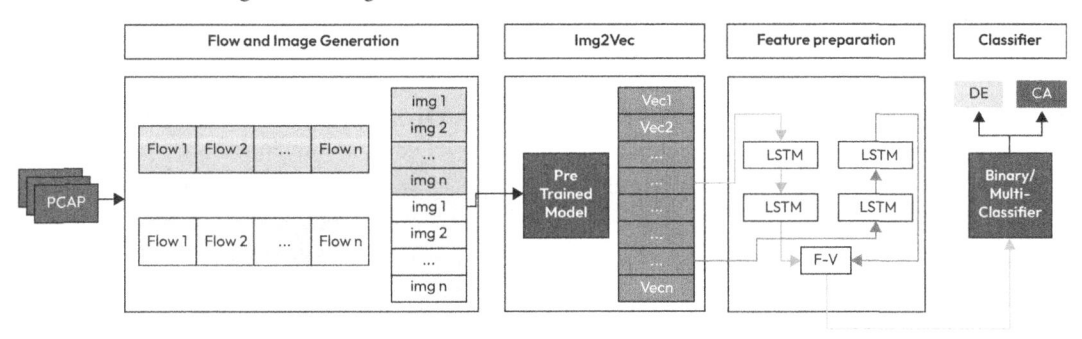

Figure 6.5 – Feature Preparation: Visual network traffic processing and malicious traffic detection

This process consists of the following:

1. **Packet capture (PCAP)** file pre-processing into an image
2. Img2vec – turning the image into a vector
3. LSTM layers – sequence modeling
4. Classification

In the next subsection, we'll describe the pre-processing steps for network traffic data.

Pre-processing

The purpose of the pre-processing stage in our case is to capture the network traffic and convert the corresponding files to images. As shown in *Figure 6.5*, capturing the network traffic with tools such as wireshark and other sniffers, and saving it to PCAP files is the first step. Then we use our own tools to convert the PCAP files to images. In the realm of network traffic analysis, three distinct granularities are recognized: raw packet level, flow level, and session level. For our analytical procedure, we employ the SplitCap tool, a utility designed to segment extensive PCAP files into multiple delineated files, each corresponding to a singular **Transmission Control Protocol (TCP)** or **User Datagram Protocol (UDP)** session. SplitCap operates as a command-line utility, deriving its packet parsing library from

NetworkMiner. While SplitCap possesses the capability to filter expansive PCAPs based on port or IP criteria, in this particular context, we prioritize filtration using TCP/UDP parameters.

In our research, the primary focus is directed toward the analysis of network flows. Each raw packet is characterized by a 5-tuple comprising the source IP, source port, destination IP, destination port, and protocol types (e.g., TCP, UDP). Notably, a network flow conglomerates various packets sharing an identical 5-tuple. It is imperative to highlight that all raw packets within a singular network flow are methodically sequenced in chronological order.

The methodology for transmuting the PCAP file into an image is delineated ahead. Commencing with the segmentation of network flows from the raw network packets, these flows subsequently undergo a transformation into two-dimensional images, analogous to the representation in *Figure 6.5*. Employing normalization techniques—specifically trimming and padding—ensures a uniform size across all network flows. Should a network flow exceed 784 bytes, it is truncated to this threshold. Conversely, flows that do not meet the 784-byte criterion are padded using the 0x00 byte until they attain the stipulated length. These standardized files are then transposed into two-dimensional grayscale images wherein each byte corresponds to a pixel value—for instance, 0x80 symbolizes a gray shade, and 0xff represents white.

Concurrent with this process, classification labels are assigned, demarcating distinct network traffic categories. Within the scope of our research, five labels are designated, accounting for four distinct malware families and a benign category. This taxonomy aligns coherently with our malware classification objective. Indirect model pre-training is accomplished based on the aforementioned categorization. To elucidate further, for the malware classification task, samples are labeled under five categories (comprising four malware categories and one benign). Conversely, for the malware detection facet, samples are bifurcated into two classes (benign and malicious), denoting either malware or benign entities.

In conclusion, we use the Wireshark tool to capture the network traffic, SplitCap to split the PCAP file, and TensorFlow to train and test our AI models.

Table 6.1 summarizes all tools we use in the visual network traffic analysis use case:

Tool	Description	Version	Comments
Wireshark	Network traffic analyzer	2.2.4	macOS 13.4.1/ Ubuntu 22.04
SplitCap	Splits large PCAP files into multiple files	0.20.0	Windows 10/Ubuntu 22.04
TensorFlow	End-to-end open source platform for machine learning	2.13.0	macOS 13.4.1/ Ubuntu 22.04

Table 6.1 – Summary of tools in visual network traffic analysis

Model training and testing

For model training and testing, we use TensorFlow as the framework for our AI project in this section. With TensorFlow, we can create our neural network to learn visual network traffic features, validate the trained neural network, and of course, test or evaluate the trained neural network.

After introducing the stages of these AI projects and the tools used in those stages, we'll introduce the aforementioned tools in detail as follows.

Wireshark functions as a network traffic analyzer, colloquially termed a "sniffer," and is compatible with operating systems including Linux, macOS, *BSD, other Unix and Unix-like systems, and Windows. As a freely accessible and open source packet analyzer, it serves myriad purposes, including network troubleshooting, in-depth analysis, the facilitation of software and communication protocol development, and pedagogical applications. The software empowers users to capture packets, frames, and bitstreams from **network interface controllers** (**NICs**) operating in promiscuous mode. This enables the visualization of all traffic traversing the NIC interface, encompassing unicast traffic not specifically directed to the NIC's MAC address. Underpinning Wireshark's functionality is the Qt graphical user interface library and packet capture and filtering libraries—`libpcap` and `npcap`. Comprehensive guidance on employing Wireshark for live network data capture can be accessed by visiting the Wireshark homepage at https://www.wireshark.org/docs/wsug_html/#ChapterCapture. On macOS, we installed the macOS 4.0.7 stable version on ARM architecture. To successfully run Wireshark, we still need to install ChmodBPF in order to capture network traffic from NIC. The steps are as follows:

1. Download Wireshark from https://www.wireshark.org/download.html.
2. Select the macOS on ARM for your laptop (depending on the specific architecture, please select a suitable version).

94   Workflow, Tools, and Libraries in AI Projects

3. Install the download file and select **Wireshark >>> Applications**, which is illustrated in *Figure 6.6*:

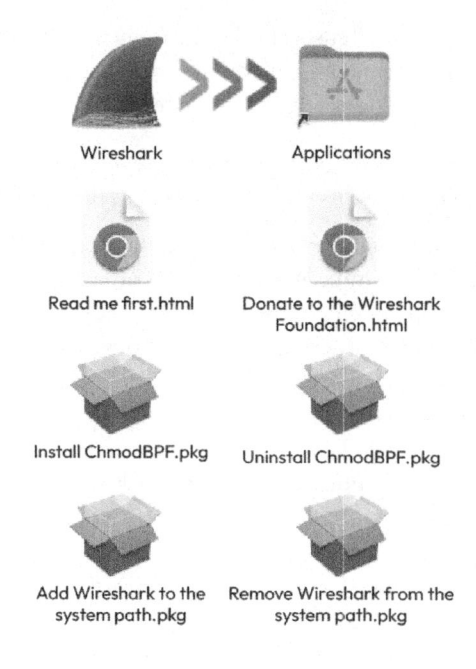

Figure 6.6 – Wireshark installation

4. Then start the newly installed Wireshark application, and if you see an image like that in *Figure 6.7*, that means the installation of Wireshark is successfully done:

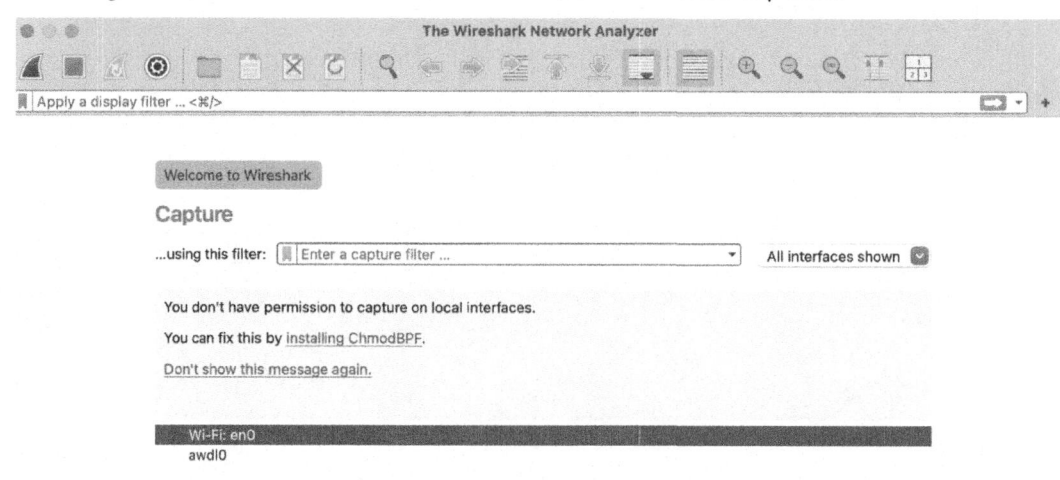

Figure 6.7 – Starting Wireshark

After installing Wireshark, let us try to use it to capture network traffic. The following figure shows the network traffic captured from the macOS Wi-Fi en0 card:

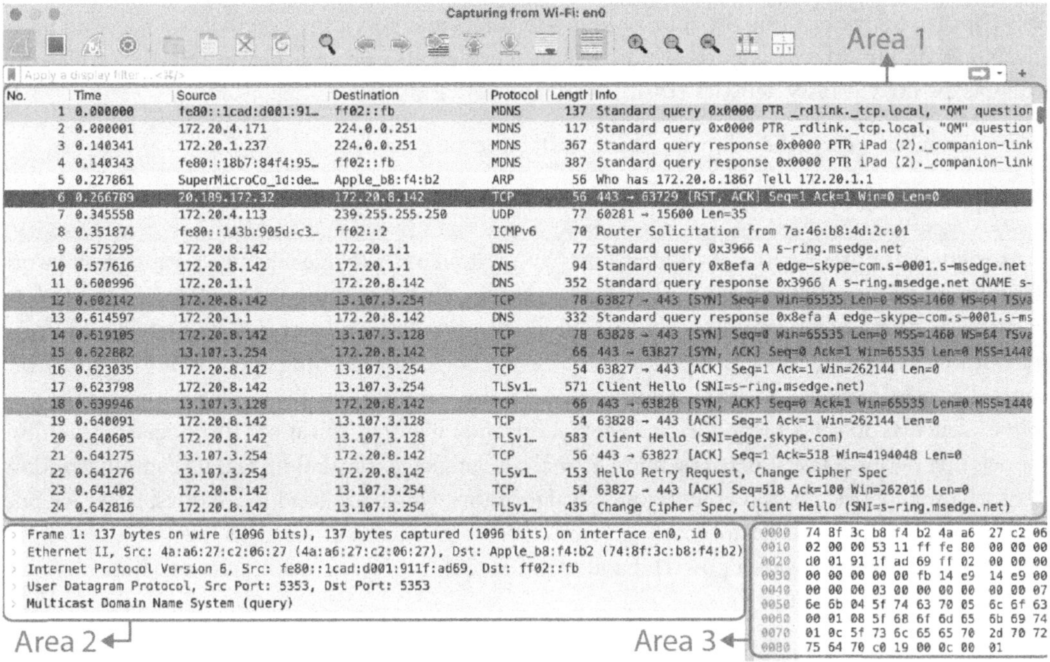

Figure 6.8 – PCAP file in Wireshark and its introduction

Area 1 shows the general information from Wireshark and each row shows the flow level information including the time, source IP address, destination IP address, the type of protocol, the length of the package, and the detailed information. Area 2 shows the captured flow information, including the information from the transmission layer to the data link layer such as the Transmission Control Protocol, Internet Protocol version 4, Ethernet II, and data frame. Area 3 extends the content in area 2 and shows the detailed data. In our case, we extend the IPv4 data in area 3.

SplitCap is a network analysis tool primarily used for handling large PCAP files. While it's not a Python library such as `scikit-learn` or `tensorflow`, it is a command-line tool that helps in splitting and filtering PCAP files based on various criteria, making it easier to analyze network traffic. It can be used to split PCAP files, splitting large PCAP files into smaller ones based on IP addresses, protocols, or other criteria. This is useful when dealing with huge capture files that are difficult to analyze as a whole. SplitCap as a tool can filter out specific types of traffic such as HTTP traffic, VoIP traffic, or traffic between certain IP addresses. Network forensics, performance analysis, and security monitoring are all use cases for SplitCap. While SplitCap is a powerful tool for network analysis, it is generally used in conjunction with other tools such as Wireshark for in-depth analysis and visualization of network traffic.

The detailed usage of SplitCap is shown at the following link: https://www.netresec.com/?page=SplitCap

SplitCap is used by executing the command in a terminal and providing arguments, such as an input file and an output directory. Furthermore, you can provide the desired way to split the traffic—for example, by flow, WLAN network (bssid), host, or MAC address.

In our use cases, we use the following command to get the pcap file split by flow:

```
$ SplitCap -r example.pcap -s flow
```

TensorFlow (https://www.tensorflow.org/) is an open source machine learning framework developed by Google. It is used for building, training, and deploying machine learning models, particularly deep learning models such as neural networks. TensorFlow provides a flexible platform for research and production, enabling users to develop machine learning applications ranging from simple linear regression models to complex deep learning architectures. TensorFlow is commonly used for tasks such as object detection, facial recognition, and image classification. It's utilized for building models that perform tasks such as sentiment analysis, language translation, and text summarization. TensorFlow supports reinforcement learning algorithms, which are used in robotics, game AI, and more, and it can be used for forecasting, anomaly detection, and other tasks involving sequential data. Overall, TensorFlow is a powerful and versatile tool for anyone working in machine learning, from beginners to experts.

This section mainly focuses on visual network traffic analysis and the final purpose is to classify the traffic as positive (benign) and negative (malware). Here, we give the classification **Modified National Institute of Standards and Technology** (**MNIST**) example with TensorFlow to show how to use it to train and test a visual task:

```
import tensorflow as tf
import tensorflow_datasets as tfds

(pcap_data_train, pcap_data_test), pcap_ds_info = tfds.load(
    'pcap_mnist', split= ['train', 'test'],
    shuffle_files=True, as_supervised=True,
    with_info=True,
)

def pcap_normalize_data(pcap_image, pcap_label):
    return tf.cast(pcap_image, tf. float32)/255., pcap_label

pcap_data_train = pcap_data_train.map(
    pcap_normalize_data, num_parallel_calls=tf.data.AUTOTUNE)

pcap_data_test = pcap_data_test.map(normalize_data,
    num_parallel_calls=tf.data.AUTOTUNE)
```

```
model = tf.keras.models.Sequential(
    [tf.keras.layers.Flatten(input_ shape= (28,28)),
    tf.keras.layers.Dense(128, activation='relu'),
    tf.keras.layers.Dense(10)]
)

model.compile(optimizer=tf.keras.optimizers.Adam(0.001),
    loss=tf.keras.losses.SparseCategoricalCrossentropy(
        from_logits=True),
    metrics=[tf.keras.metrics.SparseCategoricalAccuracy()],)

model.fit(pcap_data_train, epochs=6)
model.predict(pcap_data_test)
```

Lines 1~2 are used to import the necessary libraries. In our case, we import the `tensorflow` and `tensorflow_datasets` libraries. Line 4 is loading the `pacp_mnist` dataset and splitting it into the training and testing datasets. Lines 6~7 show a defining function to normalize the image's value to the 0~1 range. Lines 9~10 show the procedure to use the defined normalized function to process the original images loaded by the `load` function in line 4. Line 13 defines a neural network with a flatten layer, a dense layer with the `relu` function as the activation function. The last layer is a 10-output dense layer, which is used to classify all images into 10-category. Line 15 uses the compile function to configure the model for training. In our case, we use Adam optimizer with a learning rate of 0.001, `SparseCategoricalCrossentropy` as the loss function, and sparse categorical accuracy as our metric. Lines 16~17 are used to train and test the AI models.

Libraries of visual network traffic analysis

Similar to the tools in the preceding section, in this section, we'll describe all libraries in our visual network traffic analysis work in detail and explain them. *The following Table 6.2* summarizes all libraries we use in the visual network traffic analysis use case:

Library	Description	Version	Comments
Keras	Deep learning API written in Python	2.13.4	TensorFlow 2.13.0
scikit-learn	Python module for machine learning built on top of SciPy	1.3.0	TensorFlow 2.13.0
NumPy	Fundamental package for scientific computing with Python	1.25.2	TensorFlow 2.13.0
Matplotlib	Python plotting package	3.7.2	TensorFlow 2.13.0
Pillow	Python image library	10.0.0	TensorFlow 2.13.0

Table 6.2 – Summary of libraries in visual network traffic analysis

Keras serves as a high-level deep learning API, meticulously crafted in Python and operating atop the machine learning framework, TensorFlow. Its inception was driven by an imperative to expedite experimental processes while simultaneously ensuring an enhanced developer experience. Fundamentally, Keras aims to furnish developers with an efficient API, streamlining the deployment of machine-learning-integrated applications. The hallmark attributes of Keras encompass its simplicity, adaptability, and potency. The term "simplicity" in this context underscores Keras's ability to diminish the cognitive burden on developers, thereby enabling them to concentrate on salient aspects of the problem, optimizing for user-friendliness, debuggability, maintainability, and deployment efficacy.

The scikit-learn library is a popular open source machine learning library in Python. It provides simple and efficient tools for data mining, data analysis, and machine learning. It is built on top of other Python libraries such as NumPy, SciPy, and Matplotlib, making it a powerful and flexible tool for building machine learning models. Typical use cases include classification (spam detection, image recognition, and sentiment analysis), regression problems (predicting house prices and stock market forecasting), clustering (customer segmentation and document clustering), and anomaly detection (fraud detection, network security, and phishing detection).

NumPy seamlessly integrates the computational capabilities intrinsic to languages such as C and Fortran into the Python environment. This integration, while powerful, also epitomizes simplicity, as solutions articulated through NumPy often manifest as lucid and sophisticated. Serving as an indispensable library for scientific computing within Python, NumPy proffers an exhaustive suite of mathematical functionalities, encompassing linear algebra routines, random number generation, Fourier transformations, and beyond. It is designed to accommodate a diverse spectrum of hardware configurations and computational paradigms, exhibiting compatibility with distributed systems, **Graphics Processing Unit (GPU)** architectures, and sparse array libraries. At its nucleus, NumPy is predicated on adeptly optimized C code, harmoniously fusing Python's inherent adaptability with the agility of compiled languages. The elevated syntactical structure of NumPy ensures its accessibility, rendering it conducive for developers irrespective of their experiential background. Adhering to the liberal tenets of the BSD license, NumPy's open source stature is underscored by its development and sustenance on GitHub, driven by a dynamic, agile, and heterogeneous collective.

Matplotlib stands for an expansive library tailored to curating static, animated, and interactive visual representations within the Python programming milieu. Its versatility encompasses facilitating rudimentary tasks while also provisioning avenues for intricate operations. With Matplotlib, users can engender plots of publication-worthy caliber, craft interactive figures with capabilities such as zooming, panning, and real-time updating, and also refine the visual aesthetics and spatial layout. Additionally, it seamlessly integrates with platforms such as JupyterLab and diverse **graphical user interfaces (GUIs)**. Its applicability extends to Python scripts, Python/IPython interactive environments, web application servers, and an array of GUI toolkits. Originally conceptualized and developed by John D. Hunter, Matplotlib has since garnered a robust development community. It is disseminated under the auspices of a BSD-style license.

Pillow, colloquially known as **PIL**, known as the **Python Imaging Library**, serving to augment the Python interpreter with image processing proficiencies. This esteemed library proffers expansive file format compatibility, a judiciously optimized internal representation, and commendably potent image processing utilities. The crux of this image library is meticulously crafted to interface with data delineated in select foundational pixel formats. It facilitates a comprehensive spectrum of generic image processing functionalities, thereby establishing a robust foundational base. Notably, in juxtaposition with other libraries elucidated in this discourse, Pillow does not wholly embrace an open source ethos. The custodians of Pillow, in conjunction with numerous other package curators, collaborate with Tidelift to proffer commercialized support tailored to enterprise-level applications, concurrently ensuring assiduous maintenance for the open source dependencies integral to application development.

So far, we have introduced the fundamental concept, libraries, and tools used to convert network traffic to images. Let us move on to network traffic analysis in the next section.

An example of visual network traffic analysis

In this section, we'll use all the preceding tools and libraries to construct a visual network traffic project. As introduced in the *Workflow of AI projects* section, we'll divide our example into three parts.

The first part is the dataset pre-processing, which includes original data generation, visual network package generation, and so on. The following Python code snippet shows the code for this process:

```
# MNIST_Preparing_session_png.py

def getMatrixfrom_pcap(filename,width):
    with open (filename, 'rb') as f:
        content = f.read()
    hexst = binascii.hexlify(content)
    fh = numpy.array([int (hexst[i: i+2],16)
        for i in range (0, len(hexst), 2)])
    rn = len(fh)/width
    fh = numpy.reshape(fh[:rn*width], (-1, width))
    fh = numpy. uint8(fh)
    return fh
```

The `MNIST_Preparing_session_png.py` file is used to convert the network flows to images. The basic idea of this conversion is presented in the *Pre-processing* section . We use the NumPy library to define the array, reshape the array, and perform other related functions. Lines 2~3 open the split network flow files using `rb` mode and the read content is then converted to hex form by using the `binascii.hexlify()` function. The current hex form content will be padded and trimmed to 784 bits by lines 4~7. Lines 8~9 are to store the converted hex form content for further usage. The whole code will be found under the `Chapter07/01/MNIST_Preparing_session_png.py` file.

Line 1 is the start of an iteration, which is used to handle all images converted from the network flow. Lines 2 and 3 define arrays to store the image and its corresponding label. Lines 5~10 find the files ending with .png and append them into a list. Line 12 shuffles the files in the list and prepares for the further dataset-splitting process.

This code snippet can be found at the following link: https://github.com/PegX/Falcon/blob/master/4_Png2Mnist.py

```
for name in Names:
    data_image = array('B')
    data_label = array('B')

    FileList = []
    for dirname in os.listdir(name[0]):
        path = os.path.join(name[0],dirname)
        for filename in os.listdir(path):
            if filename.endswith(".png"):
                FileList.append(
                    os.path.join(name[0],dirname,filename))

    shuffle(FileList)

    for filename in FileList:
        label = int(filename.split('\\')[2])
        Im = Image.open(filename)
        pixel = Im.load()
        width, height = Im.size
        for x in range(0, width):
            for y in range(0, height):
                data_image.append(pixel[y,x])
        data_label.append(label) # labels start (one unsigned byte each)
    hexval = "{0:#0{1}x}".format(len(FileList),6) # number of files in HEX
    hexval = '0x' + hexval[2:].zfill(8)

    header = array('B')
    header.extend([0,0,8,1])
    header.append(int('0x'+hexval[2:][0:2],16))
    header.append(int('0x'+hexval[2:][2:4],16))
    header.append(int('0x'+hexval[2:][4:6],16))
    header.append(int('0x'+hexval[2:][6:8],16))
    data_label = header + data_label
```

```
        if max([width,height]) <= 256:
            header.extend([0,0,0, width,0,0,0, height])
        else:
            raise ValueError(
                'Image exceeds maximum size: 256x256 pixels');

        header [3] = 3 # Changing MSB for image data (0x 00000803)
        data_image = header + data_image
        output_file = open (name [1] +'-images-idx3-ubyte', 'wb')
        data_image.tofile(output_file)
        output_file.close()
        output_file = open (name [1] +'-labels-idx1-ubyte', 'wb')
        data_label.tofile(output_file)
        output_file.close()
```

Lines 14~22 present the process of reading the image and its corresponding label into the memory by using the Pillow library. In our case, we use the Image.open () and load () functions to image size attribute to store the images and collect them into the dataset.

Lines 25~45 store all read images and labels to four files. There are image arrays for training and testing, and also the training and testing labels. The second part is the model training. With the split training dataset and the defined neural network, we train the AI models with multiple rounds. The last part is the model testing or prediction.

The following link contains example code for training the model: https://www.tensorflow.org/datasets/keras_example

We use a sequential model with a ReLU activation function, such as the following:

```
tf.keras.models.Sequential(
    [tf.keras.layers.Flatten(input_shape= (28,28)),
    tf.keras.layers.Dense(128, activation='relu'),
    tf.keras.layers.Dense(10)]
)
```

The following line is for compiling the model and declaring the loss function and performance metrics for training:

```
model.compile(
    optimizer=tf.keras.optimizers.Adam(0.001),
    loss=tf.keras.losses.SparseCategoricalCrossentropy(
        from_logits=True),
    metrics=[tf.keras.metrics.SparseCategoricalAccuracy()],
)
```

The following two lines are used to train and test the model, respectively:

```
model.fit(ds_train, epochs=6, validation_data=ds_test,)
```

So far, we have finished the tool and library introduction for converting the network traffic to an image and doing malware detection on the image.

Tools and libraries for malware detection

In this section we describe tools and libraries used in AI projects for malware detection, in particular in Android applications. Having in mind the popularity of AI applications in computer vision and NLP, we will present content on instruction-embedding-based malware detection.

Background of malware detection

In this section, we'll primarily discuss Android malware detection as a specific case of malware detection and introduce tools and libraries focusing on Android application analysis. The same methodologies should be easily extended to other platforms such as Windows, iOS, Linux, HarmonyOS, and macOS.

Android applications predominantly utilize the Java programming language, subsequently compiled into **Dalvik Executable** (**DEX**) bytecode, and are executed within either the Dalvik virtual machine or the **Android Runtime** (**ART**). Each application within the Android ecosystem comprises DEX files, the `AndroidManifest.xml` file, Android resource files, and other pertinent configurations.

Undertaking static analysis of Android applications facilitates an intricate comprehension of the code's architecture without necessitating its execution. Such an analysis is instrumental to ascertaining that the code conforms to established industry benchmarks. A quintessential static analysis procedure commences by transposing the code of the scrutinized application into abstract models. One pivotal technique employed is the call graph analysis or API call graph analysis, which elucidates the trajectory of function transfers from the initiating callers to the respective callees. This method delineates every conceivable execution path of the program statically. However, in practical code evaluation, the call graph tends to be over-approximated, attributable to challenges such as points-to conundrums and undecidable target complications.

The area of Android malware detection has attracted several security researchers, and several works in Android malware's static analysis field are presented with different features, such as those of permissions, network traffic, program structure, and so on. Permission-based malware detection is based on extracting features from the permissions of Android apps. Together with the API calls, features of app permissions, these features are extracted by parsing data from the manifest file and source code of each mobile application.

Using API calls to detect Android malware also attracts much attention, especially when combined with the permission-based method. This type of detection methodology identifies the subset of Android APIs that have effective features and classifies Android applications as benign or malicious applications. In general, two ranked lists of Android APIs can be constructed. The first one is `benign_API_list`, which contains top-ranked APIs used in benign apps, while the other, `malicious_API_list`, includes the malicious apps' commonly used APIs [14]. After the training of using lots of training samples, the pattern from the two lists keeps the features of both benign and malicious apps.

Numerous studies leverage the n-gram opcode methodology for Android malware detection. A subset of these investigations introduces innovative Android malware detection systems utilizing CNNs. Such networks are adept at autonomously discerning malware features from unprocessed opcode sequences, obviating the need for manually curated malware feature engineering. Concurrently, alternative research explores the efficacy of n-gram opcode frequencies in discerning Android malware and questions whether specific malware families exhibit heightened or diminished susceptibility to this approach. The Adagio system presents a malware detection mechanism rooted in kernel-hashing on the function call graph. This mechanism hinges on proficient embeddings of function call graphs, drawing inspiration from a linear-time graph kernel and utilizing a distinct feature map.

As well as the preceding static-analysis-based malware detection method, there are many dynamic-analysis-based methods. Among those methods, many of them are implemented by analyzing the log file. Take **EnDroid** as an example. This tool is used for recording dynamic behavioral features of Android applications—for instance, system-level behavior and application-level malicious behavior, such as the stealing of personal information, premium service subscription, and malicious service communication. By analyzing the recorded runtime data, EnDroid adopts a feature selection algorithm and can distinguish malicious applications from benign applications with an ensemble learning algorithm. In addition to the log-file-based malware detection systems, which do not use the features from the Android code itself, several works are using the network traffic feature to detect Android malware behaviors.

In the next subsection, we'll enumerate tools for malware detection that are commonly used in cybersecurity.

Tools for malware detection

In this section, we will introduce several tools for malware detection, which include `Androidguard`, `APKTool`, `VirusTotal`, and so on.

Here is a list of tools for malware detection on various platforms:

Library	Description
Androguard	Full Python tool to analyze Android files
Apktool	Tool for reverse engineering Android APK files
Angr	Open source binary analysis platform for Python

Library	Description
VirusTotal	Tool to aggregate many antivirus products and online scan engines, allowing users to check for viruses
CopperDroid	Dynamic analysis tool that focuses on identifying malware in Android apps
SandDroid	Dynamic analysis tool

Table 6.3 – Tools for malware detection and their descriptions

We will enumerate these commonly used tools for malware detection, particularly concentrating on Android malware detection, although some tools (such as VirusTotal and Splunk) can also be used for Windows and Linux malware detection.

Androguard is a comprehensive Python tool tailored to interacting with Android files and is exclusively compatible with Python 3. Users have the option to employ Androguard's command-line interface or its graphical frontend, or integrate Androguard as a library within bespoke tools and scripts. The Androguard suite encompasses various specialized sub-tools. For instance, to decode AndroidManifest.xml or resources.arsc, one can utilize androguard axml and androguard arsc, respectively. To extract information about the associated certificates, androguard sign is available. For those seeking to construct call graphs, androguard cg is recommended, whereas androguard decompile is suitable for generating **control flow graphs** (**CFGs**).

Figure 6.8 shows the basic commands of Androguard and their usage for various purposes:

```
(ai_security_books) → 01 androguard
Usage: androguard [OPTIONS] COMMAND [ARGS]...

  Androguard is a full Python tool to play with Android files.

Options:
  --version              Show the version and exit.
  --verbose, --debug     Print more
  --quiet                Print less (only warnings and above)
  --silent               Print no log messages
  --help                 Show this message and exit.
Commands:
  analyze      Open a IPython Shell and start reverse engineering.
  apkid        Return the packageName/versionCode/versionName per APK as...
  arsc         Decode resources.arsc either directly from a given file or...
  axml         Parse the AndroidManifest.xml.
  cg           Create a call graph and export it into a graph format.
  decompile    Decompile an APK and create Control Flow Graphs.
  disassemble  Disassemble Dalvik Code with size SIZE starting from an...
  gui          Androguard GUI
  sign         Return the fingerprint(s) of all certificates inside an APK.
```

Figure 6.9 – Commands to use Androguard

We can use the `decompile` command to decompile the APK file, create CFGs for each method, and export the CFGs as images:

```
androguard decompile -o outputfolder -f png -i someapp.apk --limit
"^Lcom/elite/.*"
```

Besides the preceding `decompile` command, we can also use the `AnalyzeAPK` command to analyze the Android DEX file step by step. For example, consider the following:

```
(ai_security_books) → 01 python3.9 -m IPython
Python 3.9.17 (main, Jul  5 2023, 15:35:09)
Type 'copyright', 'credits' or 'license' for more information
IPython 8.14.0 -- An enhanced Interactive Python. Type '?' for help.

In [1]: from androguard.core.bytecodes.apk import APK

In [2]: from androguard.misc import AnalyzeAPK

In [3]: a,d,dx = AnalyzeAPK("./3c9ecaf0697c14d0cd48bd478dcdebab.apk")

In [4]: a
Out[4]: <androguard.core.bytecodes.apk.APK at 0x10413c490>

In [5]: d
Out[5]: [<androguard.core.bytecodes.dvm.DalvikVMFormat at 0x106b4e250>]

In [6]: dx
Out[6]: <analysis.Analysis VMs: 1, Classes: 512, Strings: 801>
```

Figure 6.10 – Commands to analyze an APK file

With the preceding steps, there are three objects (a, the APK file, d, the **Dalvik Virtual Machine** (**DVM**) class, and also the target dx class) we get by calling the `AnalyzeAPK` function. These three objects stand for the APK object as a symbol, the `DalvikFormat` object as d, and the `Analysis` object as dx. After getting the APK object and `DalvikFormat` object, more information about the Android application can be fetched by these corresponding functions, such as `get_permissions()`, `get_activities()`, and so on:

```
In [7]: a.get_permissions()
Out[7]:
['android.permission.READ_SMS',
 'android.permission.CLEAR_APP_CACHE',
 'android.permission.ACCESS_NETWORK_STATE',
 'android.permission.WRITE_SMS',
 'android.permission.ACCESS_WIFI_STATE',
 'android.permission.READ_PHONE_STATE',
 'android.permission.DISABLE_KEYGUARD',
 'android.permission.BLUETOOTH_ADMIN',
 'android.permission.VIBRATE',
 'android.permission.KILL_BACKGROUND_PROCESSES',
 'android.permission.WRITE_EXTERNAL_STORAGE',
 'android.permission.RESTART_PACKAGES',
 'android.permission.BLUETOOTH',
 'android.permission.INTERNET',
 'android.permission.RECEIVE_BOOT_COMPLETED',
 'android.permission.WAKE_LOCK',
 'android.permission.SEND_SMS',
 'android.permission.WRITE_SETTINGS',
 'android.permission.GET_PACKAGE_SIZE',
 'android.permission.RECEIVE_SMS',
 'android.permission.GET_TASKS',
 'android.permission.CHANGE_WIFI_STATE']
```

```
In [8]: a.get_activities()
Out[8]:
['com.extend.battery..Splash',
 'com.extend.battery..TabHandler',
 'com.extend.battery..TaskKillerActivity',
 'com.extend.battery..UninstallerActivity',
 'com.extend.battery..SecurityAuditActivity',
 'com.extend.battery..InstallerActivity',
 'com.extend.battery..FeaturesActivity',
 'com.extend.battery..BoosterActivity',
 'com.extend.battery..BatteryActivity',
 'com.extend.battery..RatingActivity',
 'com.itframework.installer.util.InstallNonMarketFromUrlActivity',
 'com.itframework.installer.util.NonMarketDialogActivity']
```

Figure 6.11 – More information about the outputs of functions for getting the activities and permissions of an Android application

The results of the `get_permissions` function are frequently used to classify applications into malware and benign entities.

APKTool is a tool for reverse engineering Android apps. Those apps could be third-party, closed-source, binary APKs. It makes working with apps easier because of its project-like file structure and the automation of some repetitive tasks, such as building APKs. APKTool can decode resources to their nearly original form and rebuild them after making some modifications. It makes it possible to debug smali (assembler for Android's **Virtual Machine** (**VM**)) code step by step.

Angr is a tool for analyzing binaries and includes static and dynamic symbolic ("concolic") analysis. Since there are more and more Android applications that include native code in themselves, and 57% of malicious Android behaviors are hidden in the native code, analyzing the native code of Android applications is therefore extremely necessary.

Angr can be used to recover the CFG to do symbolic execution, automatic building of the **return-oriented programming** (**ROP**) chain using angrop, automatic hardening binary using patcherex, automatic exploit generation using rex, and so on. It can be loaded as a library for Python. *Figure 6.11* shows an example of getting the CFGs of a binary file:

```
>>> import angr
>>> proj = angr.Project('./fauxware')
>>> cfg = proj.analyses.CFG()
>>> dict(proj.kb.functions)
{4195552L: <Function _init (0x4004e0)>,
 4195600L: <Function plt.puts (0x400510)>,
 4195616L: <Function plt.printf (0x400520)>,
 4195632L: <Function plt.read (0x400530)>,
 4195648L: <Function plt.__libc_start_main (0x400540)>,
 4195664L: <Function plt.strcmp (0x400550)>,
 4195680L: <Function plt.open (0x400560)>,
 4195696L: <Function plt.exit (0x400570)>,
 4195712L: <Function _start (0x400580)>,
 4195756L: <Function call_gmon_start (0x4005ac)>,
 4195904L: <Function frame_dummy (0x400640)>,
 4195940L: <Function authenticate (0x400664)>,
 4196077L: <Function accepted (0x4006ed)>,
 4196093L: <Function rejected (0x4006fd)>,
 4196125L: <Function main (0x40071d)>,
 4196320L: <Function __libc_csu_init (0x4007e0)>,
 4196480L: <Function __do_global_ctors_aux (0x400880)>}
```

Figure 6.12 – Example usage of the Angr tool

VirusTotal (https://www.virustotal.com/) is a tool that enables the analysis of files and URLs and provides information about hashes, domains, and threat intelligence reports. This tool aggregates outputs from various antivirus engines to provide a more accurate assessment and generate fewer false positives. Users can upload files for analysis or input URLs or hashes. VirusTotal can also do a dynamic analysis of malware behavior by executing it in a sandbox. One important feature is the VirusTotal API, which enables users to query the VirusTotal services automatically.

So far, the aforementioned tools mostly focus on static analysis, while VirusTotal uses the Cuckoo sandbox for the dynamic analysis of malware. For Android malware detection, there are two famous tools, CopperDroid and SandDroid, that should also be mentioned.

CopperDroid is a dynamic analysis tool that focuses on identifying malware in Android apps. It tracks API calls and system events to determine whether an app exhibits suspicious or malicious behaviors. It is an automatic **virtual machine introspection** (**VMI**)-based dynamic analysis system to reconstruct the behaviors of Android application introspection. Interesting OS-specific and high-level Android-specific behaviors are identified by CopperDroid, and it reconstructs these behaviors by observing and dissecting system calls. Meanwhile, the multitude of alternations the **Android runtime** (**ART**) is subjected to over its life cycle is resisted [13].

SandDroid is another dynamic analysis tool. It is an automatic Android application analysis system that combines static and dynamic techniques. It can be used to identify potentially harmful behaviors and analyze app permissions and activities. Additionally, SandDroid also can be used to detect privacy leakage and vulnerabilities. It was evaluated by Cyber Security News and appeared as one of the *10 Best Mobile App Security Scanners in 2023*.

In the next subsection, we'll present libraries that will be needed for the ML-based malware detection use case.

Libraries for malware detection

In this section, we will introduce several libraries that are commonly used in malware detection. The following table contains the libraries that will be used in the subsequent code snippets dedicated to ML-based malware detection:

Library	Description	Version	Comments
Graphviz	Open source graph visualization software library	2.2.4	TensorFlow 2.0.1
NetworkX	Python library designed for the creation, manipulation, and analysis of complex networks or graphs	0.20.0	TensorFlow 2.0.1
Androguard	Android apps analysis library	3.16.4	TensorFlow 2.0.1
Tqdm	Python package for tracking the progress of iterative processes	3.1.1	TensorFlow 2.0.1
Collections	Provides specialized container datatypes beyond the built-in datatypes, such as lists, tuples, and dictionaries	3.9.4	Python3.9
Zipfile	Common archive and compression standard	3.9.4	Python3.9

Table 6.4 – Libraries for ML-based malware detection

Let us discuss each of these libraries as follows.

Graphviz is an open source graph visualization software library that allows you to create diagrams and visual representations of graphs and networks. It provides a set of tools and APIs for generating various types of diagrams, including directed and undirected graphs, flowcharts, organizational charts, and more. Graphviz is widely used in software engineering, data visualization, network analysis, and other fields where visualizing relationships and structures is important.

To use Graphviz, you typically write DOT (which is a graph description language) language code to define the graph's structure and attributes. Then, you pass this code to the Graphviz tools, which generate the desired visual representation based on the layout algorithm and output format you specify. Keep in mind that while Graphviz is a powerful tool for generating graph visualizations, it might have a learning curve if you're new to the DOT language or graph visualization concepts. However, once you grasp these fundamentals, Graphviz can be a valuable asset for creating informative and visually appealing diagrams.

NetworkX is a Python library designed for the creation, manipulation, and analysis of complex networks or graphs. It provides a wide range of tools and functions to work with graphs, which consist of nodes (vertices) and edges (connections) between those nodes. NetworkX is particularly useful for tasks involving network analysis, graph algorithms, visualization, and modeling. NetworkX is widely used in various fields, including social network analysis, biology, transportation systems, computer science, and more. It's a versatile tool for studying the structure and behavior of networks in a computational context. In our case, we need NetworkX to store the nodes and edges for the control flow graph or program dependence graph. The nodes, in our case, are functions and their content, and the edges are function calls and program dependence relationships.

The **tqdm** library is a Python package that provides a simple and customizable progress bar for tracking the progress of iterative processes, such as loops or function calls. The name tqdm stands for *taqaddum* in Arabic, which means progress. When you're running code that involves iterations or lengthy computations, it can be helpful to have a visual indication of how much progress has been made. The tqdm library offers an easy way to add progress bars to your loops or functions, making it easier to monitor the execution progress. The tqdm library provides various customization options, such as setting the progress bar style, showing the estimated time remaining, controlling the appearance of the progress bar, and more. It can be used in various scenarios where you want to provide better feedback to users about the status of a time-consuming operation. In our case, the tqdm library is used to show the progress of handling lots of APK files.

The **Collections** library is a built-in module in Python that provides specialized container datatypes beyond the built-in datatypes such as lists, tuples, and dictionaries. It contains several classes that offer enhanced functionality and performance for common use cases. This module is part of the Python standard library, so you don't need to install it separately—it's available with any standard Python installation. The collections library provides these classes and a few other utility functions that can significantly improve your code's efficiency, readability, and functionality when working with certain data patterns.

The **Zipfile** library is a built-in module in Python that provides tools for creating, extracting, and manipulating ZIP archives. A ZIP archive is a common file format used for compressing and packaging multiple files and directories into a single archive file, making it easier to transport and share groups of files. The zipfile module allows you to work with ZIP archives in your Python scripts and programs. The zipfile library is part of the Python standard library, so you don't need to install it separately. It's a useful tool when you need to work with compressed archive files in your Python projects.

In the next subsection, we'll provide an example of an ML-based approach to Android malware detection by embedding a CFG.

An example of Android malware detection

In this section, with the aforementioned content, we'll illustrate a graph-based Android malware detection system.

The whole detection process starts from the command to prepare the original data. In our case, the following is the command to invoke the Androguard analysis framework:

```
$ python3.9 apk_2_graph.py --help
```

With this command, we can see how to use the Androguard tool and several examples.

The following figure shows a screenshot with the arguments of the apk_2_graph.py tool:

```python
29  if __name__ == "__main__":
30      parser = argparse.ArgumentParser(description='Detecting and classifying Android malware for graph embedding')
31
32      parser.add_argument("-d", "--dir", default="",
33                          help="Load APK/DEX files from this directory.")
34
35      parser.add_argument("-o", "--out", default="data/fcg",
36                          help="Select output directory for generated graphs.\
37                          If no directory is given, they will be written\
38                          to the data/fcg directory.")
39
40      parser.add_argument("-i", "--inst", default="data/inst",
41                          help="Select output directory for extracted instruction.\
42                          If no directory is given, they will be written\
43                          to the data/inst directory.")       xupeng, 24 minutes ago • add apk_graph_corpus
44
45      parser.add_argument("-s", "--size", default="64",
46                          help="Indicate the size of instruction/opcode embedding")
47
48      parser.add_argument("-e", "--function_embedding", default="mean",
49              help="Indicate the algorithm of function embedding: mean, sif, rnn")
50
51      parser.add_argument("-g", "--generate_corpus", default="False",
52              help="Indicate to generate opcode corpus or not")
53
54      fcga = parser.add_argument_group('CALL GRAPHS ANALYSIS')
55      fcga.add_argument("-f", "--fcgraphs", action="store_true",
56                          help="Extract function call graphs from all APK/DEX files\
57                          in the given directory.")
58
59      args = parser.parse_args()
60
61
62      mode = ""
63      if args.fcgraphs:
64          args.out = os.path.realpath(args.out)
65          mode='FCG'
66
67      if mode:
68          print_logo()
69          process_dir(args.dir, args.out, args.inst,args.size,args.function_embedding,args.generate_corpus,mode)
70
71      else:
72          exit()
73
```

Figure 6.13 – Code snippet from the apk_2_graph.py tool

The interesting part of *Figure 6.12* is line 69, which invokes the whole graph extraction process [15]. The `process_dir` function is used to convert a series of APKs into graph objects. It loads all APKs in a `dir` subtree and creates graph objects that are pickled for later processing and learning.

The next code snippet contains a function for embedding a **function call graph** (FCG):

```
def graphConstructing(self,inst_dir,size,fe):
    self.inst_function = inst_dir+"/"+self. filename+".inst"
    embedding_name = "embedding_matrix_opcode_"+size+".npy"
    print(embedding_name)
    self.embedding = np.load(embedding_name)
    self.function_embedding = fe
    self.size = size
    self.FE = FE (self.function_embedding,self.size)
    self.index = []
    self. instr2index = []
    print ("FE is created")
    with open("instruction.txt","r") as fd:
        inst2index = fd.readlines()
        for i in range(len(inst2index)):
            self.index.append(i)
            self. instr2index.append(inst2index [i] [: -1])
    self. index2instuction = dict(zip (self. instr2index, self.index))
    self.g = self.build_fcg()
    fd.close()
    print ("prepare to return")
    return self.g
```

Line 1 uses the FCG mode as the default working model, which indicates the generated graph as the type of FCG. Until line 16, the other lines are mainly to prepare the files list for processing by the Androguard analysis framework. Lines 16~18 show the usage of the `progressbar` library to record the percentage of the workflow.

Line 17 is used to build up the FCG. This function is to prepare the instruction, replacing the Dalvik instruction with a mathematics representation).

Line 20 returns app's function graph, and then this Python function ends.

The next screenshot shows a function for building an FCG using the `networkx` library in Python:

```python
 92    def build_fcg(self):
 93        """ Using NX and Androguard, build a directed graph NX object so that:
 94            - node names are method names as: class name, method name and
 95              descriptor
 96            - each node has a label that encodes the method behavior
 97        """
 98
 99        fcg = nx.DiGraph()
100        #f = open(self.inst_function,"w")
101        methods = self.d.get_methods()
102        print("Method length\n",len(methods))
103
104        for method in methods:
105            node_name = self.get_method_label(method)
106            # find calls from this method
107            children = []
108            for cob in method.XREFto.items:
109                remote_method = cob[0]
110                children.append(self.get_method_label(remote_method))
111
112            # find all instructions in method and encode using coloring
113            instructions = []
114
115            for i in method.get_instructions():
116                instructions.append(i.get_name())
117                #f.write(i.get_name()+" ")
118
119            if len(instructions) == 0:
120                #encoded_label2 =  np.zeros(self.size)
121                pass
122            else:
123                encoded_label2 = self.functionNode2vec(self.embedding,self.index2instuction,self.instr2index,instructions)
124
125            #print("finishing the function embedding")
126            fcg.add_node(node_name, label=encoded_label2)
127            fcg.add_edges_from([(node_name, child) for child in children])
128
129        print("graph returning")
130        return fcg
```

Figure 6.14 – Function for building an FCG data structure using the NetworkX library

Figure 6.13 shows all the code to generate a graph on the method level. As line 101 shows, with the `g.get_method()` function, we can obtain all the methods of an Android app and find out all the instructions of one method by using `get_instructions()`. Finally, we use the empty graph structure that is generated by the `networkx` library in line 99 to store the information in lines 126-127. So far, we have successfully converted an Android app to a graph function using vector representation.

Summary

In this chapter, you have been introduced to a typical workflow of AI projects. We showed the variants when the AI models would be either trained from scratch or pre-trained and possibly fine-tuned. Furthermore, we explained the approach to turning an AI model into a product. We also explained in a cybersecurity scenario, where the application of AI methods looks different, especially because of feature extraction. Thereafter, we explained the tools and libraries for malware and network traffic analysis to help you extract features in your own application scenario.

In the next section, we'll present a deep dive into a particular topic of cybersecurity where AI methods are used—malware and network intrusion detection and analysis.

References

- [1] An, Jiafu, Wenzhi Ding, and Chen Lin. 2023. "*ChatGPT: Tackle the Growing Carbon Footprint of Generative AI.*" Nature 615 (7953): 586–86. `https://doi.org/10.1038/d41586-023-00843-2`.

- [2] Radford, Alec, Jeffrey Wu, Rewon Child, David Luan, Dario Amodei, and Ilya Sutskever. 2019. "*Language Models Are Unsupervised Multitask Learners.*" `https://cdn.openai.com/better-language-models/language_models_are_unsupervised_multitask_learners.pdf`.

- [3] Brown, Tom B., Benjamin Mann, Nick Ryder, Melanie Subbiah, Jared Kaplan, Prafulla Dhariwal, Arvind Neelakantan, et al. 2020. "*Language Models Are Few-Shot Learners.*" Arxiv.org 4 (May). `https://arxiv.org/abs/2005.14165`.

- [4] Zeng, Wei, Xiaozhe Ren, Teng Su, Hui Wang, Yi Liao, Zhiwei Wang, Xin Jiang, et al. 2021. "*PanGu-$\Alpha$: Large-Scale Autoregressive Pretrained Chinese Language Models with Auto-Parallel Computation.*" ArXiv.org. April 26, 2021. `https://doi.org/10.48550/arXiv.2104.12369`.

- [5] Zhang, Susan, Stephen Roller, Naman Goyal, Mikel Artetxe, Moya Chen, Shuohui Chen, Christopher Dewan, et al. 2022. *OPT: Open Pre-Trained Transformer Language Models.* ArXiv (Cornell University), May. `https://doi.org/10.48550/arxiv.2205.01068`.

- [6] Touvron, Hugo, Thibaut Lavril, Gautier Izacard, Xavier Martinet, Marie-Anne Lachaux, Timothée Lacroix, Baptiste Rozière, et al. 2023. "*LLaMA: Open and Efficient Foundation Language Models.*" ArXiv (Cornell University), February. `https://doi.org/10.48550/arxiv.2302.13971`.

- [7] Tenney, Ian, Dipanjan Das, and Ellie Pavlick. 2019. "*BERT Rediscovers the Classical NLP Pipeline.*" Proceedings of the 57th Annual Meeting of the Association for Computational Linguistics. 0.18653/v1/p19-1452/p19-1452.

- [8] Feng, Zhangyin, Daya Guo, Duyu Tang, Nan Duan, Xiaocheng Feng, Ming Gong, Linjun Shou, et al. 2020. "*CodeBERT: A Pre-Trained Model for Programming and Natural Languages.*" Empirical Methods in Natural Language Processing, February. https://doi.org/10.18653/v1/2020.findings-emnlp.139.

- [9] Rothman, Denis. 2021. Transformers for Natural Language Processing. Packt Publishing Ltd.

- [10] Mastropaolo, Antonio, Simone Scalabrino, Nathan Cooper, David Nader Palacio, Denys Poshyvanyk, Rocco Oliveto, and Gabriele Bavota. 2021. "*Studying the Usage of Text-To-Text Transfer Transformer to Support Code-Related Tasks.*" IEEE Xplore. May 1, 2021. https://doi.org/10.1109/ICSE43902.2021.00041.

- [11] Tan, Yaoyuan Vincent, and Jason Roy. 2019. "*Bayesian Additive Regression Trees and the General BART Model.*" Statistics in Medicine 38 (25): 5048–69. org/10.1002/sim.83472/sim.8347.

- [12] Oguzhan Topsakal, and Tahir Cetin Akinci. 2023. "*Creating Large Language Model Applications Utilizing LangChain: A Primer on Developing LLM Apps Fast.*" International Conference on Applied Engineering and Natural Sciences 1 (1): 1050–56. https://doi.org/10.59287/icaens.1127.

- [13] Tam, Kimberly, Salahuddin J. Khan, Aristide Fattori, and Lorenzo Cavallaro. 2015. "*CopperDroid: Automatic Reconstruction of Android Malware Behaviors.*" Proceedings 2015 Network and Distributed System Security Symposium. https://doi.org/10.14722/ndss.2015.23145.

- [14] Delosières, Laurent, and David García. "*Infrastructure for Detecting Android Malware.*" Information Sciences and Systems 2013: Proceedings of the 28th International Symposium on Computer and Information Sciences. Springer International Publishing, 2013.

- [15] Mohammad Reza Norouzian, Peng Xu, Claudia Eckert, and Apostolis Zarras. 2021. "*Hybroid: Toward Android Malware Detection and Categorization with Program Code and Network Traffic.*" Zenodo (CERN European Organization for Nuclear Research), September. https://doi.org/10.5281/zenodo.6389838.

- [16] Prasse, Paul, Lukáš Machlica, Tomáš Pevný, Jiří Havelka, and Tobias Scheffer. 2017. "*Malware Detection by Analysing Network Traffic with Neural Networks.*" IEEE Xplore. May 1, 2017. https://doi.org/10.1109/SPW.2017.8.

Part 3: Applications of AI in Cybersecurity

After providing a strong theoretical background in the previous part, this part gives you more practical skills by dedicating chapters to concrete scenarios in cybersecurity and providing examples with coding exercises. This part equips you with ideas and implementation expertise that you can use in your organization. We are provided with a wide range of cybersecurity applications, from malware and network intrusion detection to fraud, phishing, and spam detection, as well as defending industrial control systems. We include a dedicated chapter on LLMs, a growing field in AI, highlighting their applications in cybersecurity.

This part has the following chapters:

- *Chapter 7, Malware and Network Intrusion Detection and Analysis*
- *Chapter 8, User and Entity Behavior Analysis*
- *Chapter 9, Fraud, Spam, and Phishing Detection*
- *Chapter 10, User Authentication and Access Control*
- *Chapter 11, Threat Intelligence*
- *Chapter 12, Anomaly Detection in Industrial Control Systems*
- *Chapter 13, Large Language Models and Cybersecurity*

7
Malware and Network Intrusion Detection and Analysis

Over the years, malware has evolved significantly, emerging as a pervasive global threat to our digital existence. The daily influx of malware has surged exponentially, posing a formidable challenge to security companies. These companies are tasked with scrutinizing hundreds of thousands of malicious samples daily, a workload that undeniably impacts their operational efficiency. This volume sometimes surpasses an astounding 1 million distinct files each day. Simultaneously, the classification of malware has attained heightened significance. New instances of malware incorporate increasingly sophisticated techniques designed to outsmart signature-based detectors and operate discreetly for extended durations. As a result, security analysts find themselves in a race against time to detect malware, whether by scrutinizing the infected host or by monitoring the network through which the malware transmits data.

On top of that, malware designed to exploit network vulnerabilities operates stealthily, often eluding traditional security measures. These programs leverage intricate techniques, such as zero-day exploits and polymorphic code, to avoid detection by conventional signature-based defenses. As security companies grapple with the relentless onslaught of such network-exploiting malware, the stakes are higher than ever, necessitating continuous advancements in cybersecurity protocols.

AI introduces a novel perspective on malware detection, transitioning from delivering a binary verdict to providing intricate explanations. Its proficiency in discerning malicious scripts, especially those that have been obfuscated, surpasses the detection rates achieved by conventional methods in isolation. Notably, AI exhibits heightened efficacy in identifying malware and network intrusions, thereby improving exploit identification compared to reliance solely on traditional tools.

In this chapter, we're going to cover the following main topics:

- Overcoming traditional difficulties
- Proper datasets for creating an AI model

- Exercise 1 – malware detection
- Exercise 2 – intrusion detection
- Moving from detection to classification

By the end of this chapter, you will have gained an understanding of the fundamental concepts regarding malware and network intrusion detection and analysis. Additionally, you will have discovered sources for publicly available datasets that can be utilized to train and assess your models. The exercises presented in this chapter will be the foundation for many real-world problems.

Technical requirements

This chapter has a practical component and requires the following Python libraries:

- `numpy`
- `pandas`
- `scikit-learn`
- `pickle`

For the best learning performance, it's recommended to use notebook libraries and platforms, such as the following:

- `Jupyter`
- `Google Colab`
- `Databricks`

Overcoming traditional difficulties

Traditional malware detection techniques can no longer adequately protect against the malware employed by threat actors. Malware writers often use antivirus evasion techniques, such as obfuscation and sleep functions, to evade malware detection tools such as antivirus software. This is where AI comes in. AI for malware detection offers several advantages over traditional malware detection techniques (signatures and heuristics) and propels the performance of malware detection tools if used correctly. Some advantages include the following:

- **Adaptability to new threats**: While traditional techniques rely on predefined patterns or rules, making them less effective against new and evolving threats, AI can adapt and learn from new data. This makes AI more capable of identifying novel and sophisticated malware that may not have been encountered before.

- **Feature learning**: Standard techniques depend on known patterns and predefined rules, making them less effective in detecting subtle or polymorphic malware. In contrast, AI can automatically learn relevant features from the data, capturing subtle patterns and variations in malware behavior that may be difficult to express with manual rules.

- **Reduced false positives**: Traditional approaches may produce false positives when legitimate programs exhibit behavior similar to malware. However, when properly trained, AI can distinguish between normal and malicious behavior more accurately, leading to fewer false positives and reducing the burden on security analysts.

- **Anomaly detection**: Classic methods often struggle to identify subtle anomalies or deviations from normal behavior, whereas ML models excel at anomaly detection by learning the expected behavior of a system and identifying deviations from it. This helps AI uncover previously unknown threats.

- **Scalability**: Signature databases can become large and cumbersome, affecting system performance and requiring frequent updates. Once appropriately trained, AI typically has a smaller footprint and can more easily be deployed at scale, making it well suited for real-time analysis and large-scale security deployments.

- **Continuous learning**: While signature databases need regular updates to keep up with new threats, AI can continuously learn and acclimate to changes in the threat landscape, providing a more dynamic and proactive approach to malware detection.

- **Behavioral analysis enhancement**: Behavioral analysis may struggle with identifying subtle or context-dependent malicious behavior. AI can complement behavioral analysis by learning from diverse datasets, improving the detection of nuanced malicious activities.

- **Reduced human intervention**: Security analysts often need to manually update signatures and rules compared to AI, which can automate the learning process, reducing the need for constant manual intervention and allowing security teams to focus on more complex tasks.

In this section, we have seen how AI can overcome traditional difficulties in malware detection. However, one important aspect when talking about creating an AI model is knowing how to find the proper data to train it. In the following section, we'll explore this aspect.

Proper datasets for creating an AI model

Many datasets exist for training and testing malware classification and network intrusion detection algorithms. The following compilation offers a thorough and diverse list of such datasets to aid researchers and practitioners. You can use these or any other datasets to train your AI models.

Malware analysis

Let's first see some datasets that can be used for malware analysis:

- **Kaggle**: Renowned for hosting data science competitions, Kaggle offers numerous datasets pertinent to malware detection.
- **Microsoft Malware Classification Challenge**: Microsoft has previously organized challenges focusing on malware classification, with datasets from these events potentially being accessible.
- **The Malware Dataset Repository by UCI**: Maintained by the University of California, Irvine (UCI), this repository houses diverse malware datasets. Navigate through the UCI Machine Learning Repository to discover datasets concerning cybersecurity and malware detection.
- **The NSL-KDD dataset**: Recognized as a prominent resource for intrusion detection, the NSL-KDD dataset encompasses both normal and attack traffic. Though not exclusively tailored to malware, it serves broader cybersecurity objectives.
- **Microsoft Research datasets**: Microsoft Research offers datasets dedicated to malware detection and cybersecurity research.

Network intrusion detection

As for malware analysis, there exist datasets for network intrusion detection. Those datasets have been used in plenty of research papers and scientific projects. Here are some listed:

- **NSL-KDD**: As an updated iteration of the original KDD Cup 1999 dataset, the NSL-KDD dataset rectifies several shortcomings of its predecessor. It furnishes a more authentic depiction of network traffic, encompassing both normal and attack instances.
- **UNSW-NB15**: A comprehensive resource for network-based intrusion detection, the UNSW-NB15 dataset presents a broad spectrum of attack scenarios. It incorporates features such as payload data, statistical attributes, and protocol-specific characteristics.
- **Canadian Institute for Cybersecurity Intrusion Detection Systems 2017 (CICIDS2017)**: The CICIDS2017 dataset offers a contemporary collection of diverse attacks within a realistic network environment. It encompasses both raw packet data and flow-based features.
- **Aegean Wi-Fi Intrusion Dataset (AWID)**: Tailored for Wi-Fi intrusion detection, the AWID dataset encompasses various attack modalities within a Wi-Fi network. It stands as an apt resource for exploring wireless network security.
- **Kyoto 2006+ dataset**: Derived from Kyoto University's honeypots, this dataset presents network traffic data that includes diverse attack scenarios. It serves as a prevalent choice for the assessment of intrusion detection systems.

- **CSE-CIC-IDS2018**: Engineered to assess intrusion detection systems within a realistic network setting, the CSE-CIC-IDS2018 dataset integrates both benign and malicious traffic, spanning many attack types.
- **IoTID20**: This introduces a novel IoT botnet dataset. This dataset boasts a very comprehensive array of network and flow-based features. These flow-based features are instrumental in the analysis and evaluation of flow-based intrusion detection systems.

We'll now see how we can detect malware with AI in practice.

Exercise 1 – malware detection

In this exercise, we will consider a dataset of executable files and attempt to find the malicious ones among them. We will achieve this by leveraging the Random Forest algorithm. Before delving into the code section by section, let's elucidate our rationale for selecting the Random Forest algorithm. Random Forest, an ensemble learning algorithm, constructs numerous decision trees and amalgamates their predictions to enhance accuracy and mitigate overfitting. Renowned for its robustness in ML applications for malware detection, Random Forest handles extensive datasets—a critical attribute in ML implementation—while offering commendable generalization and resistance to overfitting. Its consistent performance across diverse datasets underscores its preferability over alternative algorithms in malware detection tasks. However, optimal algorithm selection for malware detection hinges on dataset characteristics, task-specific requirements, and available computational resources. In practice, antivirus solutions often leverage ensemble models comprising a blend of algorithms to bolster detection efficacy.

The dataset employed in the current implementation originates from Kaggle, with a size of approximately 6.40 MB. Despite its modest size, this dataset boasts meticulously selected features, setting it apart from others. In real-world scenarios, a custom dataset akin to the current one, albeit larger, would be developed. For now, let's dissect our implementation section-wise.

The initial step entails importing essential libraries. `numpy` and `pandas` facilitate data manipulation, while `pickle` aids in model serialization and saving the model in a designated format. `scikit-learn`, encompassing functionalities such as `train_test_split`, `RandomForestClassifier`, and `classification_report`, is employed for various ML tasks such as dataset partitioning, model training, and performance assessment. Here is an example code:

```
import numpy as np
import pandas as pd
import pickle
from sklearn.model_selection import train_test_split
from sklearn.ensemble import RandomForestClassifier
from sklearn.metrics import classification_report
```

We now load the `dataset_train.csv` dataset in CSV format, converting it into a `pandas` DataFrame labeled as data. Dataframe facilitate efficient data handling. Following this, in the subsequent two lines, we print essential information about the dataset, including its columns or features—commonly referred to as such—along with details such as rows, size, and so on (one of the lines is commented out to minimize end user output). Subsequently, we instantiate a new DataFrame named `df` with identical content to the preceding DataFrame (no alterations made). The subsequent step involves feature selection during the preprocessing phase. Finally, the last two lines segregate the dataset features to be utilized (`X`) and the corresponding labels (`y`) from the DataFrame. Specifically, the features selected entail dropping certain columns, namely `Name`, `Machine`, `TimeDateStamp`, and `Malware`, as they were deemed uninformative during model training due to various reasons (e.g., static values and redundancy). Notably, feature selection was conducted before implementation, considering that the dataset already encompasses specific features contributing to the model's high performance. Hence, only four features were omitted during this process. The `Malware` column is designated as the `y` variable. `Malware` merely represents the classification of a sample, characterized by values 0 and 1, where 1 denotes malicious and 0 benign samples. Refer to the code block here:

```
data = pd.read_csv('dataset/dataset_train.csv')
#print(data.info())
print(data.head(10))
df = pd.DataFrame(data)
X = df.drop(['Name', 'Machine', 'TimeDateStamp', 'Malware'],
    axis=1)
y = df['Malware']
```

In the subsequent section, we partition the dataset into training and testing subsets utilizing the `train_test_split` function. Our selection was to opt for an 80-20 (0.2) division into training and testing data. However, this ratio can be adjusted, varying from 70-30 to 85-15, or even to 90-10. Through experimentation, we found that an 80-20 split yielded the most favorable results. To ensure reproducibility of the model results, the `random_state` parameter is set to 42. Additionally, the `stratify` parameter safeguards the preservation of the target variable's distribution in both training and testing datasets. In the subsequent line, we simply print the count of features utilized:

```
X_train, X_test, y_train, y_test = train_test_split(
    X, y, test_size=0.2, random_state=42, stratify=y)
print(f'Number of features used from the dataset is {
    X_train.shape[1]}')
```

We train the model by initializing a Random Forest classifier called `rfc`. This classifier comprises 100 trees, specified through the `n_estimators` parameter, and a designated random seed set via the `random_state` parameter. Additionally, we incorporate **out-of-bag** (**OOB**) scoring and restrict the maximum depth of each tree to 16 using the `max_depth` parameter. Subsequently, the model undergoes training utilizing the training data:

```
rfc = RandomForestClassifier(n_estimators=100, random_state=0,
```

```
    oob_score = True, max_depth = 16)
rfc.fit(X_train, y_train)
```

The first line in the following code section involves generating predictions based on the test set. As a reminder, we previously partitioned the data into an 80-20 ratio. In the second line, we create a classification report. This report encompasses various metrics, including accuracy, precision, recall, and F1-score, among others, for both `Benign` and `Malware` labels as shown here:

```
y_pred = rfc.predict(X_test)
print(classification_report(y_test, y_pred,
    target_names=['Benign', 'Malware']))
```

We now load the test dataset stored in CSV format from the `dataset_test.csv` file into a DataFrame. It's important to note that this dataset comprises unlabeled data containing samples without explicit classifications as malicious or benign. Subsequently, following the loading process, we drop any features not utilized. Finally, leveraging the trained Random Forest model, we make predictions to classify the samples as malicious or benign:

```
test_data = pd.read_csv("dataset/dataset_test.csv")
X_pred = test_data.drop(['Name', 'Machine', 'TimeDateStamp'],
    axis=1)
predictions = rfc.predict(X_pred)
```

In the preceding code section, we established a new DataFrame labeled `results`. This DataFrame encompasses the filenames of the samples alongside their corresponding model predictions, which are categorized as either `malicious` or `benign`. After that, the results are printed for reference. Finally, the trained Random Forest model is serialized using `pickle` and stored in a file named `RF_model.pkl`. This enables the model to be reloaded later for reuse without necessitating retraining. The `load` method from the `pickle` library facilitates loading a model saved in a `pickle` file for subsequent use. This is shown with the following code:

```
results = pd.DataFrame({'File Name': test_data['Name'],
    'Model Prediction': predictions})
print(results)
pkl_filename = "RF_model.pkl"
with open(pkl_filename, 'wb') as file:
    pickle.dump(rfc, file)
```

The following is a sample of the program execution:

```
              precision    recall  f1-score   support

      Benign       0.99      0.96      0.98      1003
     Malware       0.99      1.00      0.99      2920

    accuracy                           0.99      3923
   macro avg       0.99      0.98      0.98      3923
weighted avg       0.99      0.99      0.99      3923

                                         File Name  Model Prediction
0                                    Skype-8.10.0.9.exe                 0
1                                   vlc-3.0.2-win64.exe                 0
2                                        stinger32.exe                 1
3                                   SpotifyFullSetup.exe                 0
4                                     uftp_english.exe                 1
5   161a59f2525518f799c63f916c80fe85f50c5b09c74dc2...                 1
6   eaa478e65696ad5cbdb42c1b4bd6954f2a876fdde2e519...                 1
7                                    reverse_shell.exe                 1
8   873b9eaef6ea5ed6126086594529a3395bdbc5d63c97d8...                 1
9                                 ScratchInstaller1.4.exe                 1
10  69eb27dd3bbf5077dcd795872535b89af9a898254b90ad...                 1
11  3334686141a400bb522824fa6f7faf30614372fe11837a...                 1
12  3ec4cb928846f8298e5a13b3e96bfc2a709cb3b005a31e...                 1
13  252f705dc15d7a305afd3e0619fa014c10b679248f71b7...                 1
14                                        wordweb8.exe                 1
15  c89f1e55b418a4447394994498971c6e6f3848bfe39ef9...                 1
16                                    winrar-x64-550.exe                 0
```

Figure 7.1 – Execution of the Random Forest implementation

As illustrated here, we've achieved exceptional accuracy and precision exceeding 99%, along with commendable results across other metrics. While the size of a larger dataset may introduce some variability in the results, we can anticipate minimal loss or deviation compared to the outcomes observed thus far. It's worth noting that the dataset utilized in this demonstration is a demo dataset of less than 10 MB. In a practical scenario, we would generate another dataset with identical features but of a larger size. To accomplish this, we would initially gather Windows-executable samples from specific platforms, such as https://virusshare.com/, utilizing the platform's specific API. After amassing a diverse array of samples, we would employ a Python program to extract pertinent features.

Exercise 2 – network intrusion detection

Applying AI to intrusion detection offers analogous benefits to using AI for malware detection. In the following exercise, we will attempt to detect malicious traffic. For this purpose, we will use **support vector machines** (**SVMs**) to construct a model for intrusion detection. SVMs possess several advantages in the realm of intrusion detection systems. They excel in high-dimensional spaces, rendering them suitable for environments where feature spaces are intricate and contain many dimensions. Furthermore, SVMs exhibit resistance to overfitting compared to some alternative algorithms, a trait particularly advantageous when dealing with limited labeled data—a common occurrence in intrusion detection scenarios. Additionally, SVMs demonstrate tolerance to irrelevant features. This is a crucial aspect for intrusion detection, where certain features may hold minimal significance in attack detection. Moreover, they adeptly handle imbalanced datasets, a paramount consideration as intrusion detection datasets often feature limited instances representing attacks.

The dataset employed in this implementation is the NSL-KDD dataset, an updated iteration of the original KDD Cup 1999 dataset. While the KDD Cup 1999 dataset was extensively used to evaluate intrusion detection systems, it presented shortcomings. NSL-KDD rectifies these issues, offering a more realistic depiction of network traffic and a more challenging evaluation environment for intrusion detection systems. Other suitable datasets include UNSW-NB15, CICIDS2017, Kyoto 2006+, and CSE-CIC-IDS2018. The NSL-KDD dataset has an approximate size of 20 MB.

Similarly to the previous exercise, we will begin by importing the requisite libraries. These include standard libraries such as `numpy` and `pandas` for numerical operations and data handling, `pickle` for model serialization, and `scikit-learn` for preprocessing, model training, and evaluation. Additionally, in this scenario, we import **support vector classification** (**SVC**), an implementation of SVM tailored for classification tasks in `scikit-learn`. This is shown here:

```
import numpy as np
import pandas as pd
from sklearn.preprocessing import LabelEncoder, StandardScaler
from sklearn.model_selection import train_test_split
from sklearn.svm import SVC
from sklearn.metrics import accuracy_score, classification_report
import pickle
```

The ensuing source code section entails loading the dataset and conducting preprocessing steps to facilitate compatibility with SVMs and enhance performance. Initially, we will define the columns or features of the dataset. Since the `KDDTrain+.txt` dataset lacks explicit column headers, we will manually specify the appropriate columns based on the dataset description, the content, and our domain knowledge. Of paramount importance is the `label` column utilized for classification, distinguishing between `normal` and `abnormal` instances. Notably, the dataset constitutes a multiclass classification problem, featuring numerous options for the `label` column such as `normal traffic`, `DoS attack`, and `brute force`. To accommodate a binary classification problem, we transform the `label` column, accordingly, categorizing instances as `normal` or `abnormal`.

After delineating the columns, we will load the dataset and convert it into a DataFrame using `pandas`. Subsequently, we will discard columns that are either unusable or that could potentially degrade the model's performance. For instance, the `num_outbound_cmds` column holds static values devoid of informational utility. We will then employ the `LabelEncoder` function to convert categorical variables such as `protocol_type`, `service`, and `flag` into numerical form, thereby enabling compatibility with SVMs.

Lastly, we will convert the `label` column into binary values, denoted as `normal` and `abnormal`, and subsequently transform them into numerical values 0 and 1, respectively, to facilitate SVM usage. Additionally, we will print the count of specific values, such as 50,000 instances of normal traffic and 40,000 of abnormal traffic, for further insight into the dataset's distribution:

```
columns =(['duration','protocol_type','service','flag',
    'src_bytes','dst_bytes','land','wrong_fragment','urgent',
    'hot','num_failed_logins','logged_in','num_compromised',
    'root_ shell','su_attempted','num_root', 'num_file_creations',
    'num_ shells','num_access_files', 'num_outbound_cmds',
    'is_host_login','is_ guest_login', 'count','srv_count',
    'serror_rate','srv_serror_rate', 'rerror_rate', 'srv_rerror_rate',
    'same_srv_rate',  'diff_srv_rate','srv_ diff_host_rate',
    'dst_host_count', 'dst_host_srv_count',
    'dst_host_same_srv_rate', 'dst_host_diff_srv_rate',
    'dst_host_same_src_port_rate', 'dst_host_srv_diff_host_rate',
    'dst_host_serror_rate', 'dst_host_srv_serror_rate',
    'dst_host_rerror_rate', 'dst_host_srv_rerror_rate',
    'label','difficulty_level'])
train_data = pd.read_csv(
    'dataset/KDDTrain+.txt', header=None, names=columns)
train_data.drop(
    ['difficulty_level','num_outbound_cmds'],
    axis=1,
    inplace=True)
le = LabelEncoder()
train_data['protocol_type'] = le.fit_transform(
    train_data['protocol_type'])
train_data['service'] = le.fit_transform(
    train_data['service'])
train_data['flag'] = le.fit_transform(train_data['flag'])
train_data['label'] = train_data['label'].apply(
    lambda x: 'normal' if x == 'normal' else 'abnormal')
print(train_data['label'].value_counts())
train_data['label'] = train_data['label'].apply(
    lambda x: 0 if x == 'normal' else 1)
```

Then in the following code, the dataset is divided into features (X) and labels (y). Subsequently, the dataset undergoes partitioning into training and testing subsets utilizing a 70-30 split. We can omit this step and employ a separate test dataset, such as the `KDDTest+` dataset, exclusively for testing purposes, while utilizing the `KDDTrain+` dataset solely for model training:

```
X = train_data.drop(['label'], axis=1)
y = train_data['label']
```

```
X_train, X_test, y_train, y_test = train_test_split(
    X, y, test_size=0.30, random_state=40)
```

Then we will standardize the features using the `StandardScaler` function for scaling:

```
scaler = StandardScaler()
X_train = scaler.fit_transform(X_train)
X_test = scaler.transform(X_test)
```

Next, we will train the model using SVM by initializing an SVM model with a **radial basis function (RBF)** kernel. Henceforth, the model is trained on the standardized training data. The choice of kernel holds significant importance in SVMs, as different kernels exhibit varying performances and compatibility with diverse techniques and datasets. In our scenario, the `rbf` kernel demonstrated superior performance to other tested SVM kernels:

```
svm_model = SVC(kernel='rbf')
svm_model.fit(X_train, y_train)
```

We will now make predictions on the test set utilizing our trained model. We will print a variety of performance metrics, including accuracy for both the training and testing phases, as well as overall model accuracy. A comprehensive classification report is also generated, encompassing additional metrics such as precision, recall, and F1-score. Finally, we will save our trained model using the `pickle` library for future use:

```
y_pred = svm_model.predict(X_test)

print('Training accuracy: ', svm_model.score(X_train, y_train))
print('Testing accuracy: ', svm_model.score(X_test, y_test))
print('SVM Model accuracy: ',
    np.round(accuracy_score(y_test,y_pred),6))
print('Classification Report:\n',
    classification_report(
        y_test,y_pred,target_names=['Normal','Abnormal']
    )
)

pkl_filename = "SVM_model.pkl"
with open(pkl_filename, 'wb') as file:
    pickle.dump(svm_model, file)
```

The following is a sample of our program execution:

```
Training accuracy:   0.9922092060647986
Testing accuracy:    0.9916119813717189
SVM Model accuracy:  0.991612
Classification Report:
              precision    recall  f1-score   support

      Normal       0.99      0.99      0.99     20067
    Abnormal       0.99      0.99      0.99     17725

    accuracy                           0.99     37792
   macro avg       0.99      0.99      0.99     37792
weighted avg       0.99      0.99      0.99     37792
```

Figure 7.2 – Execution of SVM model implementation

As shown in the preceding figure, our performance metrics showcase a remarkable 99% accuracy. However, this was attained solely through dataset splitting and not validation on a separate test dataset.

Next, we will provide source code for loading a comprehensive dataset for testing our implementation. It's crucial to acknowledge that since we have already partitioned the training dataset into train and test subsets, we should exclusively utilize the ensuing test dataset once we remove the train-test partitioning code. Failure to do so may adversely impact the model's performance.

```
print('Starting Testing Process using the Test Dataset....')

test_data = pd.read_csv('dataset/KDDTest+.txt',
    header=None, names=columns)
test_data.drop(['difficulty_level', 'num_outbound_cmds'],
    axis=1, inplace=True)

le = LabelEncoder()
test_data['protocol_type'] = le.fit_transform(
    test_data['protocol_type'])
test_data['service'] = le.fit_transform(test_data['service'])
test_data['flag'] = le.fit_transform(test_data['flag'])
test_data['label'] = test_data['label'].apply(
    lambda x: 0 if x == 'normal' else 1)

X_test = test_data.drop(['label'], axis=1)
y_test = test_data['label']
X_test = scaler.transform(X_test)

y_pred = svm_model.predict(X_test)
```

```
print('Testing Phase accuracy: ', accuracy_score(y_test, y_pred))
print('SVM Model accuracy: ', accuracy_score(y_test, y_pred))
print('Classification Report:\n',
    classification_report(y_test, y_pred,
        target_names=["Normal", "Abnormal"]))
```

Here is the performance of the implementation using the test dataset:

```
Training accuracy:  0.9922092060647986
Testing accuracy:   0.9916119813717189
SVM Model accuracy: 0.991612
Classification Report:
              precision    recall  f1-score   support

      Normal       0.99      0.99      0.99     20067
    Abnormal       0.99      0.99      0.99     17725

    accuracy                           0.99     37792
   macro avg       0.99      0.99      0.99     37792
weighted avg       0.99      0.99      0.99     37792

Starting Testing Process using the Test Dataset....
Testing Phase accuracy: 0.7819819020581973
SVM Model accuracy:     0.7819819020581973
Classification Report:
              precision    recall  f1-score   support

      Normal       0.67      0.98      0.79      9711
    Abnormal       0.98      0.63      0.77     12833

    accuracy                           0.78     22544
   macro avg       0.82      0.81      0.78     22544
weighted avg       0.84      0.78      0.78     22544
```

Figure 7.3 – Execution of SVM model implementation

This highlights the performance of our model in real-world scenarios, showcasing an accuracy of approximately 80%. The observed deviation in performance signals the presence of overfitting, a common occurrence under such circumstances. Notably, feature selection was not conducted on the dataset, allowing certain features to contribute to overfitting. Additionally, the dataset itself is relatively small. Using an RBF kernel, which is susceptible to overfitting on small datasets, further exacerbates this issue.

Moreover, no fine-tuning was performed on the kernel itself, compounding the problem. However, in the final implementation, these issues will be addressed. By integrating feature selection alongside a different algorithm, such as **recurrent neural networks** (**RNNs**), and refining the model, we can anticipate achieving a performance exceeding 95% in real-world environments.

Deep learning techniques, notably neural networks, can discern intricate patterns and relationships in data. They excel in learning from large datasets and are adaptable to structured and unstructured data. Neural networks have consistently demonstrated high performance in the domain of intrusion detection. **Convolutional neural networks (CNNs)** are particularly effective in image-based intrusion detection. In contrast, RNNs excel in handling sequential data.

Moving from detection to classification

The transition from malware detection to malware classification represents a significant evolution in the sophistication and granularity of the analysis performed on potentially harmful software. In the realm of malware detection, the primary goal is to identify whether a given piece of software exhibits malicious behavior or not. This typically involves analyzing features extracted from binaries, system calls, network traffic, or other sources to apply a binary decision—benign or malicious. Algorithms used for malware detection focus on distinguishing between these two classes, often employing techniques such as anomaly detection or pattern recognition to flag suspicious activity.

On the other hand, malware classification delves deeper into the categorization and characterization of malicious software, aiming to classify malware into different types or families based on their behavioral patterns, code structures, or other attributes. Unlike detection, classification involves multiple classes or categories of malware, each representing different types of threats or attack vectors. ML algorithms for malware classification not only need to differentiate between benign and malicious software but must also categorize the detected malware into specific groups, such as trojans, ransomware, worms, or viruses, among others.

This shift from detection to classification introduces several challenges and opportunities. With classification, there is a greater emphasis on feature engineering to capture the nuances and variations across different malware families. Additionally, algorithms must handle the complexities of multi-class classification, including class imbalance, overlapping features, and hierarchical relationships between malware types. However, the payoff is a more comprehensive understanding of the malware landscape, enabling security practitioners to develop targeted defenses, prioritize threats, and respond more effectively to evolving cybersecurity threats. Overall, the move from malware detection to classification represents a maturation of ML techniques in cybersecurity, empowering defenders with more nuanced and actionable insights into the ever-evolving threat landscape.

Frequently, the algorithms employed for classification are akin to those utilized for detection purposes. Renowned algorithms such as Random Forest, SVMs, gradient boosting machines, or K-nearest neighbors, as well as ensemble methods such as AdaBoost, nagging, or stacking, are commonly employed for classification tasks. These methodologies are equally applicable when discussing network traffic classification, adhering to the same principles as other classification tasks.

Summary

Throughout this chapter, we embarked on a comprehensive journey through malware and network intrusion detection and classification. We delved deep into the advantages of AI compared to traditional methodologies. We then ventured into publicly available datasets, recognizing them as invaluable resources for training and validating detection models. By leveraging these datasets, we equipped ourselves with the tools to develop and fine-tune our models effectively, ensuring their efficacy in real-world scenarios.

Two hands-on exercises provided a practical avenue for applying theoretical knowledge, allowing us to actively engage in the model development process. Through these exercises, we honed our skills in crafting tailored solutions for detecting malware and identifying malicious network traffic, bridging the gap between theory and practice. Finally, we discussed the transition from detection to classification.

In summary, this chapter has equipped us with a comprehensive understanding of the intricacies involved in malware and network intrusion detection and classification. In the next chapter, we'll step away from malicious programs and cover the concepts of user and entity behavior.

8
User and Entity Behavior Analysis

Traditional enterprise cybersecurity relies on a **Security Information and Event Management** (**SIEM**) system to gather data needed for detecting and triaging security-related incidents. Based on this data, security experts create rules using their knowledge of known attacker tactics and techniques. They create a playbook where those rules are stored and a series of steps are defined on what actions are performed when data is found that matches those rules. This type of defensive system can detect a majority of threats that traditionally affect IT systems. However, especially in recent years, this kind of defense is becoming insufficient, as the attacks are getting more sophisticated and the threat actors find a way to evade defenses.

User Entity and Behavioral Analysis (**UEBA**) is a concept that enables us to provide solutions to these problems. Under this term, we consider a set of techniques for characterizing the behavior of users and entities (computers, routers, switches, printers, etc.) in a computer network in order to capture regular behavioral patterns. These patterns will comprise a baseline against which we can compare the behavior of a user or an entity. This gives us a context that can enable a better assessment of behavioral anomalies that pose a security risk. Furthermore, it enables us to detect attacker behavior when the attacker circumvents a rule-based defense. This is because the attacker's behavior is typically outside of the ordinary behavior of a particular user or entity. The assumption is that by having enough data and context, we can detect this anomalous behavior and give an assessment of the risk it poses to our computer network.

In this chapter, we will cover the following topics:

- Shortcomings of traditional tools
- Leveraging AI for UEBA (theoretical section)
- Exercise – UEBA anomaly detection (practical section)

Technical requirements

This chapter has a practical component and requires the following Python libraries:

- `numpy`
- `pandas`
- `scikit-learn`
- `ipaddress`

For the best learning performance, it's recommended to use "notebook" libraries and platforms, such as the following:

- Jupyter
- Google Colab
- Databricks

Shortcomings of traditional tools

Traditional signature-based defenses are becoming insufficient for the following reasons:

- **Evolution of attacker techniques**: Security rules are static by nature, and as the software evolves, new vulnerabilities are found and new attacker techniques arise that can be used to circumvent the defenses. This requires the security operations teams to constantly change the security rules to keep their false positive and false negative rate at a minimum, often with limited resources.
- **Insider threats**: Static rules can often be circumvented by insiders who have knowledge and are part of the organization of the systems (for example, disgruntled employees) or attackers with stolen credentials. They can adjust their behavior to look similar to the behavior of a valid user and stay under the radar while performing their attacker activities.
- **Context is needed**: Many times, attacker behavior can look like normal behavior without having a full context of it. Particularly if the attacker gets to a service account or a privileged account, their behavior (even though it could trigger some rules) can be ignored or whitelisted. This is because, in IT systems, there are often accounts that have higher privileges to enable them to perform administrative tasks efficiently. Furthermore, techniques such as **lateral movement** are difficult to detect with static rules, as they need to be differentiated from normal remote logins.

In the following sections, we will describe AI methods as an alternative that addresses these problems.

Leveraging AI for UEBA

There were multiple heuristic techniques used to capture the patterns of user and entity behavior. However, the success of **Artificial Intelligence (AI)** in other problems of pattern recognition has motivated its use in computer security, particularly in UEBA. The idea is to use the data gathered in the SIEM tool, vectorize it, and define features that we will use to train a machine learning model. This model can be trained in an unsupervised manner, or we can also encode the domain knowledge of experts into labels to be able to use supervised machine learning methods.

In the following section, we will describe the common features and feature extraction techniques for UEBA.

UEBA features

The behavior of a user or entity can be characterized by extracting the following feature categories:

- **Access patterns**: Usually, normal user accounts in a network are used to log in to a subset of physical or virtual devices. Furthermore, the use of privileges should be regulated in a way that each privilege has a purpose and will be used to perform a particular job or activity. Therefore, it should be possible to capture this pattern or any outlier from this access pattern. Anomalous use of accounts or privileges can point to misuse of those accounts or privileges or actual attacker activity.
- **Network connections**: The SIEM software allows us to centralize the logging of network connections made by users on particular hosts. This means that we can follow how often a user on a host connects to a particular server, on which port, and at what time of the day.
- **Process execution**: Endpoint agents should enable us to capture which process has generated a notable event that could cause an incident. This process could be originating from malware, or from an attacker using an executable already present on the victim's machine. Knowing the context around this is crucial for determining the risk of an event and detecting possible attacker activity.
- **Data traffic**: There are multiple tools that can enable tracking of how data is transferred between devices in the network. For instance, an endpoint agent can detect traffic to and from a device. Network security devices, such as firewalls or routers, can also capture and even inspect the network traffic between devices. Furthermore, software for **Data Loss Prevention (DLP)** actively monitors the transfer of data to be able to detect whether sensitive data is being exfiltrated outside of the corporate network. All these categories of software can be enabled to transfer logs into the SIEM or a data warehouse/data lake. These logs can be used as input data into the UEBA model.

- **Spatio-temporal features**: An often-important aspect of the context for a security-relevant event is the location and time of the event. For instance, if a device owned by a person working in one location is seen in a login event from an entirely different continent, this is an indicator of a potential security risk. Furthermore, if a login event is occurring at a time outside of the usual working time interval, this makes the event even more suspicious.

Feature extraction

Feature extraction can be done in various ways in order to find an appropriate numerical representation of data. This is done to create input for a machine learning model. In some cases, the data is initially numerical, for example, the number of network connections or the number of megabytes transferred. However, sometimes, the data is textual or has a custom structure. Some of the most common feature extraction methods are as follows:

- **One-hot encoding**: This feature extraction method is appropriate and often used for categorical data. If the data has values from N categories, we construct a feature vector of length N. For each data point, we construct a vector with all values set to zero, except for the value in the *i-th* index of the feature vector, where *i* is the index of the category assigned to this data point. The value of this index is set to one, which means that this point belongs to the indexed category.

- **Frequency counting (bag of words)**: This method is similar to one-hot encoding, in that we create a feature vector of length N. However, instead of just assigning the value of *1* in the appropriate index, we count the occurrence of events in every category and assign the number of occurrences in the index of each category. For instance, we can measure the number of connections from a device grouped by port. If the ports used for remote connection are, say, 389, 464, and 3389, then we have three categories ($N=3$).

- **Term Frequency/Inverse Document Frequency (TF/IDF)**: This method is common in **Natural Language Processing** (**NLP**), but can be used in any data that has a structure similar to text documents. For instance, it can be used in any free text fields such as contents of network packets or source code. With the following formula, we count the number of occurrences of every term (*TF*):

$$TF(t,d) = \frac{c_{t,d}}{\sum_{\tau \in d} c_{\tau,d}}$$

However, we also want to take into account that some terms appear often in every document, so we actually want to count the relative frequency to characterize a feature vector. Therefore, we also count the IDF, which is the inverse of the fraction of documents that contain a particular term:

$$IDF(t, D) = \log \frac{|D|}{|d, d \in D, t \in d|}$$

The TF/IDF value is therefore determined with a product of TF and IDF of a term. If we have N possible terms, then we have a feature vector of length N, similar to one-hot encoding and frequency counting.

- **Time series features**: In time series, the features need to take into account the temporal dependencies in the data. There are multiple methods to obtain this; here are some of the common approaches:

  - **Fourier transform**: Fourier transform is used to change the input data from the time domain to the frequency domain. We consider the underlying function that governs the changes in data to be a sum of sinusoidal components of different frequencies. Fourier transform enables us to uncover these frequencies and characterize the intensities on multiple frequencies as features.

  - **Wavelet transform**: Wavelet transform is an alternative to Fourier transform that is more appropriate for non-stationary time series. This means that it can capture patterns in frequency that can change through time.

  - **Time series attributes**: As a feature, we can also use the timestamp values, such as time epoch or lag, or statistical properties such as mean or standard deviation. Furthermore, we can use the attributes of time series, such as autocorrelations.

In this section, we have briefly described the AI approach to UEBA. In the following section, we will try a practical application through an exercise resembling a real UEBA problem.

Exercise – UEBA anomaly detection

In this exercise, we consider a dataset of network activity of an account and a device in the network. We extract features from this simplified network log and determine whether there is any anomalous traffic that could point to stolen credentials, policy violations, or anything similar.

We use a dataset that represents a simplified version of network traffic data that could be seen in the logs of an endpoint agent or a network device. We load it as follows:

```
df_ueba = pd.read_csv('ueba.csv')
```

User and Entity Behavior Analysis

The contents of the UEBA dataset are shown in the following figure:

	Username	HostIP	RemoteIP	RemotePort	Date/Time	HostLocation	RemoteIPLocation	ProcessImage
0	John.Doe	192.168.1.2	10.0.1.4	389	23.12.2023 08:10:00	USA	USA	LDAP_Client.exe
1	John.Doe	192.168.1.2	10.0.1.4	464	23.12.2023 08:21:21	USA	USA	LDAP_Client.exe
2	John.Doe	192.168.1.2	10.0.1.4	636	23.12.2023 08:22:23	USA	USA	LDAP_Client.exe
3	John.Doe	192.168.1.2	10.0.1.4	464	23.12.2023 08:10:30	USA	USA	LDAP_Client.exe
4	John.Doe	192.168.1.2	10.0.1.5	3389	23.12.2023 09:01:25	USA	France	RDP_Client.exe
...	...	...	...	...	...	...	...	...
76	John.Doe	192.168.1.3	10.0.0.4	8080	23.12.2023 17:02:04	USA	USA	HTTP_Client.exe
77	John.Doe	192.168.1.3	137.176.116.172	443	23.12.2023 17:02:15	USA	USA	HTTP_Client.exe

Figure 8.1 – DataFrame with UEBA data

We can observe that the dataset includes a snippet of network traffic for one day from the device of a user named John Doe.

From the figure, the columns represent the following:

- Username: The username of the device user
- HostIP: The IP address of the host device
- RemoteIP: The IP address of the device to which the user is connecting
- RemotePort: The port that is used to connect to the remote device
- Date/Time: The date and time of the network connection
- HostLocation: The location of the host that the user is using
- RemoteIPLocation: The location of the remote host
- ProcessImage: The name of the executable that is making the network connection on a host device

The goal of the exercise is to explore this dataset, define features, and detect whether there are any anomalies that we should be concerned about in terms of security risk. Now, let's delve into this table deeper.

The dataset consists of one table with multiple columns that describe the network activity. We can notice that there are multiple groups of columns:

- Communication identifiers relate to the IP and port columns
- Spatio-temporal refers to `Date/Time`, `HostLocation`, and `RemoteIPLocation`
- Context is `Username` and `ProcessImage`

One easy way to get a quick idea of the values in columns is to use the `describe()` function from the `pandas` library:

```
df_ueba.describe(include='all')
```

This function shows an output like the following:

	Username	HostIP	RemoteIP	RemotePort	Date/Time	HostLocation	RemoteIPLocation	ProcessImage
count	81	81	81	81.000000	81	81	81	81
unique	1	2	18	NaN	81	1	5	7
top	John.Doe	192.168.1.2	75.63.15.74	NaN	23.12.2023 08:10:00	USA	USA	HTTP_Client.exe
freq	81	46	8	NaN	1	81	71	58
mean	NaN	NaN	NaN	1587.691358	NaN	NaN	NaN	NaN
std	NaN	NaN	NaN	2841.218707	NaN	NaN	NaN	NaN
min	NaN	NaN	NaN	22.000000	NaN	NaN	NaN	NaN
25%	NaN	NaN	NaN	80.000000	NaN	NaN	NaN	NaN
50%	NaN	NaN	NaN	80.000000	NaN	NaN	NaN	NaN
75%	NaN	NaN	NaN	465.000000	NaN	NaN	NaN	NaN
max	NaN	NaN	NaN	8080.000000	NaN	NaN	NaN	NaN

Figure 8.2 – DataFrame with UEBA data profile

This shows that there are two constant columns – `Username` and `HostLocation`. The reason for this is that this dataset follows the activity on one host, which was only used by one user in this instance. Furthermore, there is one column that only has two possible values – `HostIP`. This is because the IP of a host has changed during the data collection, which can happen if the IP address is not set to be static. However, the activity is performed on the same host.

We have one `Date/Time` column, which would be useful if we had multiple days of data, but keeping in mind that our dataset contains only one day, we won't use it in anomaly detection. If we had multiple days of data, we could add it as a feature to check whether the event is happening in an anomalous time period (for example, outside of working hours).

Now, we need to create features. Here, we mainly use numerical feature options – integer numbers and categorical features – out of a discrete number of categories.

The following columns are left as potential features:

- `HostIP`: Numerical feature
- `RemoteIP`: Numerical feature
- `RemotePort`: Categorical feature
- `HostLocation`: Categorical feature
- `RemoteIPLocation`: Categorical feature
- `ProcessImage`: Categorical feature

We decided to create numerical features out of IP addresses because it's simple to convert IP addresses to integers using existing Python libraries (for instance, `ipaddress`). As an example, we use the following function:

```
import ipaddress
def iptointeger(ip):
    return int(ipaddress.ip_address(ip))
```

Other features, such as locations and `ProcessImage`, will be categorical since they contain values from a discrete set and the number of possible values is not so large to make the feature vector very high dimensional.

We create a new DataFrame to store the features:

```
ueba_features = pd.DataFrame() # creation of a dataframe
# creation of a HostIP numerical feature
ueba_features['HostIP'] = df_ueba['HostIP'].apply(iptointeger)
# creation of the RemoteIP numerical feature
ueba_features['RemoteIP'] = df_ueba['RemoteIP'].apply(iptointeger)
```

For the categorical features, we use the `OneHotEncoder` function from the scikit-learn library:

```
from sklearn.preprocessing import OneHotEncoder
encoder = OneHotEncoder(
    sparse_output=False).set_output(transform="pandas")
categorical_encoded = encoder.fit_transform(
    df_ueba[['RemotePort', 'HostLocation',
        'RemoteIPLocation', 'ProcessImage']])
```

We assign the categorical features back to the feature DataFrame:

```
ueba_features[categorical_encoded.columns] = categorical_encoded
```

In order to equalize the influence of features, we scale them to values between 0 and 1. For this, we can use the `MinMaxScaler` function from scikit-learn:

```
min_max_scaler = MinMaxScaler()
ueba_features_scaled = min_max_scaler.fit_transform(ueba_features)
```

Once we have converted the data into numerical features, this is what the feature matrix looks like:

```
array([[0.00000000e+00, 6.86376355e-08, 0.00000000e+00, ...,
        0.00000000e+00, 0.00000000e+00, 0.00000000e+00],
       [0.00000000e+00, 6.86376355e-08, 0.00000000e+00, ...,
        0.00000000e+00, 0.00000000e+00, 0.00000000e+00],
       [0.00000000e+00, 6.86376355e-08, 0.00000000e+00, ...,
        0.00000000e+00, 0.00000000e+00, 0.00000000e+00],
       ...,
       [1.00000000e+00, 5.74376526e-01, 0.00000000e+00, ...,
        0.00000000e+00, 0.00000000e+00, 0.00000000e+00],
       [1.00000000e+00, 5.74376526e-01, 0.00000000e+00, ...,
        0.00000000e+00, 0.00000000e+00, 0.00000000e+00],
       [1.00000000e+00, 5.74376526e-01, 0.00000000e+00, ...,
        0.00000000e+00, 0.00000000e+00, 0.00000000e+00]])
```

Figure 8.3 – Display of the feature matrix of the UEBA dataset

There are multiple AI methods we could use for anomaly detection. Since we don't have any labels for our data, we choose a method from the unsupervised anomaly detection subset. This time, we pick `IsolationForest`, as it is a robust method and available in the scikit-learn library:

```
from sklearn.ensemble import IsolationForest
```

Then, we instantiate the `IsolationForest` model and train it on our dataset:

```
isolationforest = IsolationForest(n_estimators=100, contamination=0.05)
anomaly_labels = isolationforest.fit_predict(ueba_features_scaled)
```

The `decision` function gives us a form of anomaly score. The lower the score, the more anomalous the data point is:

```
anomaly_scores = isolationforest.decision_function(
    ueba_features_scaled)
ueba_features['anomaly_labels'] = anomaly_labels
ueba_features['anomaly_scores'] = anomaly_scores
```

After training the model and testing it on our dataset, we can plot the decision function values as follows:

```
ueba_features['anomaly_scores'].plot(kind='bar')
```

Here is the output plot:

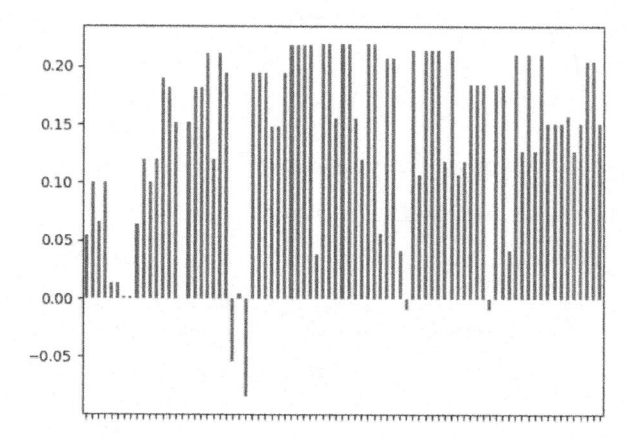

Figure 8.4 – Decision function values for rows in the UEBA dataset

The plot shows two very significant outliers with negative decision functions. This shows the strong confidence of the anomaly detector that these two data points are anomalous. In fact, we can show all points where the anomaly label is -1. Refer to the following screenshot, created by displaying the contents of the `ueba_features` DataFrame:

l_Client.exe	ProcessImage_RDP_Client.exe	ProcessImage_SSH_Client.exe	ProcessImage_custom_script.py	ProcessImage_fast_script.py	anomaly_labels	anomaly_scores
0.0	0.0	0.0	1.0	0.0	-1	-0.053908
0.0	0.0	0.0	0.0	1.0	-1	-0.083715
0.0	0.0	1.0	0.0	0.0	-1	-0.009324
0.0	0.0	1.0	0.0	0.0	-1	-0.009324

Figure 8.5 – First part of the feature set of anomalous points

ocation_USA	RemoteIPLocation_China	RemoteIPLocation_France	RemoteIPLocation_Portugal	RemoteIPLocation_UK	RemoteIPLocation_USA	anomaly_labels	anomaly_scores
1.0	1.0	0.0	0.0	0.0	0.0	-1	-0.053908
1.0	0.0	0.0	1.0	0.0	0.0	-1	-0.083715
1.0	0.0	0.0	0.0	1.0	0.0	-1	-0.009324
1.0	0.0	0.0	0.0	1.0	0.0	-1	-0.009324

Figure 8.6 – Last part of the feature set of anomalous points

Here, we can see that the locations for these points are also different. For instance, here we can see China and Portugal as locations of the destination IP, which is different from more common locations in this dataset such as the UK or USA.

Apart from true positives, we can also see two other points that look like outliers. These are false positives and are barely over the zero line in terms of anomaly scores. However, they are pronounced as outliers, partially due to the *contamination* hyperparameter that determines the percentage of outliers expected in the dataset.

One of the general problems with unsupervised anomaly detection is the presence of false positives. The reason for this is that in every outlier detection method, the decision boundary depends on hyperparameters, and it's difficult to know the values of these hyperparameters in advance. For instance, in `IsolationForest`, one such hyperparameter is *contamination*. We can estimate this value based on previous domain knowledge about the data or based on statistical testing (for example, *cross-validation*) to minimize the false positive and false negative rate. However, we should still be ready to postprocess the data to detect and remove false detections.

Other use cases

UEBA can be used in multiple other use cases in order to detect attacker activity. Some examples are as follows:

- **Detection of lateral movement**: An attacker could be trying to move between different machines to find a server with valuable information
- **Data leakage detection**: An attacker might be trying to exfiltrate highly restricted data from the organization
- **Abuse of privileged access**: An attacker might use a privileged account to access sensitive information
- **Reconnaissance activities by attackers**: An attacker might investigate the network to try to find vulnerabilities or servers with valuable data
- **Detection of security misconfigurations**: There could be accidental misconfigurations that could be exploited by an external attacker or an insider threat

These are examples of activity that can be detected using the UEBA approach to anomalous network traffic compared to the baseline network activity of hosts in the organization.

Summary

This chapter describes UEBA as a use case for AI methods applied to cybersecurity. You have learned how to design and implement an AI approach to problems in UEBA. Furthermore, by working through a concrete problem on a UEBA dataset and in an exercise, you were able to practice the AI practical skills you learned in previous chapters. The workflow that you have practiced ranges from data preprocessing, machine learning model training, and results analysis.

In the next chapter, you will work on another group of cybersecurity problems that benefit from AI methods. These are fraud, spam, and phishing detection.

9

Fraud, Spam, and Phishing Detection

Fraud, phishing, and spam detection are critical areas in cybersecurity, aimed at identifying and preventing malicious activities. Fraud detection involves identifying deceptive actions aimed at gaining financial or personal benefits through dishonest means. Fraud detection is particularly important in industries such as banking, e-commerce, and insurance. Phishing detection aims to identify and prevent attempts to acquire sensitive information such as usernames, passwords, and credit card details by masquerading as a trustworthy entity. Meanwhile, spam detection focuses on identifying and filtering out unwanted and unsolicited messages, typically in email communication. Spam can include advertisements, malware, and phishing attempts.

In this chapter, alongside covering fraud, phishing, and spam detection as typical examples of anomaly detection, we also introduce methods for using collaborative machine learning to achieve privacy-preserving anomaly detection. We present fundamental contents and examples of fraud, phishing, and spam detection, then discuss **multi-party computation** (**MPC**)[1] and federated learning-based detection methods for collaborative machine learning. This chapter will help us with the following:

- Introducing fraud, phishing, and spam detection methods
- Understand phishing detection with a practical example

Introducing fraud, phishing, and spam detection methods

Fraud, phishing, and spam detection in cybersecurity are distinct areas, each focusing on identifying and mitigating different types of malicious activities.

Let us explore each of these in the following sections.

Fraud detection

Similar to malware and phishing issues, online fraud and credit card fraud are also critical problems in contemporary cybersecurity. The issues can cause huge losses for banks, consumers, insurance companies, and other organizations (in both the public and private sectors). Therefore, fraud detection increasingly attracts attention, and it indicates the classification of malicious transactions from legitimate ones. Please find references [2,3,4] in the *References* section for more details. Traditionally, fraud analyzers use expert-driven predictive models, which are based on the rules standardized by experts. One example of a rule-based model would be a model where a transaction will be denied if the previous one was issued five minutes ago at a store 500 kilometers away. Currently, with the development of machine learning and deep learning, the data-driven predictive models in the fraud detection field attract more interest. Using machine learning and deep learning techniques in fraud detection has several significant advantages, such as analyzing multidimensional features, updating models dynamically, and making use of large amounts of data. Similar to malware detection and phishing detection, fraud detection is also a binary classification problem in machine learning.

Typical examples of fraud detection include credit card fraud, insurance claim fraud, and identity theft. The scope of fraud detection often involves financial transactions and personal data integrity. Keeping detection systems updated to handle new types of fraud requires continuous research and development, especially in sectors such as banking and e-commerce, which involve processing a massive number of transactions daily. Analyzing such large volumes in real time without affecting performance is technically challenging and resource-intensive.

Fraud detection is a complex and challenging field that requires a multi-faceted approach to address various issues effectively. Organizations must stay vigilant and adaptive, employing a combination of advanced technologies, skilled personnel, and robust processes to combat evolving fraud threats. Tools such as SAS Fraud Management, FICO Falcon, and specialized fraud detection systems are used in banking, insurance, and e-commerce. For the fraud detection task, we use the Kaggle dataset directly (`https://www.kaggle.com/datasets/mlg-ulb/creditcardfraud`). It has 30 features. The principal components are obtained with PCA, except for time and amount.

Phishing detection

Phishing detection [5,6,7,8] is one of the primary topics of anomaly detection in the cybersecurity field. Phishing is increasingly becoming the most severe threat in contemporary society. It is typically associated with the motive of stealing the sensitive information (such as personal identification numbers, usernames, and passwords) of internet users by sending spam messages or mimicking famous websites. Traditionally, to detect and prevent phishing, several blacklists and whitelist filter-based methods were introduced. Recently, some machine learning-based automatic classifier works have also appeared in the phishing detection area.

Typical examples of phishing cases and their corresponding detection methods include phishing email detection and phishing website detection. A phishing email normally contains a link that appears to lead to a legitimate bank website but actually redirects to a malicious site. An email might claim to be from PayPal, asking the recipient to verify their account by clicking on a link. A phishing site might mimic the login page of a popular social media platform. The detection methods for these types of phishing include URL checking, keyword analysis, heuristic analysis, sender reputation, form field analysis, visual similarity check, SSL certificate checks, and machine learning. For instance, we can check whether the URL in the link matches the purpose of the link. For example, if the sender claims it's an e-mail for sharing company information, the link should be to a URL with a domain belonging to that company.

Spam detection

The purpose of spam detection [9,10,11,12] is to identify and filter out unsolicited and unwanted messages, primarily to reduce clutter and prevent potentially malicious content from reaching users.

The scope of spam primarily focuses on email communications but is also applicable to SMS, social media, and other messaging platforms. Typical examples include unwanted email advertisements, bulk marketing emails, irrelevant promotional messages, and so on.

In the next subsection, we move to a practical example in the area of phishing detection.

Understanding phishing detection with a practical example

Phishing detection involves a multi-faceted approach that combines various techniques and tools to effectively identify and mitigate phishing attempts. Continuous advancements in technology, along with user education, are essential to keep up with the evolving tactics used by phishers. Examples of phishing detection tools and services include Google Safe Browsing, Microsoft Office 365 Advanced Threat Protection: PhishTank, and **Anti-Phishing Working Group** (**APWG**). These tools and services analyze and share phishing attack data to help in detection and prevention service that warns users and administrators about unsafe websites, offer email filtering and URL analysis to detect phishing attempts, and report phishing URLs to public databases.

Feature categorization	Feature names	Comments
URL format	url_have_ip url-length stortening service has_atCharacter double slash prefix_suffix sub_domains https_token fac_icon	• Has ip address or not in urls • The length of urls is over 54 characters or not • The urls is in a short name of famous website or not, such as goo. or .gl for google • Has "@" in urls or not • Has "//" or not • Has" https?://- + -/" or not • Has sub domain in urls • Has "https" or not • Has head link or link or href in urls or not
URL objects	request URL anchorURL links Tags	• The percentage of "img", "audio", "embed", "iframe" is over 61% • The percentage of anchor link "#" in urls is over 97% of links • The percentage of "link" and "script" is over 90%
HTML and JS	redirectURL statusBar right Click popularWindow	• Has redirected URL or not • Has "onmouseover" in response or not • Has "event.button" in response or not • Has "alert()" in response or not
Domain	domainLife dnsRecord pageLinks	• The creating time is less then 12 months or not • The nds record is smaller than 1 year • Has more than two <a href=> in the url's page

Table 9.1 – Features for phishing detection

We normally consider the manually indicated features for the phishing detection task because phishing detection is relatively simpler than malware detection. We take the URL format, URL objects, HTML, JS, and domain information as our feature information and get a 20-element vector. The list of vector elements is enumerated in *Table 9.1*.

After getting the vector representation, we use logistic regression as our local classification to do the local training. We take the following equation as our objective function:

$$l_X(\theta) = \frac{1}{n} \sum_{i \in X} \log(1 + e^{-y_i \theta^T x_i})$$

Here, θ is the model parameter of the logistic regression and X is the learning sample with n sample-label pairs (). We use the **stochastic gradient descent** (**SGD**) method using the following equation to train our model:

$$\nabla_{l_X(\theta)} = \frac{1}{X'} \sum_{i \in X'} (\frac{1}{1 + e^{-y_i \theta^T x_i}} - 1) y_i x_i$$

The datasets for phishing detection are collected from PhishTank [43] and TLDs [44]:

- PhishTank is a collaborative dataset of phishing on the internet. We collect 25,000 items as our malicious URL lists.
- With TLD, we collect a benign list of TLDs in Alexa's top one million domain list, which contains one million entries of the most popular websites worldwide.

Using the previously listed datasets, as well as the logistic regression, decision tree, factorization, max-voting, naive Bayes, and **Random Forest** (**RF**) machine learning algorithms, we finally get the result shown in *Figure 9.1*. This contains the ROC performance metric and precision-recall curve for various classifiers. We use these metrics to show the performance of the classifier for various classification threshold values.

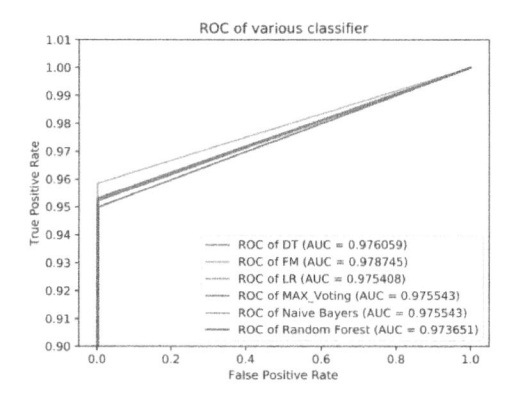

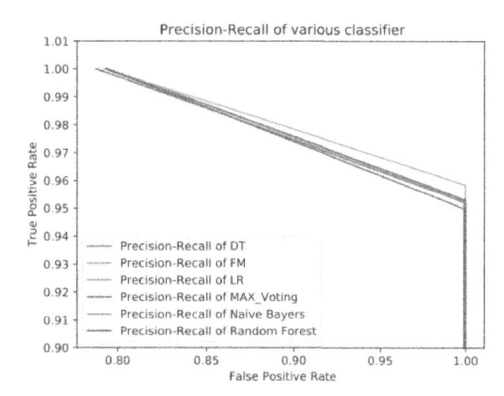

Figure 9.1 – Performance of the phishing classifier

For phishing detection, our benchmark model is trained with 20 features. In *Figure 9.1*, we can see that the best ROC is from the FM classifier, which is 97.87%. Other classifiers get around 97.5% AUC, except for the RF, classifier which gets 97.36% AUC.

In the next subsection, we describe the topic of collaborative anomaly detection, which we believe is of increasing importance in spam and phishing detection, as well as in cybersecurity in general.

Introducing the collaborative anomaly detection

Collaborative machine learning and deep learning increasingly require more attention from various fields, such as health-sensitive prediction, product recommendation, and bank credit ranking systems. However, user privacy issues constrain the development of collaborative machine learning and deep learning systems. As a classic example, anomaly detection systems in cybersecurity, such as malicious code detection, phishing detection, and fraud detection, also suffer from this weakness.

With the increasing number of accidents that involve large companies compromising on data security and privacy, on the one hand, the topic of data privacy and security has become a worldwide issue. Additionally, more and more countries and organizations have released standards and specifications to set the bar for data usage. Take the **General Data Protection Regulation** (**GDPR**) as an example. It was implemented by the **European Union** (**EU**) on May 25, 2018, aiming to protect users' privacy and data security. GDPR requires companies and organizations to present the user agreement to their users and delete users' data without any data leakage risks. Similar to the EU's GDPR, China's **Cybersecurity Law** and **General Principles of the Civil Law**, which were released in 2017, require internet-based companies to not leak or tamper with the personal information they collect and transact the collected data to third parties. Although these regulations will help build a more civil society, new challenges are introduced to the data transaction procedures commonly used today in machine learning and deep learning.

On the other hand, collaborative machine learning and deep learning increasingly attract attention from various fields, such as product recommendation systems [1,13], personal text classification [14, 15], and fraud detection [16,17]. Similar to incremental learning [18,19,20], collaborative machine learning needs several participants working together and training the model with private datasets or model parameters. However, one of the critical problems of the collaborative machine learning system is data privacy. For various organizations or companies, private data is one of the most important assets. Although nearly all of them know that data sharing can improve the trained model's performance, none of them want to share those private data assets with others, especially competitors. Take anomaly detection in cybersecurity as an example. Both FireEye and VirusTotal, famous cybersecurity companies, could detect a specific program's malicious code or behaviors by various features, such as networking traffic, run-time resource consumption, and static API call characteristics. However, none of them want to share their whole datasets with their competitors.

In the anomaly detection field, although there are many datasets, we cannot get enough anomaly samples. Data imbalance in machine learning refers to the situation where the number of instances in different classes of a dataset is significantly disproportionate. This is common in classification problems where one class might be much more prevalent than others. Data imbalance is a significant challenge in machine learning that can lead to biased models and misleading performance metrics. For example, in a medical diagnosis dataset, there might be many more healthy cases than cases with a specific disease. In the anomaly detection area, data imbalance is also a challenging problem. Similar to medical diagnoses, most items in the dataset (sometimes even more than 95%) are benign, and just a few of them are anomalies. Therefore, knowing how to effectively use valuable samples without losing data privacy from various vectors and organizations is a challenge. Using a collaborative machine learning method can partially overcome this challenge.

In order to do collaborative machine learning without the risk of private data leakage, we need some privacy-preserving solutions in the contemporary anomaly detection field. From the technologies standpoints, privacy-preserving machine learning provides many solutions, such as **multi-party computation** (MPC) [21,22,23], **homomorphic encryption** (HE) [24,25,26], **differential privacy** (DP) [27, 28], **trusted execution environment** (TEE) and **federal learning** (FL) [29,30]. DP is a robust and mathematical definition of privacy in statistical and machine learning analysis. According to this mathematical definition, DP is a criterion of privacy protection, which many tools that are designed for analyzing sensitive personal information have been devised to satisfy [28]. DP provides a provable guarantee of privacy protection against a wide range of privacy attacks, such as differentiating and reconstruction attacks.

HE [24,25,26] is a form of encryption that allows computation on ciphertext, generating an encrypted result that, when decrypted, matches the result of the operations as if they had been performed on the plaintext. HE enables protocol parties (such as cloud providers) to apply functions on encrypted data without revealing the values of data. HE can be used for privacy-preserving outsourced storage and computation. The three types of HE are **Partially Homomorphic Encryption (PHE)**, **Somewhat Homomorphic Encryption (SHE)**, and **Fully Homomorphic Encryption (FHE)**.

PHE [30,31] allows only select mathematical functions to be performed on encrypted values, which means that only one operation, either addition or multiplication, can be performed an unlimited number of times on the ciphertext. A SHE scheme is one that supports select operations (either addition or multiplication) up to a specified complexity, but these operations can only be performed a set number of times. FHE requires both addition and mathematical multiplication functions to be performed on encrypted values. In a real HE-based system, the SHE scheme accompanies secure MPC and plays the role of triplet generation.

As a distributed training model, FL increasingly attracts academic and industrial partners. The original FL is designed to collaborate with Android users to improve its user experience on the device [29,30]. However, it is developed as a privacy-preserving machine learning technique that trains a model by having multiple decentralized edge devices or servers holding local data without exchanging it. In other words, it enables multiple actors to build a standard, robust machine learning/deep learning model without sharing data. **Federated stochastic gradient descent (FedSGD)** [29] is another instance of a federated model, where collaboration is achieved by using a random fraction C of the clients and then using all the data on this client. The gradients are averaged by the central server proportionally to the number of training samples on each client and used to make a gradient descent step. **Federated averaging (FedAvg)** is a generalization of FedSGD, allowing local nodes to perform more than one batch update on local data and exchange the updated weights rather than the gradients.

Secure Multi-Party Computation (sMPC) is a fundamental technique in theoretic cryptography. Usually, there are parties in the MPC protocol, and each of them keeps its private data . The MPC's purpose is to jointly calculate a function using the distributed data from parties

. Meanwhile, the participants do not want to leak their own data to others. There are several theoretical methods to achieve MPC goals. **FHE**, **garbled circuit (GC)**, and **secure sharing (SS)** are among the most popular ones. Although FHE and GC are easy to implement, FHE is very expensive in computation. GCs are heavily dependent on network transfer, making them unsuitable to support large-scale data mining tasks.

In contrast to the FHE and GC schemes, the SS-based schemes attract more attention due to their relatively low overhead under certain assumptions. Thus, most practical MPC systems leverage an SS scheme [21,22,12].

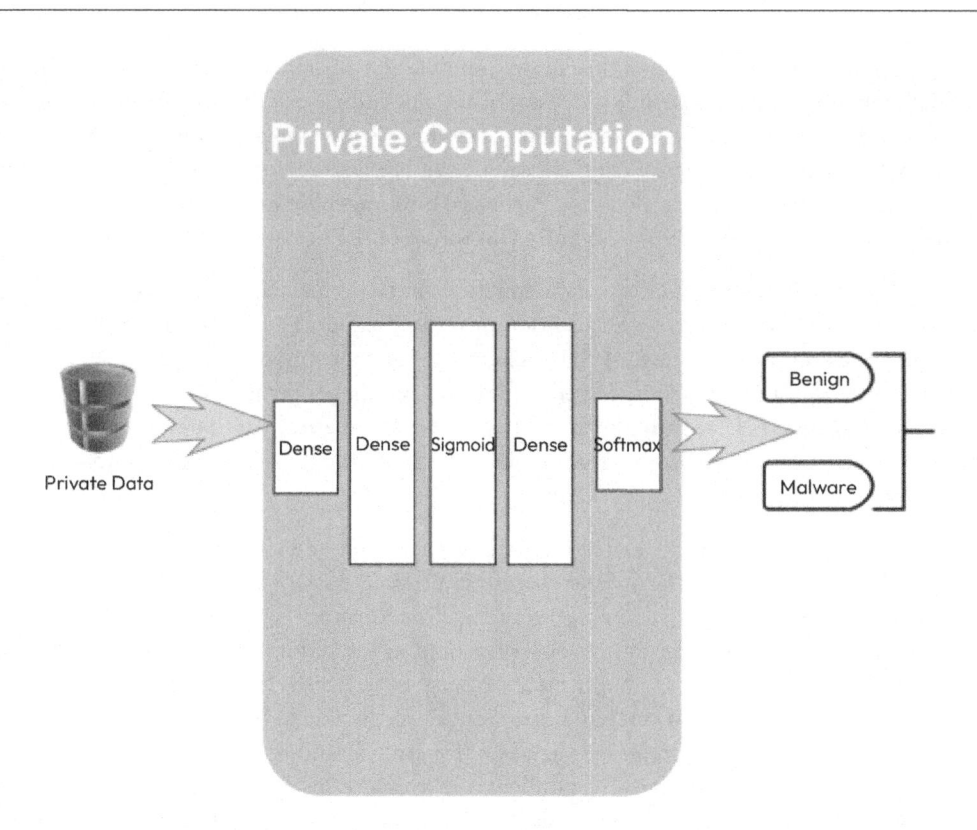

Figure 9.2 – Architecture of collaborate/private computation

In the next subsection, we dive into the design and implementation of collaborative anomaly detection.

Designing and implementing collaborative anomaly detection systems

As we discussed earlier, there are several methods to implement collaborative anomaly detection systems. It is an effective way to solve the data imbalance problem, alongside the classical methods for balancing datasets, such as undersampling and oversampling. Our work aims to design and implement a privacy-preserving machine learning system in the cybersecurity field. That means the primary task is the machine learning-based system. Then, the next task is to consider the privacy-preserving scheme. The architecture of the collaborative anomaly detection is shown in *Figure 9.2*. We will present federated learning-based and MPC-based collaborative anomaly detection in the following sections.

FL-based collaborative anomaly detection systems

Federated anomaly detection can be classified into horizontal and vertical systems. We show them in the following sections in detail.

Horizontal federated anomaly detection

Horizontal federated anomaly detection is presented in *Figure 9.3*. In this system, *k* participants with the same data structure collaboratively learn the anomaly detection model with the assistance of a cloud server.

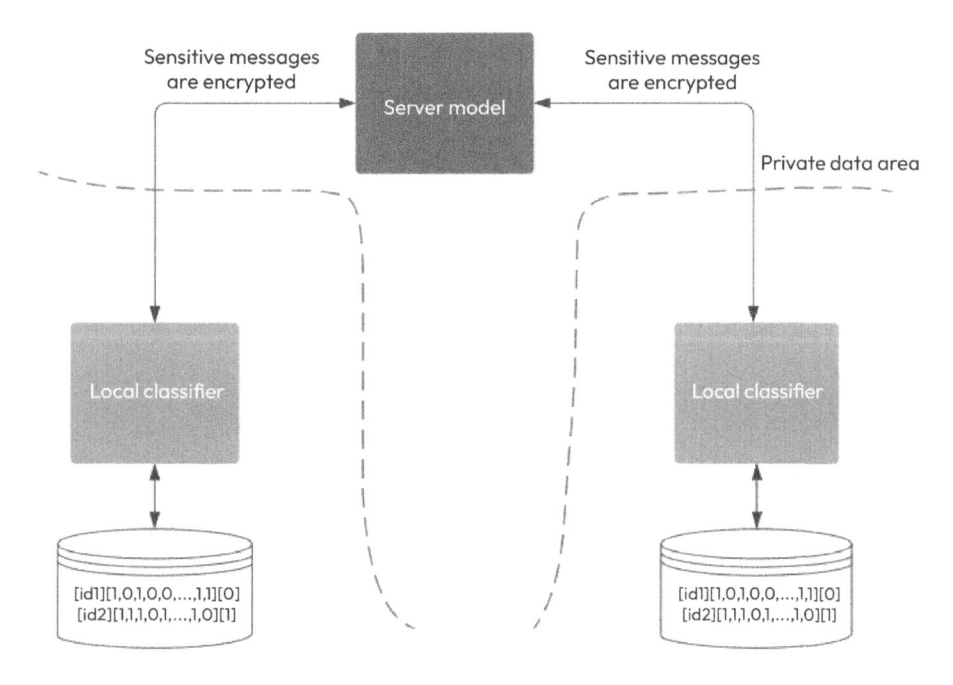

Figure 9.3 – Horizontal phishing detection

The training process for horizontal federated malware detection contains the following six steps:

1. **Initialization**: The server generates the key pairs (*Spr*, *Spu*) using an RSA algorithm and sends them to *k* client participants. Meanwhile, *k* client participants run the same RSA algorithm on a local machine and get *k* key pairs (*Cipr*, *Sipu*), i ∈ k.
2. **Local training**: *k* local participants train their own model.
3. **Model sending**: *k* local participants send the encrypted and masked local models to the server.
4. **Model aggregation**: The server performs secure aggregation without any learning information from *k* local participants.
5. **Model distributing**: The server distributes the aggregated results to *k* local participants.
6. **Local updating**: *k* local participants update their own model with the decrypted gradients.

In order to finish the entire training step and make the trained model converge, the six preceding steps continue through multiple iterations until the loss function converges.

Vertical federated anomaly detection

Similar to vertical federated learning in general, vertical federated anomaly detection contains two parts. The first one is the encrypted entity alignment, and the other one is encrypted model training [76, 204].

After accomplishing the encrypted entity alignment, these common entities' data can be used to train phishing detection algorithm.

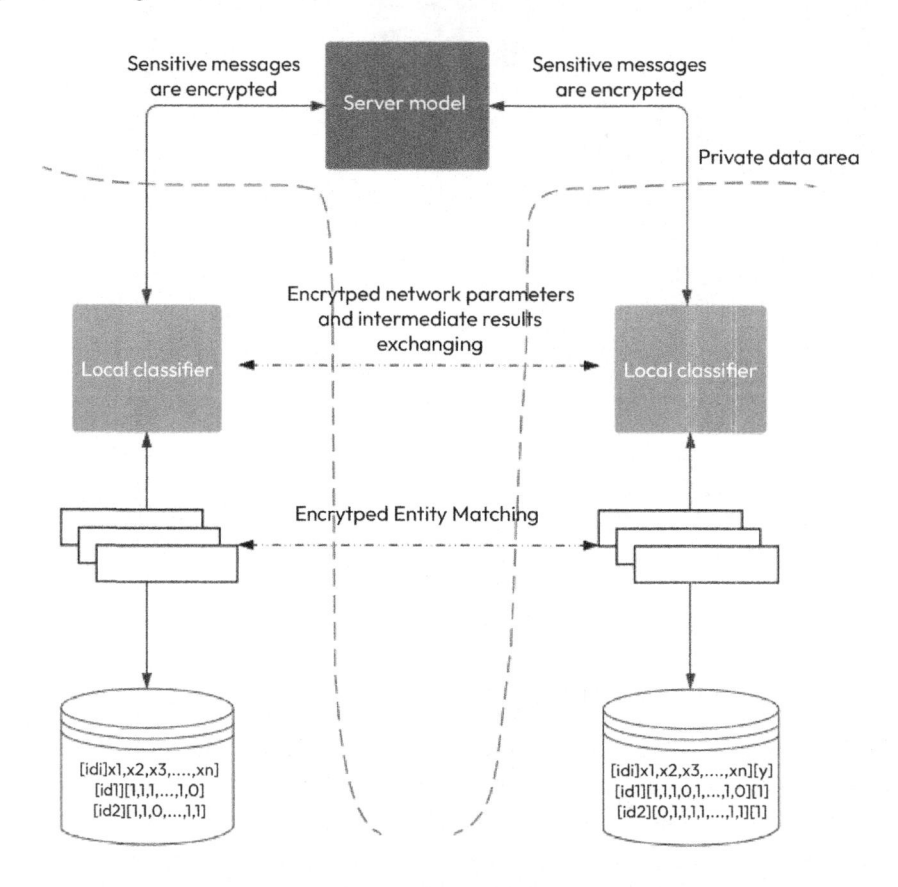

Figure 9.4 – Vertical Phishing Detection

The training task contains the seven steps shown in *Figure 9.4*, as explained here:

1. **Initialization**: Generate key pairs both for the server (Spr, Spu) and k client participants (Cipr, Cipui, i ∈ k) by the RSA algorithm. The server and k client participants broadcast their public keys to other participants.

2. **Encrypted entity matching**: Identify the same entities for several client participants collaboratively training the model.

3. **Encrypted parameters and results exchanging**: Exchange the encrypted parameters and intermediate results among k client participants.
4. **Local training**: k local participants train their own model.
5. **Model sending**: k local participants send the local model parameters to the server with encryption.
6. **Model distribution**: The server distributes the encrypted aggregated model parameters to k client participants.
7. **Local updating**: k local participants update the local models.

Similar to horizontal anomaly detection, vertical anomaly detection iterates through seven steps to converge the model. We present encrypted entity matching, encrypted parameters and results exchanging, local training, and model distribution in detail in *Figure 9.4*.

In this section, we described how to create anomaly detection models in a collaborative manner by taking security and privacy into account.

Summary

Now that you have finished this chapter, we hope you understand what fraud, phishing, and spam detection are. We also explored how to construct simple phishing detection systems with traditional machine learning algorithms. These examples are also very easy to extend to other deep learning-based methods. On the other hand, we hope you now understand the data imbalance issues in several AI applications and why we need collaborative anomaly detection systems to consider the performance and privacy issues. These topics were covered in this chapter. We exemplified federated learning as an example of collaborative anomaly detection in this chapter as well. It is also very easy to extend this to MPC-based or TEE-based collaborative anomaly detection.

In the next chapter, we will focus on user authentication and access control.

References

- [1] E. Shmueli and T. Tassa, "Secure multi-party protocols for item-based collaborative filtering," in *Proceedings of the eleventh ACM conference on recommender systems*, 2017, pp. 89–97.
- [2] Varmedja, D., Karanovic, M., Sladojevic, S., Arsenovic, M., & Anderla, A. (2019, March). "Credit card fraud detection-machine learning methods". In *2019 18th International Symposium INFOTEH-JAHORINA (INFOTEH)* (pp. 1-5). IEEE.
- [3] Raghavan, P., & El Gayar, N. (2019, December). Fraud detection using machine learning and deep learning. In *2019 international conference on computational intelligence and knowledge economy (ICCIKE)* (pp. 334-339). IEEE.

- [4] Omar, S. J., Fred, K., & Swaib, K. K. (2018, May). A state-of-the-art review of machine learning techniques for fraud detection research. In *Proceedings of the 2018 International Conference on Software Engineering in Africa* (pp. 11-19).
- [5] Moghimi, M., & Varjani, A. Y. (2016). New rule-based phishing detection method. *Expert systems with applications*, 53, pp. 231-242.
- [6] Zuraiq, A. A., & Alkasassbeh, M. (2019, October). Phishing detection approaches. In *2019 2nd International Conference on new Trends in Computing Sciences (ICTCS)* (pp. 1-6). IEEE.
- [7] Orunsolu, A. A., Sodiya, A. S., & Akinwale, A. T. (2022). A predictive model for phishing detection. *Journal of King Saud University-Computer and Information Sciences*, 34 (2), pp. 232-247.
- [8] Jain, A. K., & Gupta, B. B. (2017). Phishing detection: analysis of visual similarity based approaches. *Security and Communication Networks*, 2017 (1), 5421046.
- [9] Crawford, M., Khoshgoftaar, T. M., Prusa, J. D., Richter, A. N., & Al Najada, H. (2015). Survey of review spam detection using machine learning techniques. *Journal of Big Data*, 2, pp. 1-24.
- [10] Guzella, T. S., & Caminhas, W. M. (2009). A review of machine learning approaches to spam filtering. *Expert Systems with Applications*, 36(7), pp. 10,206-10,222.
- [11] Gupta, S. D., Saha, S., & Das, S. K. (2021, February). SMS spam detection using machine learning. In *Journal of Physics: Conference Series* (Vol. 1797, No. 1, p. 012017). IOP Publishing.
- [12] Teja Nallamothu, P., & Shais Khan, M. (2023). Machine learning for SPAM detection. *Asian Journal of Advances in Research*, 6 (1), pp. 167-179.
- [13] T. Qi, F. Wu, C. Wu, Y. Huang, and X. Xie, "Privacy-preserving news recommendation model training via federated learning," arXiv preprint arXiv:2003.09592, 2020.
- [14] Reich, D., Todoki, A., Dowsley, R., De Cock, M., & Nascimento, A. C. (2019). Privacy-Preserving Classification of Personal Text Messages with Secure Multi-Party Computation: An Application to Hate-Speech Detection. *ArXiv*. https://arxiv.org/pdf/1906.02325
- [15] Badawi, A. A., Hoang, L., Mun, C. F., Laine, K., & Aung, K. M. (2019). PrivFT: Private and Fast Text Classification with Homomorphic Encryption. *ArXiv*. https://doi.org/10.1109/ACCESS.2020.3045465
- [16] Nabil, Mahmoud, Muhammad Ismail, Mohamed Mahmoud, Waleed S. Alasmary and Erchin Serpedin. "PPETD: Privacy-Preserving Electricity Theft Detection Scheme With Load Monitoring and Billing for AMI Networks." *IEEE Access* 7 (2019): 96334-96348.
- [17] Yao, Wenyan, Na Ruan, Feifan Yu, Weijia Jia and Haojin Zhu. "Privacy-Preserving Fraud Detection via Cooperative Mobile Carriers with Improved Accuracy." *2017 14th Annual IEEE International Conference on Sensing, Communication, and Networking (SECON)* (2017): pp. 1-9.
- [18] Ross, David A., Jongwoo Lim, Ruei-Sung Lin and Ming-Hsuan Yang. "Incremental Learning for Robust Visual Tracking." *International Journal of Computer Vision* 77 (2008): pp. 125-141.

- [19] Castro, Francisco M., Manuel J., Nicolás Guil, Cordelia Schmid, and Karteek Alahari. "End-to-End Incremental Learning." *ArXiv*, (2018). Accessed September 9, 2024. /abs/1807.09536.
- [20] Wu, Yue, Yinpeng Chen, Lijuan Wang, Yuancheng Ye, Zicheng Liu, Yandong Guo, and Yun Fu. "Large Scale Incremental Learning." *ArXiv*, (2019). Accessed September 9, 2024. /abs/1905.13260.
- [21] Li, Yi, Yitao Duan, Yu Yu, Shuoyao Zhao, and Wei Xu. "PrivPy: Enabling Scalable and General Privacy-Preserving Machine Learning." *ArXiv*, (2018). Accessed September 9, 2024. /abs/1801.10117.
- [22] Mohassel, Payman and Peter Rindal. "ABY3: A Mixed Protocol Framework for Machine Learning." *Proceedings of the 2018 ACM SIGSAC Conference on Computer and Communications Security* (2018): n. pag.
- [23] Wagh, Sameer, Divya Gupta and Nishanth Chandran. "SecureNN: Efficient and Private Neural Network Training." *IACR Cryptol. ePrint Arch.* 2018 (2018): pp. 442.
- [24] Kanan, K., & Sharma, N. (2021). A review of homomorphic encryption scheme analysis of cloud computing. J. Interdiscip. Cycle Res, 13, pp. 458-467.
- [25] Hoskin, R. A. (2019). Femmephobia: The role of anti-femininity and gender policing in LGBTQ+ people's experiences of discrimination. Sex Roles, 81(11), pp. 686-703.
- [26] O'Donoghue, K., & Guerin, S. (2017). Homophobic and transphobic bullying: Barriers and supports to school intervention. Sex Education, 17(2), pp. 220-234.
- [27] Dwork, C. (2006, July). Differential privacy. In International colloquium on automata, languages, and programming (pp. 1-12). Berlin, Heidelberg: Springer Berlin Heidelberg.
- [28] Abadi, M., Chu, A., Goodfellow, I., McMahan, H. B., Mironov, I., Talwar, K., & Zhang, L. (2016, October). Deep learning with differential privacy. In Proceedings of the 2016 ACM SIGSAC conference on computer and communications security (pp. 308-318).
- [29] Yuan, H., & Ma, T. (2020). Federated accelerated stochastic gradient descent. Advances in Neural Information Processing Systems, 33, 5332-5344.
- [30] Wu, Z., Ling, Q., Chen, T., & Giannakis, G. B. (2020). Federated variance-reduced stochastic gradient descent with robustness to byzantine attacks. IEEE Transactions on Signal Processing, 68, 4583-4596.
- [31] Li, Chenglin, Keith G. Mills, Di Niu, Rui Zhu, Hongwen Zhang and Husam Kinawi. "Android Malware Detection Based on Factorization Machine." *IEEE Access* 7 (2018): 184008-184019.
- [32] Wu, Dong-Jie, Ching-Hao Mao, Te-En Wei, Hahn-Ming Lee and Kuo-Ping Wu. "DroidMat: Android Malware Detection through Manifest and API Calls Tracing." *2012 Seventh Asia Joint Conference on Information Security* (2012): pp. 62-69.

- [33] Peiravian, Naser and Xingquan Zhu. "Machine Learning for Android Malware Detection Using Permission and API Calls." *2013 IEEE 25th International Conference on Tools with Artificial Intelligence* (2013): 300-305.

- [34] Aafer, Yousra, Wenliang Du and Heng Yin. "DroidAPIMiner: Mining API-Level Features for Robust Malware Detection in Android." *Security and Privacy in Communication Networks* (2013).[15] N. Peiravian and X. Zhu, "Machine Learning for Android Malware Detection Using Permission and Api Calls," in IEEE International Conference on Tools with Artificial Intelligence (ICTAI), 2013.

- [35] Peiravian, Naser and Xingquan Zhu. "Machine Learning for Android Malware Detection Using Permission and API Calls." *2013 IEEE 25th International Conference on Tools with Artificial Intelligence* (2013): 300-305.

10
User Authentication and Access Control

In the context of information security, user authentication and access control are two fundamental concepts that help protect data and resources from unauthorized access. These mechanisms are critical for maintaining the integrity, confidentiality, and availability of systems and data within an organization or network. User authentication is the process of verifying the identity of a user who attempts to access a system. Access control is the mechanism that follows authentication, which determines and enforces what authenticated users are allowed to do.

By the end of this chapter, you'll understand not only the "how" but also the "why" behind these critical components of cybersecurity. You'll see that they're more than just technical jargon; they're the guardians of our digital universe. Let's embark on this journey to demystify these guardians—to understand how they work tirelessly behind the scenes to keep our information, our identities, and our lives secure.

We will cover these topics in the following sections in this chapter:

- Understanding user authentication and access control
- Exemplifying user authentication and access control
- Practicing user authentication and access control with Python
- AI for user authentication – face recognition

Understanding user authentication and access control

At the beginning of this chapter, we will briefly mention the fundamental purpose and techniques of user authentication and access control. Imagine a world where your identity is the gateway to everything you hold dear. Your bank accounts, personal emails, health records, and even your home security system hinge on the ability to prove you are who you claim to be. Now, envision a breach—someone else claims to be you, and suddenly, your world is turned upside down. This isn't a plot from a dystopian novel; it's a potential reality in our digital age, where user authentication and access control form the crux of information security.

User authentication—at its core—is the process of verifying that someone is who they claim to be. It is fundamentally about identity confirmation. When a user attempts to access a system, authentication is the gatekeeper. It acts as the front line of defense, ensuring that the front door of a system only opens for those with the right keys; it involves verifying credentials against a set of known values within the security infrastructure of the system. This verification could be as simple as a password (something the user knows) or as complex as biometric verification, which could involve fingerprint or retina scanning (something the user inherently *is*). This is the first line of defense. This verification can take many forms: something you know (such as a password), something you have (such as a smartphone), or something you are (such as a fingerprint). The robustness of authentication is what stands between safety and vulnerability.

Moving a step further, access control is the method by which systems determine user rights and privileges. It, conversely, decides the extent of use and visibility that authenticated users have within a system. Think of it as the various locks and doors within a secured facility. Just because one has access to the building doesn't mean they can enter every room or access all the information. Access control can be discretionary, allowing owners or administrators to set access on an individual basis, or it can be rule-based, whereby access is granted according to predefined security models. This defines the boundaries within which a user operates, preventing overreach and potential internal threats. It controls which users are allowed access to certain resources or information. It's not enough to just know who someone is; we must also ensure they have the right level of access—no more, no less—to the critical assets they seek to interact with.

In a world where digital transactions are as commonplace as breathing, the significance of these concepts cannot be overstated. Authentication and access control are not merely about safeguarding data—they protect lives, livelihoods, and the uninterrupted functionality of our society. They are the unsung heroes in a world rampant with cyber threats and relentless attackers, and a landscape that evolves faster than one can keep pace with.

Despite their differences, user authentication and access control are interconnected. They work in tandem, a seamless interplay between verifying a user and then providing them with the appropriate level of access. Successful authentication is a precursor to access control; without establishing a user's identity, the system cannot enforce access policies. This sequence is pivotal – authentication without access control is akin to having a fortified gate with no fence, and access control without authentication is like having a complex series of locked doors with no initial checkpoint.

Together, these mechanisms uphold the principles of least privilege and need-to-know – foundational concepts in information security that dictate users should have the minimum level of access necessary to perform their functions. By effectively applying both authentication and access control, organizations can protect themselves against both external threats and internal vulnerabilities, ensuring that their digital assets are accessed securely and used appropriately.

The evolution of authentication and access control from rudimentary practices to sophisticated, technologically advanced systems is a fascinating narrative that mirrors the broader history of human civilization and its ever-increasing dependency on information security. This historical trajectory demonstrates not only technological innovation but also a response to the escalating challenges posed by an increasingly interconnected world.

As we delve into this chapter, we'll traverse the evolution of authentication methods, from primitive passwords to cutting-edge biometric systems. We'll unlock the complexities of access control models, from the simplicity of discretionary access to the advanced algorithms of dynamic access controls. We will delve into their history, how they've become an inseparable part of the fabric of information security, and why, as we sail further into the digital horizon, their role only grows in stature and complexity. We'll encounter real-world scenarios that demonstrate the catastrophic consequences of failure and the triumphs of security done right.

User authentication

Authentication is a critical component of information security, ensuring that individuals, systems, or entities attempting to access a digital environment are indeed who they claim to be. This verification is based on three foundational principles: something you know, something you have, and something you are. Together, these form the cornerstone of robust authentication strategies, each adding a layer of security that protects sensitive information and systems from unauthorized access.

Let us understand each of these in the following sections.

Something you know – knowledge-based authentication

The first and most traditional form of authentication involves something that the user knows. This category includes passwords, **Personal Identification Numbers** (**PINs**), passphrases, and personal security questions. Knowledge-based authentication is widely used due to its simplicity and cost-effectiveness. Users are required to recall and provide information that ideally only they should know.

A typical example is the login process for an email account, whereby the user is prompted to enter their password. Similarly, withdrawing money from an ATM requires entering a PIN, which theoretically only the cardholder knows.

However, this method has vulnerabilities. The strength of knowledge-based authentication is directly tied to the complexity and secrecy of the information. Users often choose passwords that are easy to remember, which also makes them easy to guess or hack. Additionally, widespread practices such as reusing passwords across multiple sites significantly compromise security. To mitigate these risks, best practices suggest combining this method with others for a **Multi-Factor Authentication** (**MFA**) approach, enhancing overall security.

Something you have – possession-based authentication

The second principle involves something the user physically possesses. This form of authentication is harder to compromise remotely and includes items such as security tokens, smart cards, and mobile phones used to receive **One-Time Passwords** (**OTPs**). Devices may also be configured to act as part of the authentication process, utilizing pre-registered device fingerprints such as **Mandatory Access Control** (**MAC**) addresses or unique device IDs.

A bank might issue a security token to a customer that generates a code, which must be entered in conjunction with a password to access online banking services. Another common example is the use of a smartphone app that generates a time-sensitive OTP, or SMS-based OTPs sent to a user's phone during a transaction process.

This method enhances security by requiring physical possession of an item, making unauthorized access without possession of the device significantly more challenging. However, it also introduces the risk of loss or theft of the device, which can lead to potential security breaches if additional protective measures (such as PIN or biometric locks on the device) are not in place.

Something you are – biometric authentication

Perhaps the most secure authentication factor is based on something inherent to the user—biometric characteristics. Biometrics utilizes unique physical or behavioral traits of individuals for identification and authentication. Common examples include fingerprint scanning, facial recognition, iris scanning, and voice recognition. More advanced systems might analyze patterns in behavior such as typing rhythm or mouse movements.

Modern smartphones often come equipped with fingerprint sensors or facial recognition systems that users must authenticate to unlock their devices or make purchases. High-security facilities might use iris scanners to control access to sensitive areas, relying on the unique patterns in an individual's iris to confirm identity.

Biometric authentication offers a high level of security as these traits are extremely difficult to replicate or steal. However, concerns regarding privacy and the potential misuse of biometric data have been raised. Additionally, factors like injuries or changes in physical traits can sometimes affect the reliability of biometric authentication.

Multi-factor authentication

No single authentication method is foolproof when used in isolation. Modern security paradigms often involve integrating two or more of these authentication principles—a strategy known as MFA. By requiring multiple forms of verification, MFA significantly reduces the risk of unauthorized access, providing a layered defense that compensates for the potential weaknesses of each single authentication method.

In summary, the three basic principles of authentication form the backbone of securing access to systems and data. Understanding and implementing these principles effectively is crucial for anyone involved in securing digital environments, whether they're managing personal devices or securing enterprise-level systems. As cyber threats evolve, so must our strategies for authentication, continuously adapting to protect against an ever-changing landscape of risks.

Authentication technologies

In the digital world, authentication technologies are crucial for verifying the identity of users and securing access to systems and data. As cyber threats evolve, so too do the methods of authentication, each designed to provide different levels of security based on the needs of users and organizations. Currently, with the development of AI technology, there are also many methods to use model-based authentication. Most of them are connected with biometric authentication, such as face recognition, video recognition, and so on.

In the following sections, we will explore four central authentication technologies—passwords, tokens, biometrics, and MFA—discussing their mechanisms and applications, and assessing their strengths and weaknesses.

Passwords

Passwords are the most traditional form of authentication, relying on something the user knows. This method requires users to create and recall a string of characters, which they must provide to gain access to systems or services. The security of a password depends largely on its complexity and unpredictability.

Passwords are ubiquitous in personal and professional contexts. They are used to secure everything from email accounts and smartphones to databases and corporate networks.

Here are some strengths of passwords:

- **Ubiquity and familiarity**: Nearly everyone understands how to use passwords, and they are supported by almost all systems
- **Cost-effective**: They require no additional hardware or sophisticated software to implement

Here are some weaknesses:

- **Vulnerability to attacks**: Passwords are susceptible to a range of attacks, including brute force (trying all possible combinations), dictionary attacks (using per-compiled lists of common passwords), and phishing.
- **User management challenges**: Users often choose weak passwords or reuse them across multiple sites, increasing the risk of breaches. Managing strong, unique passwords for numerous accounts can be cumbersome without the aid of password managers.

Tokens

Tokens are physical or digital objects that users possess and present as proof of their identity. Physical tokens may be hardware devices that generate an OTP, while digital tokens could include software-based tokens or certificates stored on a device. Tokens are often used in banking for online transactions, in corporate environments for accessing secure networks, and anywhere where an additional layer of security is required over and above a username and password.

Here are some strengths of using tokens:

- **Increased security**: Tokens add a layer of security by requiring something the user physically possesses
- **Difficult to duplicate**: Physical tokens, especially those designed to be tamper-resistant, are hard to clone

Here are some of the weaknesses associated with using tokens:

- **Cost and management**: Tokens can be expensive to distribute and replace. Managing them can be logistically challenging, especially in large organizations.
- **Inconvenience**: Users must carry tokens with them, and they can be lost or stolen.

Biometric

Biometric authentication technologies use unique physical or behavioral characteristics—such as fingerprints, facial features, iris patterns, or voice—to identify individuals. Biometrics are used in various high-security settings such as government buildings, airports, and smartphones. They are also increasingly found in consumer electronics and banking applications.

The strengths of using biometrics include the following:

- **High security**: Biometric characteristics are extremely difficult to replicate or steal, making them secure identifiers
- **User convenience**: Biometrics provide a quick and easy way for users to authenticate themselves without needing to remember passwords or carry tokens

Some of the weaknesses of using biometrics are as follows:

- **Privacy concerns**: Storing and processing biometric data raises significant privacy issues, as compromise of such information can be irreversibly damaging
- **Environmental and physical limitations**: Conditions such as poor lighting for facial recognition, injuries to fingers used for prints, or voice changes can affect the reliability of biometric systems

MFA

MFA involves combining two or more independent credentials: what the user knows (password), what the user has (token), and what the user is (biometric verification). The idea is that if one factor is compromised, the others will still protect the system. MFA is critical for protecting sensitive data in industries such as finance, healthcare, and government. It is also recommended for personal use in securing important online accounts such as email and banking.

The associated strengths are as follows:

- **Enhanced security**: With the use of multiple authentication methods, MFA is able to significantly reduce the risk of unauthorized access
- **Flexibility and customizability**: Organizations can implement MFA in various ways to meet their specific security needs and user contexts

On the other hand, the associated weaknesses are as follows:

- **User friction**: MFA can add complexity and time to the authentication process, potentially frustrating users
- **Implementation complexity:** Properly implementing MFA requires careful planning, testing, and support to ensure effectiveness without overly burdening the user

User authentication is described in standards such as the NIST Digital Identity Guidelines (`https://pages.nist.gov/800-63-3/`), and its importance is emphasized by cybersecurity organizations such as the **Open Web Application Security Project (OWASP)**.

Access control

Access control is a fundamental concept in security that restricts access to resources based on specific criteria. Essentially, it determines who is allowed to access or use resources in a computing environment. There are several models of access control, each with its methodology and application context. The most commonly implemented are **Discretionary Access Control (DAC)**, MAC, **Role-Based Access Control (RBAC)**, and **Attribute-Based Access Control (ABAC)**. Understanding these models provides the groundwork for securing sensitive information and systems effectively.

Access control models

Let's explore the previously mentioned different access control models in detail.

DAC

In the **DAC** model, the control of access is left up to the discretion of the resource owner. Owners can decide on their own to whom they want to grant access and the level of access they wish to grant. This is typically managed through **Access Control Lists (ACLs)** that specify which users or groups are granted access to objects and what operations they can perform.

The associated strengths are as follows:

- **Flexibility**: DAC offers great flexibility, allowing users to easily share resources
- **Ease of use**: It is straightforward and intuitive for users to control their own resources
- **Owner control**: Resource owners have full control over their resources, which can be essential in dynamic environments

The associated weaknesses are as follows:

- **Lack of central control**: Since access is at the discretion of individual users, it can lead to inconsistencies in security policies across the organization
- **Security risks**: DAC systems are susceptible to Trojan horse attacks, as malicious programs can misuse the access rights of users to access unauthorized data
- **Scalability issues**: Managing individual permissions can become cumbersome in large organizations with thousands of users and resources

MAC

MAC is more stringent than DAC and is used primarily where confidentiality and classification are crucial. Under MAC, access to resource decisions is enforced by a central authority based on predefined security labels assigned to both resources (data) and users. Users cannot change the access controls.

Here are some of the strengths of MAC:

- **High security**: Provides a higher level of security as users cannot modify the access control settings
- **Standardized**: Enforces a uniform security policy across all controlled resources
- **Ideal for sensitive contexts**: Widely used in military and governmental contexts where data classification is critical

The weaknesses of MAC are as follows:

- **Inflexibility**: Users cannot share resources at their discretion, which can hinder collaboration and operational flexibility
- **Complex administration:** Requires careful setup and management of classifications and clearances
- **Limited by static policies**: Does not adapt dynamically to changing conditions or contexts within the system

RBAC

RBAC restricts system access based on a person's role within an organization. It simplifies access management by assigning rights and permissions to roles, rather than to individual users, and users are then made members of these roles.

Here are some of the strengths of RBAC:

- **Simplified management**: Eases the administration of permissions as changes to roles apply to all members of that role
- **Least privilege**: Facilitates adherence to the principle of least privilege, reducing the chance of excess privilege
- **Scalable**: Effective in organizations of almost any size

RBAC has the following weaknesses:

- **Role explosion**: Complex environments may lead to an excessive number of roles
- **Initial setup**: Requires careful planning and setup to define roles accurately
- **Rigid structure**: This may not cater to exceptions or special cases without creating new roles

ABAC

ABAC utilizes a range of attributes (user, resource, and environment) to make access control decisions. Policies can evaluate these attributes using logical rules to allow or deny access, making ABAC highly dynamic and context-aware.

Some of the strengths of ABAC are as follows:

- **Flexibility and granularity**: Offers highly granular access control, capable of considering a wide range of contextual factors
- **Dynamic conditions**: Adapts to changing conditions in real time
- **Comprehensive policy enforcement**: Can enforce complex policies that reflect real-world scenarios

These are some of ABAC's weaknesses:

- **Complex policy management**: Developing and managing ABAC policies can be complex and time-consuming
- **Performance concerns**: The decision-making process can be resource-intensive, impacting system performance
- **Higher implementation costs**: Typically requires more sophisticated technology and expertise to implement effectively

Access control mechanisms

In the realm of information security, access control mechanisms are vital to ensuring that the right people have the appropriate level of access to organizational resources. These mechanisms are implemented through various tools, including ACLs, policies, and comprehensive models/frameworks. Each of these plays a crucial role in defining and enforcing the boundaries within which users operate. This detailed exploration covers how these mechanisms function and interact to maintain security and operational integrity.

ACLs

An ACL is a table that informs a computer operating system about the access rights assigned to each user for specific system objects, such as file directories or individual files. Each entry in an ACL specifies a subject and an action. For instance, an ACL entry might allow user John read access to a file but deny him write access. ACLs are used extensively in various systems, from traditional file systems to modern network devices such as routers and firewalls. They work by filtering traffic based on the defined rules that apply to user requests. These lists are crucial for maintaining segregation in multi-user environments, ensuring that users do not exceed their authorized access.

Imagine a diagram showing a file with an attached list. The list entries specify that User A can read and write, User B can only read, and User C has no access.

Access control policies

Access control policies are sets of rules that determine how ACLs and other control mechanisms are applied. They define the conditions under which access can be granted and detail the capabilities of different user roles within the organization. Policies might be broad, covering general access, or very specific, detailing actions that can be performed on particular resources. These policies are implemented through administrative tools and software that manage user permissions. They help ensure compliance with security standards and legal regulations by systematically enforcing the necessary controls. Policies must be regularly reviewed and updated to adapt to new threats or changes in the organization.

Models/frameworks

Access control models provide a theoretical framework for understanding and implementing access control policies and mechanisms. Common models include the following:

- **DAC**: The resource owner decides on access
- **MAC**: Access is governed by a centralized authority based on predefined policies
- **RBAC**: Access is based on the user's role in the organization
- **ABAC**: Access decisions are based on policies that evaluate attributes

These models are used to design the access control architecture of an organization. They guide how policies are created and enforced, as well as how ACLs are structured. By choosing a particular model, an organization defines the method and philosophy of its access control.

The design and implementation of access control mechanisms are fundamental to protecting an organization's informational assets. ACLs provide the specific entries for user permissions, policies ensure these are applied consistently, and models/frameworks define the overarching method and philosophy of security. Together, these mechanisms create a robust defense.

Exemplifying the user authentication and access control

In this section, we will show several examples of the user authentication and access control field. These examples include OAuth2.0, SELinux, and face recognition.

OAuth2.0 and user authentication

OAuth 2.0 is an open standard for authorization that allows third-party services to exchange web resources on behalf of a user. It's important to clarify that OAuth 2.0 is primarily about authorization, not authentication. Authorization is the process of specifying access rights to resources, which is different from authentication, the process of verifying identity. OAuth 2.0 provides delegated access to server resources on behalf of the resource owner. It establishes a process through which users are redirected to a service provider to authenticate and then return to the original website with an authorization code that can be exchanged for an access token. This token grants specific types of access to specific resources for a defined duration.

While OAuth 2.0 is designed for authorization, it is often used in conjunction with protocols such as OpenID Connect to provide authentication services. OpenID Connect is built on top of OAuth 2.0 and adds an identity layer, enabling clients to verify the identity of the end user and obtain basic profile information in an interoperable and REST-like manner.

Here we will see two real-life examples of OAuth 2.0 being used for authentication in conjunction with additional tools or in scenarios where the distinction between authentication and authorization might blur.

Social media login (Google, Facebook, and X (formerly Twitter)

Implementing OAuth 2.0 for user authentication via social media platforms such as Google, Facebook, or X involves integrating these platforms as identity providers for your application. This process allows users to authenticate using their existing social media accounts, which simplifies the login process and can enhance user engagement and convenience. Here's a step-by-step guide on how to set this up.

Step 1 – register your application

Before you can implement OAuth 2.0, you need to register your application with the social media platform you intend to use for authentication:

- **Google**: Use the Google Developer Console to create a new project and set up an OAuth 2.0 client ID
- **Facebook**: Use the Facebook Developer portal to create a new app and get your app ID and app secret
- **X (Twitter)**: Use the Twitter Developer portal to create an app and obtain API keys and tokens

During this process, you will need to provide information about your application, such as the name, the website URL, and most importantly, the redirect URI, which is the URL to which users will be redirected after they have authenticated with the social media service.

Step 2 – implement the OAuth 2.0 flow

Integrate the OAuth 2.0 flow into your application. This typically involves the following steps:

1. **Authentication request**: Redirect the user to the authorization URL of the social media platform. The URL will include parameters such as your client ID, the requested scope (information or permissions you want access to), and your redirect URI. Here's an example of what this might look like for Google:

   ```
   https://accounts.google.com/o/oauth2/v2/auth?response_type=code&client_id=YOUR_CLIENT_ID&redirect_uri=YOUR_REDIRECT_URI&scope=openid%20email&state=SECURE_RANDOM_STATE
   ```

2. **User login and consent**: The user will be asked to log in to their social media account (if they are not already logged in) and consent to share information with your application. The *scope* of the information shared is defined by the scope parameter in the authentication request.

3. **Authorization code**: After the user consents, the social media platform redirects back to your application using the `redirect_uri` parameter provided, appending an authorization code to it.

4. **Access token**: Your application then exchanges the authorization code for an access token using the social media platform's token endpoint. This requires sending the client ID, client secret, authorization code, and redirect URI to the token endpoint:

   ```
   POST https://accounts.google.com/o/oauth2/token
   Content-Type: application/x-www-form-urlencoded
   code=AUTHORIZATION_CODE&client_id=YOUR_CLIENT_ID&client_secret=YOUR_CLIENT_SECRET&redirect_uri=YOUR_REDIRECT_URI&grant_type=authorization_code
   ```

5. **Access user information**: Using the access token, your application can now access the user's information via the social media platform's API, depending on the granted scopes. This token represents the user's authorization to access their data.

Step 3 – handle the user data

Once your application has access to the user's data (such as their email, name, and profile picture), you can use this information to either create a new user account in your application or update an existing one. Ensure you handle this data according to privacy best practices and regulations.

Step 4 – maintain security

This is how we can make it secure:

- Use HTTPS for all exchanges involving sensitive information to prevent interception
- Validate the state parameter in the redirect URI to mitigate CSRF attacks
- Regularly review and update the requested permissions to avoid asking for more data than necessary

Using OAuth 2.0 for social media login simplifies the authentication process for users and can enhance the security of your application by leveraging the robust authentication mechanisms of established social media platforms. This not only improves the user experience by reducing the burden of remembering multiple passwords but also benefits from the trust users have in their social media platforms.

Mobile app access to services

Using OAuth 2.0 to facilitate user authentication in a mobile app, especially when accessing third-party services, is a common practice that enhances both security and the user experience. The OAuth 2.0 framework helps mobile applications authenticate users and obtain permission to access services or resources on their behalf without exposing user credentials.

We will look at the steps to implement this process.

Step 1 – register your mobile app with the service provider

First, you need to register your mobile application with the third-party service provider whose resources you want to access. This is typically done through the service provider's developer portal. During registration, you will provide details such as your app name and logo, and most importantly, the redirect URI that the service provider will use to send responses back to your app. You will receive `client_id` and a `client_secret` upon registration, which are essential for the OAuth 2.0 authentication process.

Step 2 – implement the OAuth 2.0 authorization flow in your mobile app

Choose the appropriate OAuth 2.0 flow for your mobile application. For mobile apps, the recommended flow is usually the *Authorization Code Flow* with **Proof Key for Code Exchange** (**PKCE**), which enhances security by preventing interception of the authorization code:

1. Initiate the authorization request. Construct an authorization request URL and direct the user to it. This URL will include the following:

    - The `response_type` set to `code` (for Authorization Code Flow)
    - The `client_id` you received during registration

- The `redirect_uri` that points back to your mobile app
- A `scope` parameter defining the permissions your app is requesting
- A randomly generated `code_challenge`, part of PKCE, ensures that the exchange of the authorization code for the access token is secure and performed by the requester; here is an example of a URL for an authorization request:

   ```
   https://example.com/oauth/authorize?response_type=code&client_
   id=YOUR_CLIENT_ID&redirect_uri=YOUR_REDIRECT_URI&scope=profile
   email&state=XYZ&code_challenge=YOUR_GENERATED_CHALLENGE&code_
   challenge_method=S256
   ```

2. Handle the redirect. The service provider will direct the user to a login and consent page. Upon successful authentication and consent, the service provider redirects the user back to your application using the `redirect_uri`, appending an authorization code.

3. Exchange the code for an access token. Your app then exchanges the authorization code for an access token using the token endpoint of the service provider. This request will also include the `code_verifier` (the value used to generate the `code_challenge`) to complete the PKCE validation. Here is an example of a POST request for this purpose:

   ```
   POST https://example.com/oauth/token Content-Type: application/
   x-www-form-urlencoded grant_type=authorization_code&code=AUTH_
   CODE_RECEIVED&redirect_uri=YOUR_REDIRECT_URI&client_id=YOUR_
   CLIENT_ID&code_verifier=YOUR_CODE_VERIFIER
   ```

4. Retrieve user information. With the access token, your application can now access the user's information from the service provider's API, as per the allowed scopes. The access token authenticates your application's requests on behalf of the user.

Step 3 – use the access token

Store the access token securely within your application. Use this token to make API requests to the third-party service, fetching or posting data according to the scopes granted. Ensure the token is sent in the `HTTP Authorization` header as a bearer token.

Step 4 – handle token expiry and refresh

Access tokens typically have a limited lifespan. If a refresh token was provided during the initial token exchange, your app could use it to obtain a new access token without requiring the user to authenticate again. Handle token expiry gracefully in your app by checking the token's validity and refreshing it as needed.

Step 5 – ensure security best practices

Here are some best practices for security:

- Implement secure storage practices on the mobile device to protect tokens and sensitive data
- Use HTTPS for all communications between your mobile app, the user's device, and external services
- Regularly review and update the OAuth 2.0 implementation to comply with new security standards and mitigate emerging security threats

Implementing OAuth 2.0 in mobile apps for accessing third-party services enhances security by leveraging robust, standardized protocols for user authentication and authorization. It allows your app to act on behalf of the user without handling sensitive password data, thereby reducing the security burden and improving the user experience.

So far, we have finished the OAtuh2.0 introduction, and we have learned the fundamental information about its functionality and use cases. Let us move to **Security-Enhanced Linux** (**SELinux**) and access control at the kernel level in the next part.

SELinux and access control

SELinux is a security architecture integrated into various Linux distributions. It was originally developed by the United States **National Security Agency** (**NSA**) to implement MAC in the Linux kernel. SELinux enhances the existing security measures in Linux by adding an additional layer of access controls, which improves the system's ability to restrict running processes and their access to files, devices, and other resources.

SELinux operates based on policies that define how applications and processes can access resources on a system. These policies are enforced by the Linux kernel, and they are typically much more granular than the traditional user and group permissions found in Unix-like systems. SELinux policies include rules that dictate whether an action by a particular subject (a process or user) on an object (like a file, directory, or port) should be allowed or denied.

The system uses two main modes:

- **Enforcing**: SELinux actively enforces its policies, blocking unauthorized access and logging actions.
- **Permissive**: SELinux logs actions that would have been blocked if running in enforcing mode but it does not block them. This mode is useful for debugging and policy creation.

SELinux is fundamentally about access control. It labels every process and resource with security contexts that include a type, a role, and a user (not directly corresponding to Linux users). Access decisions are based on these contexts, roles, and types, not merely on the object's owner or the process' user ID. It excels at limiting how processes can interact with files, other processes, and system resources.

This capability is crucial in preventing or mitigating damage from system compromises, such as those arising from software vulnerabilities.

SELinux supports RBAC, which restricts system access based on the roles assigned to users within an organization. This feature allows administrators to define security policies that are adaptable to the specific roles and responsibilities within the organization. At the heart of SELinux is type enforcement, which dominates how policies are written. Every file, process, and resource is tagged with a type, and rules in the SELinux policy database govern interactions based on these types.

SELinux is a powerful tool for implementing MAC in Linux environments, enhancing system security by controlling what resources users and processes can access after they are authenticated. It does not replace user authentication mechanisms but complements them by providing robust access control that aligns with the principles of least privilege and separation of duties. This makes SELinux an essential component of a comprehensive security strategy in environments requiring stringent security measures.

In the following sections, we will see two practical examples of how SELinux can be used to control access in real-life scenarios and test our examples under the Ubuntu 23.04 version.

Web server isolation

In a hosting environment, multiple websites are served by a single Apache web server on a Linux system. The goal is to prevent a security breach on one website from affecting other hosted websites. SELinux can be configured to isolate each website by running each instance of the web server in its own confined domain. By applying strict SELinux policies, each web server instance is restricted to access only its own files and configurations. This containment is achieved using SELinux's **Multi-Category Security** (**MCS**) or **Multi-Level Security** (**MLS**), which assigns unique categories to each instance.

Implementing web server isolation using SELinux involves setting up a secure environment where each instance of a web server operates independently of others. This ensures that if one instance is compromised, the security breach does not affect other instances on the same system.

The following are the steps to achieve web server isolation using SELinux.

Step 1 – installing and enabling SELinux

Ensure SELinux is installed and enabled on your Linux system. You can check the current status by running the following:

```
sestatus
```

If SELinux is disabled, you'll need to enable it by editing the `/etc/selinux/config` file and setting the `SELINUX` mode to `enforcing`:

```
SELINUX=enforcing
```

Then, reboot your system for changes to take effect.

Step 2 – use different SELinux contexts

Assign unique SELinux contexts to each web server instance. This involves setting different SELinux types for each instance's document root, configuration files, and logs. For example, if you are running multiple Apache web servers, you can create and manage separate contexts for each:

1. **Create new context types**: First, define new types in your SELinux policy. You might need to use a policy management tool such as `checkmodule` and `semodule_package`, or use existing utilities to copy and modify existing types. For instance, you might use the following:

   ```
   checkmodule -M -m -o mypolicy.mod mypolicy.te
   semodule_package -o mypolicy.pp -m mypolicy.mod
   ```

   These two commands enable us to compile our policy. We will also need a `mypolicy.te` file for these commands, as this file will be used to define the policy.

   The following command enables us to install the policy:

   ```
   semodule -i mypolicy.pp
   ```

2. **Set file contexts**: Use `semanage fcontext` to apply the appropriate SELinux context to each web server's directories and files:

   ```
   semanage fcontext -a -t httpd_sys_content_t "/www/site1(/.*)?"
   semanage fcontext -a -t httpd_sys_content_t "/www/site2(/.*)?"
   restorecon -Rv /www/site1
   restorecon -Rv /www/site2
   ```

Step 3 – configuring SELinux Booleans

SELinux Booleans control certain policy features that can be toggled on or off to refine access controls. For web servers, you might need to set booleans that allow network connectivity, manage user home directories (disable in the following example), and handle logging appropriately (enable full logging in the following example):

```
setsebool -P httpd_can_network_connect on
setsebool -P httpd_enable_homedirs off
setsebool -P httpd_log_full on
```

Step 4 – testing and debugging

After configuring SELinux policies and contexts, perform the following steps:

1. Restart the web servers to apply all changes.

2. Check for **Access Vector Cache (AVC)** denials in the audit logs (`/var/log/audit/audit.log`). Use tools such as `audit2why` to understand any denials that occur and adjust your SELinux policies accordingly:

```
ausearch -m avc -ts recent | audit2why
```

Step 5 – maintaining and updating SELinux policies

As your web servers and hosting requirements evolve, continuously update and refine SELinux policies to ensure they remain effective and do not unnecessarily restrict legitimate web server operations.

This setup ensures that if one website is compromised, the malicious process is unable to access or interfere with the files or processes of other websites on the same server. This containment significantly limits the damage a compromised site can inflict. Using SELinux for web server isolation provides a robust layer of security by ensuring that each server operates within strictly enforced boundaries defined by SELinux policies. This approach significantly mitigates risks associated with web server vulnerabilities by containing potential breaches within a limited scope. It requires a detailed understanding of SELinux policy management but offers a high degree of protection in multi-hosting environments.

User authentication in multi-user environments

Using SELinux to protect user data in a multi-user environment involves configuring SELinux policies to strictly control how data is accessed and by whom. SELinux's MAC capabilities allow you to define precise access controls that are more granular and secure than traditional DAC systems.

Here are the steps to show how you can implement such protections.

Step 1 – ensuring SELinux is enabled and configured

First, verify that SELinux is active and set to enforcing mode to ensure that policies are being actively applied:

```
sestatus
```

If SELinux is not enabled, you will need to enable it by editing the `/etc/selinux/config` file to set `SELINUX=enforcing` and rebooting the system.

Step 2 – assigning SELinux user contexts

SELinux controls access based on SELinux user contexts, roles, and types. Assigning different SELinux user contexts to different system users can help segregate their access:

1. List available SELinux user contexts to find an appropriate context for each type of user:

```
semanage user -l
```

2. Map Linux users to SELinux user contexts. For example, assigning a user to a restricted SELinux user role that limits their permissions looks like this:

3. `semanage login -a -s user_u -r s0-s0:c0.c1023 username`

Step 3 – configuring SELinux policies for user data protection

Here, we develop and enforce SELinux policies that define and limit user interactions based on their roles and the sensitivity of the data as follows:

1. Define and apply file context rules to user data directories, ensuring that only authorized users can access sensitive data:

    ```
    semanage fcontext -a -t user_home_t "/home/user1(/.*)?"
    restorecon -Rv /home/user1
    ```

2. Create custom policy modules if necessary to handle complex scenarios where default policies do not suffice. For example, to restrict access between users in different departments, do the following:

    I. Write a custom SELinux policy module to control access.

    II. Compile and load the module:

    ```
    checkmodule -M -m -o custom_policy.mod custom_policy.te
    semodule_package -o custom_policy.pp -m custom_policy.mod
    semodule -i custom_policy.pp
    ```

Step 4 – using MCS or MLS

SELinux MCS/MLS can be used to further refine access controls in environments with highly sensitive data:

1. Assign unique MCS/MLS labels to different data that need to be segregated. For example, assign different sensitivity labels to documents based on their confidentiality level.

2. Apply MCS/MLS labels to user processes and data files, ensuring that only processes running with the appropriate sensitivity level can access corresponding data:

    ```
    chcon -l s0:c1,c2 /some/sensitive/file
    ```

Step 5 – monitoring and auditing

Regular monitoring and auditing are crucial to ensuring that SELinux policies are correctly enforced and that no unauthorized access occurs. Here's how:

1. Audit system logs for denied operations that might indicate attempted security breaches or misconfigurations:

    ```
    ausearch -m avc -ts recent | audit2why
    ```

2. Adjust policies based on audit findings to address any identified issues or to refine access controls further.

In the following section, we describe security enhancements in the context of database management.

Database security enhancement

A financial institution uses a Linux-based server to host its customer database, which contains sensitive personal and financial information. The security requirement is to ensure that only authorized services and users can access this database. SELinux is configured to ensure that only the specific database daemon (e.g., `mysqld` or `postgresd`) can read and write to the database files and directories. Access control policies are also put in place to restrict which system administrators or scripts can manage the database process or access the database files.

Implementing database security enhancements using SELinux involves leveraging its MAC capabilities to restrict database processes and protect database files from unauthorized access and manipulation. This setup ensures that even if an attacker compromises the database server, the damage they can inflict is minimized.

The following is a step-by-step guide on how to use SELinux to enhance database security.

Step 1 – ensuring SELinux is enabled and configured

First, make sure SELinux is enabled on your Linux system. You can check the SELinux status with the following:

```
sestatus
```

If it's not enabled, you will need to enable it by modifying the `/etc/selinux/config` file to set `SELINUX=enforcing` and then rebooting the system.

Step 2 – defining and applying appropriate SELinux contexts

SELinux uses contexts to manage access controls. Each file, process, and user has an SELinux context that includes a user, role, type, and level. You'll need to ensure that your database files, processes, and users have the correct SELinux contexts. Here is how we do it:

1. **Set the context for database files**: Use `semanage fcontext` to set the appropriate file context types for your database directories and files. For example, if you are using MySQL in the following line: `semanage fcontext -a -t mysqld_db_t "/var/lib/mysql(/.*)?" restorecon -Rv /var/lib/mysql`, you can set the context to existing files.

2. **Set the context for database executables**: Similarly, ensure the database executables have the correct context:

   ```
   semanage fcontext -a -t mysqld_exec_t "/usr/sbin/mysqld"
   restorecon -v /usr/sbin/mysqld
   ```

   This applies the proper SELinux type to the MySQL daemon

Step 3 – configuring SELinux Policies for database access

To strictly control access to the database, you may need to write or modify SELinux policies. This could involve creating custom policies if the default ones do not meet your specific requirements:

1. **Create custom modules**: If necessary, create custom SELinux policy modules. For example, you might write a policy to restrict which other services or users can interact with your database server.
2. **Compile and load the module**: Use `checkmodule`, `semodule_package`, and `semodule` to compile and load your custom module:

   ```
   checkmodule -M -m -o mydb.mod mydb.te semodule_package -o mydb.pp -m mydb.mod semodule -i mydb.pp
   ```

Step 4 – utilizing SELinux Booleans

SELinux Booleans can be toggled to refine policy enforcement without needing to modify or compile policies. Check which Booleans are relevant to your database and adjust them as needed. For example, for a MySQL database, you might need the following:

```
setsebool -P daemons_enable_cluster_mode 1 setsebool -P mysqld_disable_trans 0
```

You can list all Booleans related to MySQL using the following:

```
getsebool -a | grep mysql
```

Step 5 – regularly monitoring and auditing

Regularly check SELinux logs for denied accesses or policy violations as follows:

```
ausearch -m avc -ts recent | audit2why
```

This helps in understanding SELinux denials related to your database server and fine-tuning your policies accordingly.

Implementing database security with SELinux is a robust approach to enhancing the security of database systems on Linux. By properly configuring SELinux contexts, writing precise policies, and regularly monitoring the security logs, you can achieve a highly secure database environment that protects sensitive data from unauthorized access and minimizes the risks of data breaches.

This setup prevents unauthorized applications and users from accessing or modifying the database files, even if they gain access to the server through some other means. It helps protect against both external breaches and insider threats, ensuring that only those explicitly granted permission can interact with the database.

The following section contains a practical example in Python for setting up authentication and access control on your system.

Practicing user authentication and access control with Python

So far, we have explained user authentication and access control in theory. We have also presented several examples on OAuth2.0 and SELinux in the Ubuntu23.04 environment. In this section, we present several practical examples based on Python to further demonstrate the key concepts discussed earlier.

To try the examples, we need to have a practical component. It requires the following Python libraries:

- `Requests-oauthlib`
- `Accesscontrol`
- `Selinux`
- `Pycasbin`

Using OAuth2.0 in mobile application authentication

To implement OAuth 2.0 in a Python environment for a scenario similar to mobile application authentication, you can use the popular `requests-oauthlib` library. This library provides a high-level interface to interact with OAuth 1.0/a and OAuth 2.0 providers.

In this section, I will guide you through setting up a basic OAuth 2.0 client that could be used to authenticate a user via a third-party service (like Google) in a Python application:

1. **Install the required library**: First, you need to install the `requests-oauthlib` library. You can do this using `pip`:

   ```
   pip install requests-oauthlib
   ```

2. **Register your application**: Assuming you're using Google as an OAuth 2.0 provider, do the following:

   I. Go to the Google Developer Console at `https://console.developers.google.com/`.

   II. Create a project.

III. Go to **Credentials**, create credentials, and select **OAuth client ID**.

IV. Configure the consent screen, and set the application type to **Web application**.

V. Define the authorized redirect URIs (this should match the URI in your application where Google will send the response).

VI. Once registered, you'll receive `client_id` and `client_secret`.

3. **Create the OAuth2 session**: Here's a sample Python script that creates an OAuth2 session and guides the user through the authentication process:

```
from requests_oauthlib import OAuth2Session from oauthlib.oauth2
import MobileApplicationClient
# Replace these with the actual client ID and URI
client_id = 'YOUR_CLIENT_ID' redirect_uri = 'YOUR_REDIRECT_URI'
# This should match the one you set in the Google Console
# Define the scope of the application
scope = [
    'https://www.googleapis.com/auth/userinfo.email',
    'https://www.googleapis.com/auth/userinfo.profile' ]
# Create a client instance
client = MobileApplicationClient(client_id)
# Create an OAuth2 session
oauth = OAuth2Session(
    client=client, scope=scope, redirect_uri=redirect_uri)
# Get the authorization URL
authorization_url, state = oauth.authorization_url(
    'https://accounts.google.com/o/oauth2/auth',
    access_type="offline", prompt="select_account")
print('Please go to {} and authorize
    access.'.format(authorization_url))
# Get the authorization response from the callback URL
redirect_response = input('Paste the full redirect URL here: ')
 # Fetch the token
token = oauth.fetch_token(
    'https://accounts.google.com/o/oauth2/token',
    authorization_response=redirect_response,
    client_secret='YOUR_CLIENT_SECRET')
print('Authentication successful, here is the token: ', token)
```

From the preceding code, we can understand the following:

- **OAuth2Session setup**: We initialize an OAuth2Session with the necessary client ID, redirect URI, and scope. The scope determines what information the application is requesting access to.

- **Authorization URL**: The user needs to visit this URL to authorize access. In a real-world application, you would direct them to this URL through a web browser.
- **Handle redirect**: After the user authorizes access, they will be redirected to a URL that includes a code. This is simulated here by asking the user to paste the redirect URL.
- **Fetch the token**: Using the redirect response, the script exchanges the code for an access token. This token is used to authenticate API requests on behalf of the user.

This script provides a basic demonstration of how to implement OAuth 2.0 user authentication in a Python application. This is suitable for scenarios such as mobile app authentication, where the app is acting as an OAuth client.

Writing SELinux in Python to control the Ubuntu files

To implement an access control program on Ubuntu using SELinux and Python, you'll need to work with SELinux commands and potentially leverage Python to script these changes or monitor SELinux states. However, it's important to note that Ubuntu typically uses AppArmor as its primary MAC system rather than SELinux, although SELinux can be installed and used if desired.

Here, we will see how to set up a basic Python script that uses SELinux commands to manage file access controls, assuming SELinux is installed and managing your system. If SELinux is not already set up on your Ubuntu system, you'll first need to install it and switch from AppArmor. This setup can be complex and is generally recommended only for advanced users or specific use cases:

1. **Install SELinux (if not installed)**: If you decide to use SELinux on Ubuntu, you can install it via the terminal:

   ```
   sudo apt-get install selinux-basics selinux-policy-default
   auditd sudo selinux-activate sudo reboot
   ```

2. **Check SELinux Status**: After rebooting, ensure SELinux is active:

   ```
   getenforce
   ```

   It should return `Enforcing`. If it doesn't, you may need to troubleshoot the SELinux installation or configuration.

3. **Use Python script to manage SELinux file contexts**: Here's a basic Python script that uses the subprocess module to change SELinux contexts and thus control access to files. This script will change the SELinux type of a file, which could be used to restrict or grant access depending on your SELinux policies:

   ```
   import subprocess

   def set_selinux_context(filepath, selinux_type):
       """ Set the SELinux context for a specific file. """
   ```

```python
    try:
        command = f'semanage fcontext -a -t
            {selinux_type}"{filepath}"'
        subprocess.run(command, check=True, shell=True)
        subprocess.run(f'restorecon -v {filepath}', check=True,
            shell=True)
        print(f"SELinux context for {filepath} set to
            {selinux_type}.")
    except subprocess.CalledProcessError as e:
        print(f"Failed to set SELinux context
            for {filepath}: {str(e)}")

def get_selinux_context(filepath):
    """ Get the SELinux context for a specific file. """
    try:
        result = subprocess.run(f'ls -Z {filepath}',
        check=True, shell=True, capture_output=True, text=True)
        print(f"SELinux context
            for {filepath}: {result.stdout.strip()}")
    except subprocess.CalledProcessError as e:
        print(f"Failed to get SELinux context
            for {filepath}: {str(e)}")

# Example Usage
file_path = '/path/to/your/file'
selinux_type = 'httpd_sys_content_t'
# Example type, change as needed
set_selinux_context(file_path, selinux_type)
get_selinux_context(file_path)
```

From this code, we understand the following:

- `set_selinux_context`: This function sets the SELinux context of a specified file to restrict or allow access based on the type specified. For instance, setting a file type to `httpd_sys_content_t` might allow a web server to read the file but deny other types of access.

- `get_selinux_context`: This function fetches and prints the current SELinux context of the specified file.

4. **Customize and extend**: You can customize this script to handle different files and types based on your specific security needs. You might also integrate error handling or logging to keep track of any changes or issues.

AI for user authentication – face recognition

Face recognition is a biometric technology that identifies or verifies a person by analyzing and comparing patterns based on their facial features. Unlike traditional identification methods, such as passwords or PINs, face recognition uses the unique structure of an individual's face to recognize them. This technology involves several steps, including face detection, feature extraction, and face matching. Face detection is locating the face within an image or video. Feature extraction is analyzing facial features such as the distance between the eyes, the shape of the nose, the contour of the lips, and the structure of the jawline. Last but not least, face matching is comparing the extracted features to a database of known faces to identify or verify the person.

In user authentication, face recognition is used as a biometric method to verify a user's identity. It is increasingly being adopted for secure access control in devices, applications, and services. Instead of relying on something the user knows (such as a password) or something the user has (such as a key or token), face recognition leverages something the user inherently possesses—their facial features.

Let's consider an example of using face recognition for authenticating users to access a secure system, such as a company's internal portal. Applications for face recognition in user authentication include smartphones and laptops that can use it to unlock devices using the built-in camera. Access control systems use it to allow entry to secured areas or buildings. In the financial transactions field, face recognition is used to authorize payments or access banking services. Verifying the identity of patients is the typical use case in medical facilities.

The user takes several pictures of their face from different angles. The system processes these images, extracts facial features, and creates a unique facial template. This template is securely stored in the system's database. When the user attempts to log in, the system captures a new image of their face using a camera. The system extracts facial features from the captured image. The extracted features are then compared to the stored template. If the match score exceeds a predefined threshold, the user is authenticated and granted access.

Contrastive Language-Image Pretraining (CLIP) is a powerful model designed to understand both images and text by learning to associate images with their descriptions. While CLIP is not explicitly designed for face recognition (which is usually done with specialized models such as FaceNet, ArcFace, or VGGFace), you can use CLIP for face similarity tasks or even some basic face recognition by comparing the embeddings of images.

Here's an example of how you might use CLIP for a simple face recognition task in Python.

The following code snippet is our prerequisite for this section.

Install the required libraries:

```
# Installing torch and clip library using pip
pip install torch torchvision clip-by-openai
```

Download CLIP from OpenAI:

```
import torch
import clip
from PIL import Image

# Load the model
device = "cuda" if torch.cuda.is_available() else "cpu"
model, preprocess = clip.load("ViT-B/32", device=device)
```

Step 1 – preprocess the images

First, you'll want to load and preprocess the images of the faces you want to recognize:

```
# Load and preprocess the images
image_path1 = "path_to_face_image1.jpg"
image_path2 = "path_to_face_image2.jpg"

image1 = preprocess(Image.open(image_path1)).unsqueeze(0).to(device)
image2 = preprocess(Image.open(image_path2)).unsqueeze(0).to(device)
```

Step 2 – generate embeddings

Next, use CLIP to generate embeddings for each image:

```
# Generate image embeddings
with torch.no_grad():
    image_features1 = model.encode_image(image1)
    image_features2 = model.encode_image(image2)

# Normalize the embeddings
image_features1 /= image_features1.norm(dim=-1, keepdim=True)
image_features2 /= image_features2.norm(dim=-1, keepdim=True)
```

Step 3 – compare the embeddings

Finally, compare the embeddings of the two images. The cosine similarity between these embeddings can give you an idea of how similar the two faces are:

```
# Calculate cosine similarity
similarity = torch.cosine_similarity(image_features1, image_features2)

print(f"Similarity: {similarity.item()}")
```

Step 4 – determine whether the faces match

You can set a threshold for the similarity score to decide whether the faces match:

```
# Define a threshold
threshold = 0.8  # Adjust this value based on your needs

if similarity.item() > threshold:
    print("The faces match!")
else:
    print("The faces do not match.")
```

This example shows how to use CLIP for a simple face recognition task by comparing the embeddings of two face images. Note that CLIP is not specifically optimized for face recognition, so this approach may not be as accurate as using a model explicitly designed for this purpose. However, it can still work reasonably well for tasks where high precision is not critical.

Face recognition offers a seamless and secure method for user authentication, blending convenience with advanced security features. By leveraging powerful libraries and following best practices, developers can integrate face recognition into various applications, enhancing both user experience and system integrity.

Summary

In this chapter, you learned the basics of user authentication and access control that you can use in your AI projects. Similar to other computer systems, AI systems need to control access to their functionality because they can contain critical capabilities and highly restricted data, as you learned in this chapter. The chapter included a theoretical introduction, practical considerations, and exercises that helped you enhance your knowledge. Furthermore, it included an example use case of AI methods used for user authentication, in particular face biometrics.

The next chapter introduces AI methods to enhance another area of cybersecurity – threat intelligence.

11
Threat Intelligence

In cybersecurity, it's important to stay on top of changes in the threat landscape and investigate how threat actors behave and what are the emerging risks. This means that we cannot rely only on our detections; we need to seek information from outside sources as well to enhance our knowledge and make better decisions. Therefore, one of the most important parts of the detection and analysis of cyber threats is the analysis of various sources of information and the synthesis of knowledge from it. Apart from internal detection mechanisms, there are many public and non-public sources of information describing current cyber threats that can be used to prepare defenses, improve cybersecurity analytics, and shed light on cybersecurity risks.

In this chapter, we will cover an introduction to **threat intelligence** and explain how **artificial intelligence** (**AI**) can be helpful in retrieving information from various threat intelligence data sources. Furthermore, we include a practical exercise with an example of information retrieval from public data.

To summarize, in this chapter, we are going to cover the following topics:

- Understanding threat intelligence
- Working with AI for threat intelligence
- Exercise – retrieval of threat intelligence from X (formerly Twitter) using AI

Technical requirements

This chapter has a practical component and requires the following Python libraries:

- `numpy`
- `pandas`
- `gensim`
- `PyLDAvis`
- `collections`
- `ntlk`

For the best learning performance, it's recommended to use `notebook` libraries and platforms, like the following:

- `jupyter`
- `google-colab`
- `Databricks`

Understanding threat intelligence

Cyber threat intelligence (**CTI**) consists of the collection, analysis, correlation, and sharing of cybersecurity-related data from public and non-public data sources. The information retrieved through CTI is used not only to adapt and strengthen defenses but also to assess risk and justify funding for these improvements. Therefore, accurate and complete information retrieval is crucial for cybersecurity in enterprises and other organizations.

We can differentiate the following types of threat intelligence:

- **Tactical**: Tactical threat intelligence focuses on providing data about potential attacker activity. This means to gather data about malware, network traffic patterns, malicious URLs, phishing, and similar. The data is structured as **tactics, techniques, and procedures** (**TTPs**) and can be used to adjust the defensive security tools to provide optimal protection.

- **Strategic**: While tactical threat intelligence provides more particular and fine-grained information about attacker activity, in strategic threat intelligence, the focus is on high-level information that can be used by the cybersecurity leadership to understand risk and the cybersecurity posture. Furthermore, there is other non-technical information useful for strategic decisions, such as legal information or social media. This strategic information directly impacts business decisions.

- **Operational**: In operational threat intelligence, the idea is to complement the tactical technical information with other data about the attacker's motive and activity. Common data sources for this are the social media of the attacker, forums, chats, and so on. This data should be directly useful in operational activity, for example, to inform the people performing operational activities if there is an ongoing attack.

Organizations need all three types of threat intelligence to work together to provide proper insight into current cybersecurity posture, assess risk, and improve defenses.

CTI includes a large set of activities that should be structured optimally. Broadly speaking, there are six phases in the CTI life cycle:

1. **Planning**: Here, we gather requirements and set the roadmap for the threat intelligence process.
2. **Collection**: In this part, we collect data relevant to our requirements, from different public or non-public sources.

3. **Processing**: The data could be collected in various formats and have variable quality. Here, we process collected data to transform it into the form needed for analysis.
4. **Analysis**: In the analysis part, we take a deeper look to obtain insight from the transformed data that can be used to inform our threat investigation and the impact of threats on cybersecurity risk.
5. **Dissemination**: In this step, we share the information and insights obtained through the analysis of CTI data. The information could be shared within the organization, externally with organizations in the same area, or even publicly.
6. **Feedback**: Here, we gather feedback from the recipient of the CTI information. Furthermore, we will possibly be directed to the planning phase again if there are any further requirements.

In the next section, we will describe how AI fits into this life cycle and what value it can bring to the CTI analysts and the recipients of CTI information.

Working with AI for threat intelligence

Having in mind that *Steps 3 and 4* in the CTI life cycle are about processing and analysis of data, we can anticipate that some form of intelligent data analytics techniques are beneficial to obtain information from threat intelligence data. **Open-source intelligence** (**OSINT**) data sources, such as online publications, media, blogs, newspapers, and so on offer a large volume of mainly textual data that may contain useful information for cyber threat investigation. Having in mind the volume and variety of information, we could benefit from **natural language processing** (**NLP**) techniques and machine learning techniques to automate and accelerate information extraction.

There were multiple academic and industrial approaches to such information extraction. They are based on extracting topics or keywords related to cyber threats in general or to a specific topic of investigation. For instance, we could be investigating a particular vulnerability and attempting to extract information about this, or similar vulnerabilities that affect a particular type of devices. On the other hand, we could be extracting information about zero-day vulnerabilities in general, or about a topic such as **denial of service** attacks or **ransomware**.

To extract features from textual documents, we can employ a typical sequence of steps:

1. **Tokenization**: In this step, we parse the text data into tokens (usually words). This is a very common first step in text processing.
2. **Stemming**: This step is to remove prefixes and suffixes to get the root of the word, which enables us to look at the words with the same root to have the same meaning. For instance, the words "hacking," "hack," and "hacked" might be stemmed from a root "hack," as they have essentially the same meaning, from the NLP standpoint.
3. **Lemmatization**: We group different forms of the same word to map to the same token (e.g., different tenses), for instance, the words "is" and "be."

4. **Part of speech tagging**: In this step, we identify parts of speech for different tokens (e.g. nouns, verbs, etc.)

   These steps enable us to extract knowledge or meaning from textual documents, which is often the form of data in CTI. It is not obligatory to perform all the steps in each CTI analysis; the exact steps depend on the task and the data at hand.

5. **Machine learning**: After these processing stages, it's possible to apply AI methods to further analyze the CTI data. We can employ typical methods, which were described in *Chapter 5*, such as classification, regression, or clustering, to classify documents into predefined categories, predict some properties of documents, or cluster the documents to discover the underlying structure. Furthermore, there are other AI, or, in particular, machine learning methods that could be interesting to do even deeper analysis. One of them is **topic modeling**.

Topic modeling

Topic modeling is a set of methods for uncovering groups of words that happen to be together in documents. These groups are called **topics**. The assumption is that a set of texts comes from a discrete set of topics, and one text can belong to one or more topics.

During the training, we can determine the topics as distributions over words, and documents also belong to different topics with different probabilities. This is different than just clustering or classification because we don't just group documents into different clusters or categories.

In topic modeling, we assume that documents can belong to different "groups" at the same time, as they can contain text from multiple topics. Furthermore, the words don't necessarily point to one topic, but they can be characteristic of multiple topics at the same time. One common example of topics is in the analysis of newspaper articles, where topics would be politics, sports, business, and so on. Words belonging to sports could be "football," "game," or "match," and words belonging to politics could be "election," "president," or "debate." However, some words don't have a clear assignment. For instance, words such as "agreement" or "deal" could belong to both politics and business. One article could also encompass multiple topics, such as business and politics, or business and sports. Therefore, topic modeling needs to take account of those cases.

In the next section, we present an exercise where topic modeling is applied to the problem of information extraction from X.

Exercise – extracting CTI information from X data

Social media can be a fascinating and useful source of information for discovering new vulnerabilities and investigating trends in the threat landscape through acquiring data that are publicly released by experts or organizations. X is a way to quickly publish short text content and is often used to share news about new vulnerabilities.

However, the volume of data from X can pose a challenge, as we try to retrieve information from text and find what's important for our analysis. Tools that can help us automate this analysis and increase its accuracy in potentially noisy data are always welcome. Intelligent data processing is exactly what AI methods excel at, and given that the data is textual, we can apply methods that are used in NLP, which is a discipline that also intersects with AI.

In the next subsection, we describe the preprocessing of data as a preceding step to applying NLP methods.

Data preprocessing

In this exercise, we analyze data from posts (posts) about vulnerabilities, gathered by the **CyberTweets** project and posted on GitHub by Bezhadan et al. (Behzadan, Vahid, et al. *Corpus and Deep Learning Classifier for Collection of Cyber Threat Indicators in Twitter Stream*. 2018 IEEE International Conference on Big Data (Big Data). IEEE, 2018). We use the data available in a `tweets.bson` file from a MongoDB database snapshot.

The data is available on a public GitHub repository:

https://github.com/behzadanksu/cybertweets/tree/master/threat

You need to download the `tweets.bson` file. Since it's a `.bson` (Binary JSON) file, you need the bson library to read this file. For example, you can install it using `pip`:

```
%pip install bson
```

We decode the contents of the file and load them into a list of post (earlier called a tweet) strings using the following code:

```
Import bson
posts_text = []
with open('tweets.bson', 'rb') as f:
    content = f.read()
    base = 0
    while base < len(content):
        base, d = bson.decode_document(content, base)
        posts_text.append(d['text'])
```

We can output a sample of posts using the following code:

```
[print(t) for i,t in enumerate(posts_text) if i<num_posts]
```

This can give us the following sample:

```
Best way to build empathy is through honesty and vulnerability.
Cryptocurrency Scams Replacing Ransomware as Attackers' Fave |
https://t.co/zesoY4oLQs
Cryptocurrency Scams Replacing Ransomware as Attackers' Fave
https://t.co/2jCvL5ZvpQ
Protect your customers access Prestashop Anti DDoS Attack #Prestashop
https://t.co/xUGxiI8zw8 https://t.co/hXmPX30FmK
Data leak from Huazhu Hotels may affect 130 million customers
https://t.co/RRXVTAWnIu
Instagram App 41.1788.50991.0 #Denial Of #Service https://t.co/
pTsuQe2HT9 #PacketStorm (good slides):
```

To further analyze and process the text, we use the `nltk` and `gensim` libraries. The following code sample enables us to import the required libraries:

```
from gensim import models, corpora
from collections import defaultdict
import nltk
```

The next step is to preprocess the text. In our case, we tokenize the text by splitting the string into an array and removing stopwords and common words that likely don't contribute to the model.

Luckily, we don't need to know all the stopwords for this. We can use the `nltk` library and its `stopwords` function. This module doesn't necessarily come with the library. If you don't have this module, you can use the following line to run the download:

```
nltk.download()
```

Furthermore, you need to select `stopwords` when asked about the module to download, to download the right module.

Once you have the `stopwords` module, you can import and use it as follows:

```
from nltk.corpus import stopwords
s=set(stopwords.words('english'))
texts = [
    [word for word in document.lower().split() if word not in s]
    for document in posts_text
]
```

Furthermore, we can remove some custom stopwords or common words that we find in the text using the following code lines:

```
stoplist = set('for a of the and to in is
    you are I this that - ='.split())

texts = [
    [word for word in document if word not in stoplist]
    for document in texts
]
```

Lastly, we can remove the words that only appear once:

```
frequency = defaultdict(int)
for text in texts:
    for token in text:
        frequency[token] += 1

texts = [
    [token for token in text if frequency[token] > 1]
    for text in texts
]
```

Once the preprocessing is done and the texts are prepared, we can apply AI models for NLP. We describe our modeling approach in the next subsection.

Building the model

To build the model, we need to define the following:

- **Corpus**: A set of documents considered when training the model
- **Dictionary**: A set of tokens (words) taken into account by the model
- **Model type and hyperparameters**: The type of AI model we are using and the hyperparameters that predefine model adjustments

There are many possible ways to model a set of texts. In our case, we are trying to uncover the underlying structure of our set of documents. One good way to do that is topic modeling, described in the previous section.

One of the most common methods of topic modeling is the **latent Dirichlet allocation** (**LDA**). LDA is a probabilistic topic modeling method where we define an assumption that the content of our posts belongs to a small set of topics of a predefined size. In training time, we adjust the parameters of the topic in a way that a small set of topics can cover all the documents in our corpus. The Gensim library has an implementation of LDA, and we will use this implementation in our exercise. The idea is to uncover the structure of our set of posts, to see what kind of content we can find, without manually going through all the data.

Implementation

In the following lines, we define the dictionary and the corpus based on the preprocessed texts:

```
dictionary = corpora.Dictionary(texts)
corpus = [dictionary.doc2bow(text) for text in texts]
```

We create a model using the following lines:

```
model = models.LdaModel(
    corpus, id2word=dictionary, num_topics=num_topics)
```

This creates a model that defines the topics present in our set of texts. One important condition is that we need to predefine the number of topics. This means that we need to have a good guess of how many topics there might be or find it by experimentation. There are methods such as **hierarchical Dirichlet processes** that don't require a predefined number of topics, but they do have different hyperparameters that need to be defined before training the model.

To get an insight into the topics, we can use the `print_topics` method, like the following:

```
>>> model.print_topics(num_topics)
[(0,
  '0.028*"vulnerability" + 0.017*"ddos" + 0.008*"security" +
0.006*"attack" + 0.005*"one" + 0.005*"like" + 0.005*"businesses" +
0.004*"know" + 0.004*"&amp;" + 0.004*"fake"'),
 (1,
  '0.026*"ransomware" + 0.025*"vulnerability" + 0.010*"new" +
0.008*"buffer" + 0.008*"#ransomware" + 0.007*"#cybersecurity" +
0.007*"via" + 0.007*"overflow" + 0.006*"#infosec" + 0.005*"exploit"'),
 (2,
  '0.032*"vulnerability" + 0.014*"1,0day" + 0.009*"&lt;=" +
0.009*"alarm" + 0.009*"gsm2" + 0.009*"[hw1.4]" + 0.009*"calibration" +
0.009*"cb_xlabor-dauertest" + 0.009*"kunde:" + 0.009*"mydatasens"')]
```

From the distributions of the three topics, it seems like the first topic is largely dedicated to the **Distributed Denial of Service** (DDoS) attacks, the second topic can be attributed to ransomware and buffer overflow, and the third is mostly to ransomware and 1- or 0-day attacks.

The Gensim library also enables us to print topic distribution for documents in our corpus, or outside of it.

Here are two examples:

```
>>>print(posts_text[1])
Cryptocurrency Scams Replacing Ransomware as Attackers' Fave |
https://t.co/zesoY4oLQs

>>>model[corpus][1]

[(0, 0.047974426), (1, 0.9041977), (2, 0.047827844)]

>>>print(posts_text[3])
Protect  your customers  access Prestashop Anti DDoS Attack
#Prestashop   https://t.co/xUGxiI8zw8 https://t.co/hXmPX30FmK

>>>model[corpus][3]

[(0, 0.93299556), (1, 0.033533014), (2, 0.03347142)]
```

These two documents decisively contain one topic. The first document belongs mostly to topic 1 (second topic, ransomware), whereas the second document belongs to topic 0 (first topic, DDoS).

Visualization

Topics can be understood better using visualization. Luckily, there is the pyLDAvis library that can help us with that. This library enables interactive topic visualization and is also interoperable with Gensim. This means that we can use Gensim models as input for the visualization functions of this library.

We can use the plotting functions by executing the following code:

```
import pyLDAvis
import pyLDAvis.gensim_models

pyLDAvis.enable_notebook()
pyLDAvis.gensim_models.prepare(model, corpus, dictionary)
```

This should generate two images:

- **Intertopic distance map**: This shows the distances between topics
- **Top 30 model salient terms**: This shows terms (words) that are most important, or most common in particular topics or overall, in the corpus

The following figures show the example images for the corpus that we used in this exercise. The first figure shows the intertopic distance map:

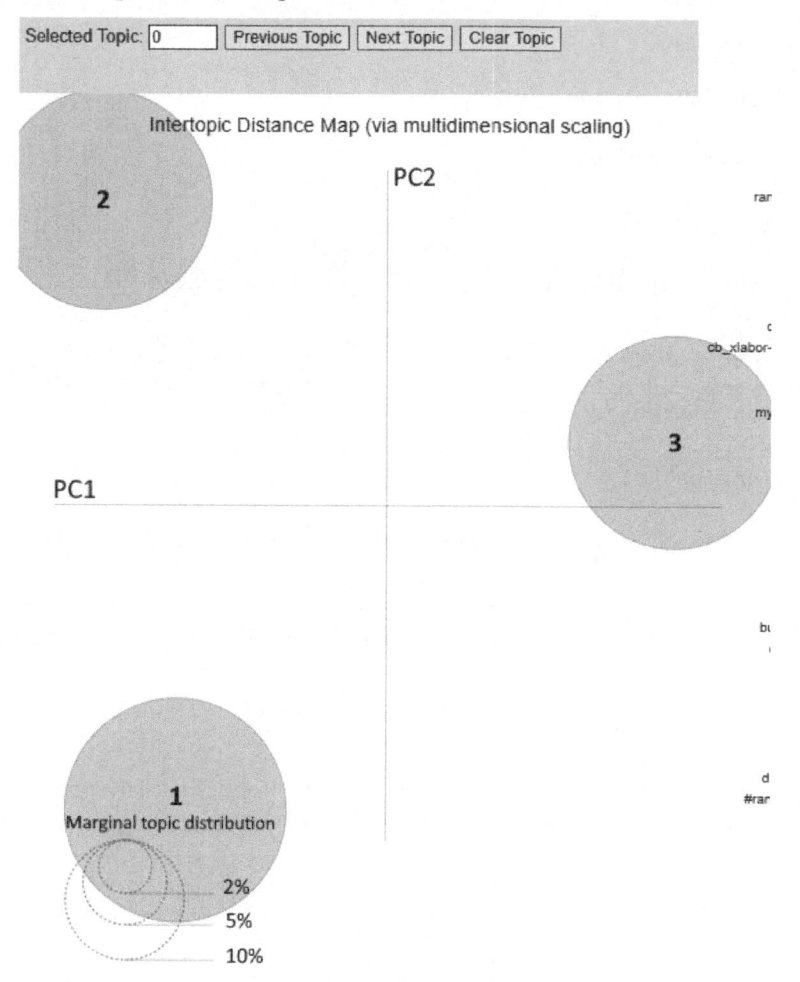

Figure 11.1 – A visualization of topic models using the pyLDAvis library – intertopic distance map

The preceding figure is an illustration of the distance between topics (i.e., how different the topics are in the distribution of words (tokens)).

The next figure shows the top 30 model salient terms.

Top-30 Most Salient Terms[1]

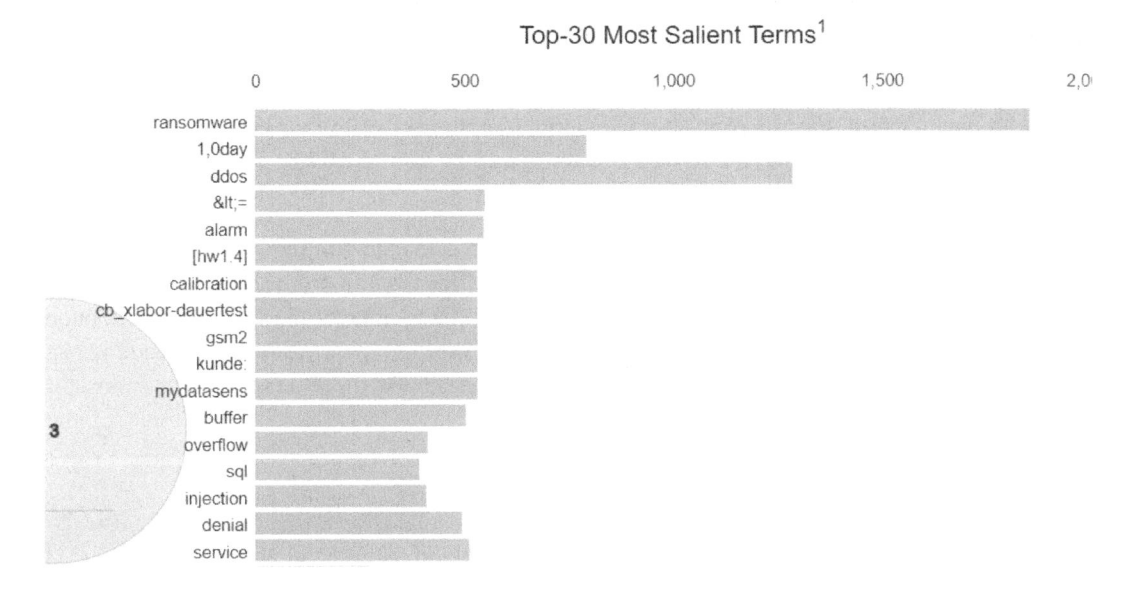

ransomware
1,0day
ddos
<=
alarm
[hw1.4]
calibration
cb_xlabor-dauertest
gsm2
kunde:
mydatasens
buffer
overflow
sql
injection
denial
service

Figure 11.2 – A visualization of topic models using the pyLDAvis library – the top 30 model salient terms

The second figure shows which words are most important in the corpus. Furthermore, in the interactive figure that would be generated, it's possible to select a specific topic and review the word distribution for that topic.

Topics can also be used further as features for clustering, regression, or classification methods. For example, if we had labeled posts by assigning importance labels, we could try to train a model to classify each post into one of the predefined classes, by using topic probabilities for each document as a feature set.

This exercise shows one example of how we can use AI methods to retrieve information from CTI data. In the next section, we summarize the content of this chapter and review what we have learned.

Expanding on use cases of AI in threat intelligence

Apart from posts, there are many other datasets where NLP is relevant. For instance, in threat intelligence, the source of information could be blogs, news posts, but also security-specific data such as IP blacklists or malware data. Aggregating this threat data using big data platforms and applying machine learning algorithms can enable us to get important information that can be intractable or just be missed by a human analyst.

For instance, Microsoft, in their blog, described their application of machine learning to get information about advanced persistent threats (`https://www.microsoft.com/en-us/security/blog/2019/08/08/from-unstructured-data-to-actionable-intelligence-using-machine-learning-for-threat-intelligence/`).

Another company called *Abnormal Security* used machine learning to analyze email threats (`https://abnormalsecurity.com/blog/using-artificial-intelligence-address-email-threats`).

In recent years, we have also seen an expansion of the use of **large language models** (**LLMs**) in many areas to improve search through textual information. This is also a promising avenue for threat intelligence analysts. For instance, they can use the power of LLMs to improve the search for information about indicators of compromise, improve the aggregation of various types of information, and interact with it more naturally. Furthermore, it can help threat analysts make more accurate conclusions, for instance, about attack attribution.

Summary

In this chapter, we were introduced to CTI and the possibility of using AI methods to improve CTI workflows by making information extraction faster and more insightful. This chapter teaches you how to recognize opportunities to apply AI in CTI problems, design the right approach, and implement it. Moreover, we saw a concrete CTI problem and were guided through an exercise using topic modeling to process CTI data.

The next chapter contains a description of a problem with increasing importance – detecting security anomalies in industrial control systems, and AI approaches to solve this problem.

References

- Qamar, Sara, Zahid Anwar, Mohammad Ashiqur Rahman, Ehab Al-Shaer, and Bei-Tseng Chu. "Data-driven analytics for cyber-threat intelligence and information sharing." *Computers & Security* 67 (2017): 35-58.
- Wheelus, Charles, Elias Bou-Harb, and Xingquan Zhu. "Towards a big data architecture for facilitating cyber threat intelligence." In *2016 8th IFIP international conference on new technologies, mobility and security (NTMS)*, pp. 1-5. IEEE, 2016.
- Chen, Haipeng, Rui Liu, Noseong Park, and V. S. Subrahmanian. "Using twitter to predict when vulnerabilities will be exploited." In *Proceedings of the 25th ACM SIGKDD international conference on knowledge discovery & data Mining*, pp. 3143-3152. 2019.
- Mittal, Sudip, Prajit Kumar Das, Varish Mulwad, Anupam Joshi, and Tim Finin. "Cybertwitter: Using twitter to generate alerts for cybersecurity threats and vulnerabilities." In 2016 *IEEE/ACM International Conference on Advances in Social Networks Analysis and Mining (ASONAM)*, pp. 860-867. IEEE, 2016.

References

- Manning, Christopher D. *Foundations of statistical natural language processing.* The MIT Press, 1999.
- Blei, David M., Andrew Y. Ng, and Michael I. Jordan. "Latent dirichlet allocation." *Journal of machine Learning research* 3, no. Jan (2003): 993-1022.
- Teh, Yee, Michael Jordan, Matthew Beal, and David Blei. "Sharing clusters among related groups: Hierarchical Dirichlet processes." *Advances in neural information processing systems* 17 (2004).
- `https://www.crowdstrike.com/cybersecurity-101/machine-learning-cybersecurity/`
- `https://www.recordedfuture.com/threat-intelligence`
- `https://www.silobreaker.com/glossary/ai-in-threat-intelligence/`
- `https://www.secureworks.com/blog/automating-threat-intel-with-machine-learning`

12
Anomaly Detection in Industrial Control Systems

As a type of control system, an **industrial control system** (**ICS**) is used to manage and integrate industrial processes and machines in various scenarios such as manufacturing, energy/water/gas provision, and other types of critical infrastructures. Typically, an ICS plays an essential role in controlling and automating complicated operations and includes several key components, such as sensors as key data acquisition components, actuators as key action components, processing units as key controlling components, and industrial communication protocols as key interfaces. As these components are vulnerable to cyberattacks, detecting the weaknesses of ICSs and preventing advanced cyberattacks are essential for security.

In this chapter, from concept to real examples, we will explain what an ICS is and give several examples, and we will also mention its current techniques and challenges. The chapter is divided into the following sections:

- The ICS and its components
- Cyberattacks on ICSs and their components
- Anomaly detection for ICSs and their components

Introducing the ICS and its components

To manage and control most of the industrial processes and machinery in various sectors, industries, and energy plants, an ICS is used to effectively monitor, automate, and optimize complex industrial operations. As mentioned earlier, besides efficiency, there are also other objectives of the ICS, such as process control automation, safety, remote monitoring and control, and high-standard quality control.

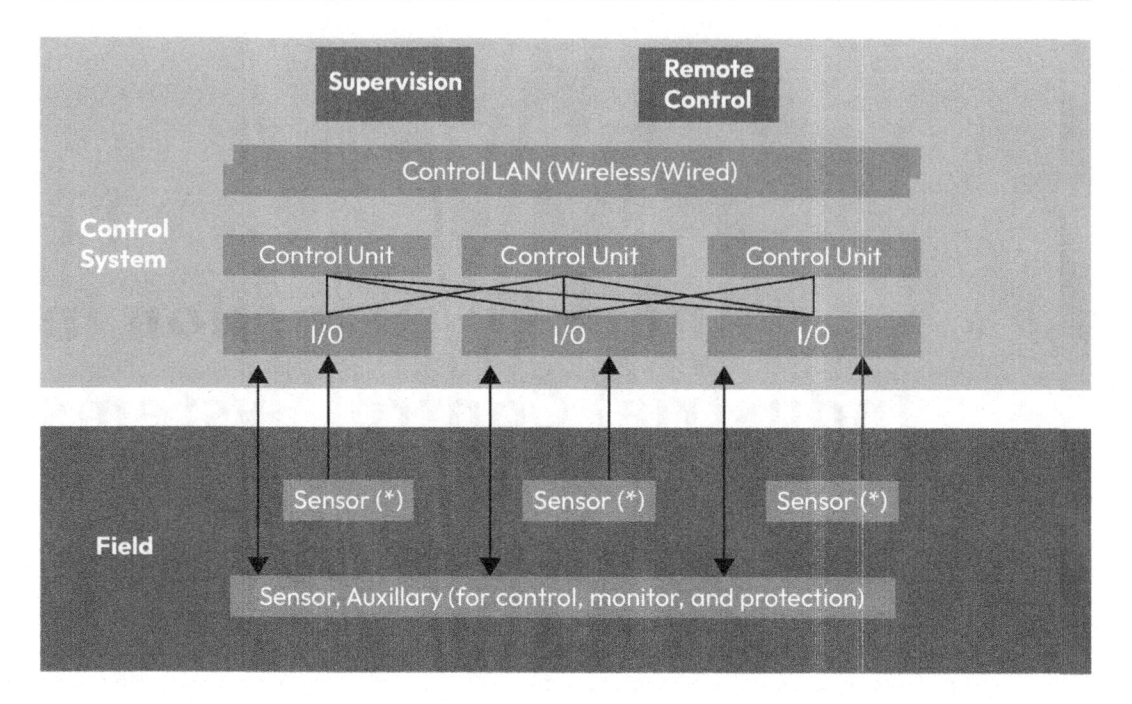

Figure 12.1 – Typical ICS at power stations

As *Figure 12.1* shows, typical ICSs normally include two parts. The first part is the control system and the other is the field. The control system includes a central control unit, operation monitoring screen, turbine building, and also a thermal power plant. There are many types of sensors in the thermal power plant to collect data and also various types of controllers receiving commands from remote stations. Central control units and **supervisory control and data acquisition** (**SCADA**) systems are used to process the raw data and present the final analysis or real-time changing tendency on the screen of the operation monitoring modules. There are various telecommunication protocols or products that are used to exchange information from one component to another. *Figure 12.1* shows a typical ICS at power stations, which includes a control system and field primarily. There are several components in the control system, such as a **human-machine interface** (**HMI**), control processor (CPU), and interconnection modules (Control LAN, I/O bus, etc.)

To classify the ICSs, there are two primary types: SCADA systems and **distributed control systems** (**DCSs**).

SCADA is used to control and monitor industrial processes, facilities, and infrastructure in various sectors such as electricity generation, water treatment, power stations, oil and gas pipelines, manufacturing, and many others. SCADA systems are crucial for industrial organizations because they help to ensure efficient, safe, and intelligent operations. Typical SCADA systems include HMIs, supervisory systems, **remote terminal units** (**RTUs**), **programmable logic controllers** (**PLCs**), communication units, and data historians. The primary functionalities of SCADA systems are monitoring the status of remote equipment and gathering data, controlling commands to remotely adjust the operations of the process, collecting data from sensors and equipment to analyze the performance of the industrial components, providing visual tools to understand the state of the related components, and also ensuring the seamless information flow between the devices and the supervisory systems. SCADA systems offer a method to maximize efficiency, enhance safety, and ensure reliability by combining the software and hardware components in modern industrial operations.

Meanwhile, DCSs are primarily designed for managing complex, large-scale industrial processes across multiple locations. The scenarios of using DCSs are similar to SCADA and include manufacturing plants, power generation, oil and gas pipelines, water treatment facilities, and other applications where process control is critical. Typical DCSs include controllers, HMIs, **input/output (I/O)** modules, engineering workstations, and communication networks. At the heart of a DCS are the controllers, which are industrial-level computers responsible for executing control logic and managing the process in real time. These controllers are distributed geographically across the plant. The key functionalities of DCSs comprise the maintenance control over continuous processes, ensuring they operate within desired parameters, the provision of comprehensive visualization capabilities allowing operators to monitor process conditions in real time, the management of sophisticated alarm systems to alert operators of process deviations, and the collection and storing process data for the historical analysis and reporting. A DCS is an integrated control system used for managing complex industrial processes within a localized area and it combines several components to ensure precise control, efficient operation, and safety of the manufacturing process.

After introducing various types of ICSs, now is the time to move to the components of ICSs. An entire ICS is normally integrated with other components around it, such as corporate IT, ERP server, production management, and corporate firewall. We define these relative components as the corporate network, which is out of this chapter's scope.

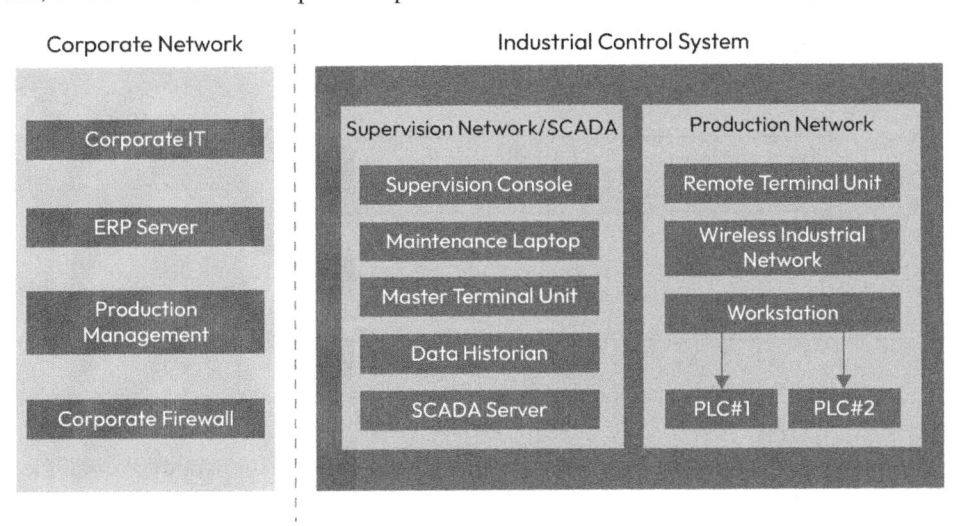

Figure 12.2 – ICSs

Figure 12.2 shows the ICS in detail. In our architecture, similar to other works, as with SCADA systems and DCSs, we divide the ICS into two parts due to the primary functionalities of these backbone components. The first one is the **supervision network** and the other is the **production network**.

Let us understand each and their functions in more detail with this table:

Parts	Names	Functionalities	Primary Element
Supervision Network/ SCADA	Supervision Console	Host the supervisory control system (software/hardware) Send commands and receive feedback	Control software
	Maintenance/ Remote Diagnostic	Identify and discover risks and abnormal behaviors. Recover to normal situation	Malware detection subsystem Flow analyzer
	Master Terminal Unit (MTU)	Delivers commands to RTUs in the field	
	Data Historian	Centralized database for logging all process information	Database
	SCADA Server	Similar to the MTU, issues commands to the RTUs	
Production network	**Remote Terminal Unit**	Control field device with command and accomplish the indicated task	Microprocessor
	Wireless industrial Network	Tele-communication connection for the workstation, PLCs, and other components	Base station Router Switcher
	Workstation	Controller for the connecting PLCs	
	PLC	Actuator	Motor Programmable controller

Table 12.1 – The primary components in the ICS with the aforementioned classification

Table 12.1 illustrates the primary components of an ICS and their functionalities. We will introduce cyberattacks in ICSs in the following section in detail.

Cyberattacks on ICSs

With the development of digitization, digital transformation, and upgrading of ICSs, more and more components in IT technologies are introduced into the ICS. Similar to the increasingly serious risks to security and privacy in the IT infrastructure such as ransomware, key data leakage, and anomaly access control, these issues are also copied into the ICS field.

We will divide our content into two parts:

- Cyberattacks on the entire ICS
- Primarily concentrate on the sub-components in the ICS, targeting the specific components, such as controller, wireless/cellular network, and ICS configuration systems

Cyberattacks on the ICS

Although there are many types of attacks targeting ICSs, some of them aim at the broader ICS network or system as a whole, potentially impacting multiple components and processes simultaneously. A few of them are listed here:

- **Ransomware attacks**: While these can target specific devices, their impact is often system-wide, encrypting files across the network and disrupting operations at a large scale.
- **Denial of service (DoS) and distributed denial of service (DDoS) attacks**: These are typically aimed at the entire network or key devices within the ICS to disrupt the availability of the system as a whole.
- **Zero-day and n-day vulnerabilities**: These are critical concepts in the cybersecurity domain, including the protection of ICSs. These vulnerabilities represent potential security holes in software or hardware that can be exploited by malicious actors to gain unauthorized access, disrupt operations, or cause damage to critical infrastructure.
- **Phishing attacks**: These attacks in the context of ICSs represent a significant cybersecurity threat, leveraging social engineering techniques to deceive individuals into divulging sensitive information, executing malicious software, or unwittingly providing access to secure systems.
- **Advanced persistent threats** (**APTs**): APTs are broad, targeted campaigns that can infiltrate the entire ICS infrastructure, remaining undetected while accessing sensitive information and disrupting operations.

Looking at some real attacks on an ICS is a good practice for presenting the aforementioned attacks.

Stuxnet is a highly sophisticated computer worm discovered in 2010 but believed to have been in development and active since at least 2005. It is notable for being one of the first cyberweapons known to have been specifically created to target and sabotage industrial systems. Stuxnet was designed to infect Windows computers and then seek out software from Siemens STEP 7 and WinCC, which are used to program ICSs that operate machinery in manufacturing and utility facilities, including nuclear power plants.

The primary target of Stuxnet was the nuclear program, specifically, the uranium enrichment facilities at Natanz. The worm was engineered to infect the PLCs used in uranium enrichment centrifuges. These PLCs were responsible for controlling the speed and operation of the centrifuges. Stuxnet operated by first spreading via USB drives and network shares to reach computers within the targeted facility, exploiting multiple zero-day vulnerabilities in Windows to propagate itself and gain administrative privileges. It exploited four zero-day vulnerabilities in Windows systems, along with several other known vulnerabilities, to spread and execute its malicious payload. These zero-day exploits allowed the worm to spread via USB drives, escalate privileges, target Siemens SCADA systems, and spread laterally across networks. Once it infected a system running the Siemens software, it would look for specific configurations that matched those of the Iranian nuclear facility's control systems.

The exact impact of Stuxnet is difficult to quantify, but it is believed to have successfully damaged approximately 1,000 of the 5,000 centrifuges Iran had operating at the time, significantly setting back the country's uranium enrichment capabilities. The worm's discovery led to widespread concern about the potential for cyberattacks to cause physical damage to critical infrastructure and opened a new chapter in cyber warfare.

The attack on the Colonial Pipeline company, which occurred on May 6, 2021, was a significant cybersecurity event that highlighted the vulnerabilities of critical infrastructure to ransomware attacks. Colonial Pipeline is one of the largest pipeline operators in the United States, providing roughly 45% of the East Coast's fuel supply, including gasoline, diesel fuel, heating oil, and jet fuel. The attack was carried out using a ransomware variant known as **DarkSide**, which is believed to be operated by a cybercriminal group with the same name. The attackers gained access to Colonial Pipeline's networks and deployed ransomware that encrypted data on the computers and servers, effectively locking the company out of its own systems.

As a precautionary measure and to contain the attack, Colonial Pipeline proactively shut down its operations. This was a significant step, given the pipeline's critical role in fuel distribution across the Eastern United States. The shutdown led to widespread concern over fuel availability, resulting in panic buying and fuel shortages in several states. The situation was exacerbated by hoarding and transportation challenges. In response to the crisis, several states declared states of emergency to address the fuel shortages and ensure continued access to essential services. Colonial Pipeline reportedly paid a ransom of approximately $4.4 million to the attackers to obtain a decryption key for the encrypted data, although the effectiveness of the decryption tool was questioned.

Another example of an attack on ICSs is the BlackEnergy attacks, which refer to a series of cyberattacks that targeted Ukraine's power generation infrastructure, among other targets, and are notable for their sophistication and the extent of their impact. These attacks, which began around 2015, were among the first publicly acknowledged incidents to successfully disrupt a nation's power grid through cyber means, marking a significant moment in the history of cybersecurity.

BlackEnergy malware was originally created and used around 2007 for relatively simple purposes, such as DDoS attacks and bank fraud. The most notable BlackEnergy attack on Ukraine's power grid occurred in December 2015. The attackers began by sending spear-phishing emails to employees of

Ukrainian power companies. These emails contained malicious attachments that, once opened, would infect the network with the BlackEnergy 3 malware. After gaining initial access, the attackers moved laterally within the networks to gain higher privileges and access control systems. During this phase, they conducted reconnaissance to understand the power grid's operations and identify key systems. The attackers took control of the systems that operated electrical substations and switched off the power to several regions, affecting tens of thousands of people. The attack was coordinated to occur simultaneously across multiple locations, maximizing its impact.

The impact of the BlackEnergy attacks was significant. The December 2015 attack left approximately 230,000 people without electricity for several hours. It was a stark demonstration of the potential real-world consequences of cyberattacks on critical infrastructure. In the aftermath of the BlackEnergy attacks, there was a significant increase in efforts to secure critical infrastructure against cyber threats. This included greater investment in cybersecurity technologies, enhanced training for staff, and improved incident response and recovery procedures. The attacks also led to increased international cooperation on cybersecurity issues, as nations recognized the global nature of the threat.

Cyberattacks on the components of ICSs

The preceding examples about the Colonial Pipeline company, Stuxnet, and BlackEnergy are three classic examples of attacks on the ICS. However, as we explained in the last section, these cyberattacks occur on the sub-systems and then extend to the whole system. We should mention that these attacks focus on exploiting vulnerabilities in specific parts of an ICS, such as individual devices or software applications, as follows:

- **Man-in-the-middle (MitM) attacks**: These can occur at any point where data is transmitted between two components within the ICS, such as between sensors and controllers or controllers and actuators. For this type of attack, these are the targeted components in the ICS:

  - **Communication links**: MitM attacks primarily target the data transmission between components. This includes communication between field devices such as sensors and actuators, and control systems such as PLCs, RTUs, and SCADA systems.

  - **Network devices**: Devices that facilitate communication within an ICS, such as switches and routers, can also be targeted to intercept or alter data in transit.

- **SQL injection**: This type of attack targets specific applications or databases within the ICS that use SQL for data management. It's more about exploiting a software vulnerability than attacking the system's infrastructure. Similar to MitM attacks, SQL injection attacks are mostly on targeted components as follows:

  - **Databases**: SQL injection attacks target databases used within the ICS for storing configuration data, event logs, or operational data. These databases are often part of larger software systems used for process management, data analysis, or reporting.

- **Web applications**: Web-based interfaces that provide access to ICS data or control capabilities can be vulnerable to SQL injection. These interfaces might be used for monitoring system status, adjusting configurations, or manual control inputs.

- **Zero-day exploits**: While these can have system-wide implications, they initially target specific vulnerabilities in software or hardware components within the ICS. The exploit's impact depends on the nature of the vulnerability and the role of the compromised component within the overall system:

  - **Software applications**: Any software component of an ICS could be targeted by a zero-day exploit. This includes operating systems, industrial applications, custom-developed software, and third-party utilities used within the ICS environment.
  - **Hardware devices**: Certain zero-day exploits may target vulnerabilities in the firmware or hardware of ICS components, such as PLCs, RTUs, network equipment, and even sensors and actuators. The specific target would depend on the nature of the vulnerability being exploited.

The aforementioned devices and their attacks are primarily categorized by the general techniques. Some of them could be found in at least one ICS component, such as communication links and their attacks relative to wireless industrial networks, SCADA servers, RTUs, and remote diagnostics. Meanwhile, others would include multiple ICS components, such as software applications and their security risks covering workstations, supervision servers, IT/OT, and PLCs.

After discussing the attack surface, their risks, and also the sensitive components, in the next section, the methodology connecting with the detection of their attack behaviors will be presented.

Detecting anomaly behaviors in ICSs

Anomaly detection in ICSs is a critical aspect of cybersecurity and operational reliability. It involves identifying patterns or activities within the system that deviate from the established normal behavior. These anomalies can indicate potential cybersecurity threats, such as cyberattacks or malware infections, as well as operational issues such as equipment failure, process deviations, or safety risks.

In the following content, we will show various techniques to detect the anomaly behaviors. We start with the classification of various detection techniques and then give examples of each.

Classification

Techniques for anomaly detection in ICSs are primarily classified into four groups: data-driven, model-based, knowledge-based, and also a combination of these. Let us understand each of these in the following sections.

Data-driven techniques

Although data-driven methods can cover a bigger scope than deep learning, here, we primarily focus on deep learning.

Deep learning utilizes neural networks with multiple layers to learn from vast amounts of data.

Deep learning models, such as **convolutional neural networks (CNNs)** and **recurrent neural networks (RNNs)**, are effective in identifying complex patterns and anomalies in time-series data, which is common in ICS monitoring. These models can be applied both to the entire system for general anomaly detection and to specific components such as sensors and actuaries for detailed monitoring. One example of this technique could be a CNN model trained on historical sensor data from a chemical processing plant to detect anomalies in temperature, pressure, and flow rates. The model learns complex patterns in the data and can identify subtle anomalies indicating equipment failure or process deviations. However, applying these methods in resource-constrained environments typical of many ICSs faces several significant limitations. ICS environments often have real-time operational constraints, limited computational power, and stringent safety requirements, which complicate the deployment of deep learning-based approaches.

Model-based techniques

After mentioning the data-driven techniques, we have to discuss the more theoretical method, which is used to solve the detection problems as the mathematical models:

- **System dynamics modeling**: This involves creating mathematical models that represent the dynamic behavior of the entire ICS or its components. By simulating the expected operation under various conditions, anomalies are detected when actual operational data deviates significantly from the model's predictions. This approach is particularly useful for complex systems where physical processes and interactions can be mathematically modeled.
- **Digital twins**: A digital twin is a virtual replica of a physical ICS or its components. By mirroring the real-time data of the system, it allows for anomaly detection through continuous comparison between the virtual model's performance and the actual system's behavior. This technique is effective for both holistic system monitoring and targeted component analysis. Let's take an example of creating a digital twin of a wind turbine to monitor its operational health. The digital twin simulates the turbine's performance under various conditions, and by comparing these simulations with actual sensor data, anomalies such as blade damage or gearbox wear can be detected.

Knowledge-based techniques

In the real industrial application scenario, the most popular technique used in real life to detect anomaly behavior is a knowledge-based one, especially a rule-based technique. This type of technique is easy and very effective:

- **Rule-based systems**: These utilize a set of predefined rules or heuristics that describe normal and abnormal conditions. These systems can quickly identify anomalies based on deviations from these rules. While effective for known scenarios and specific component monitoring, they require regular updates to remain effective against new or evolving threats. An example is implementing a rule-based system for a water treatment facility that triggers alarms when sensor readings fall outside predefined thresholds, such as chlorine levels below safety standards or unexpected changes in water flow rates, indicating potential contamination or pipe leakage.

- **Ontology-based approaches**: These use a structured framework of knowledge to understand complex relationships within the ICS and its components. By modeling the system's entities, properties, and relations, these approaches can infer anomalies through logical reasoning. This is particularly useful for detecting sophisticated cyberattacks or systemic operational issues that may not be immediately apparent. An ontology-based system for a smart grid that models relationships between electricity consumption patterns, weather conditions, and equipment status is extremely popular but also very easy to attack without a good detection method. By reasoning over this knowledge base, the system can detect anomalies such as unusual power usage, which might indicate equipment malfunction or unauthorized access.

Hybrid techniques

Combining multiple anomaly detection techniques can leverage the strengths of each approach to improve overall detection accuracy and reduce false positives. For example, a hybrid system might use machine learning to detect broad patterns of anomalies across the ICS while employing system dynamics modeling for high-fidelity monitoring of critical components. Similarly, rule-based systems can provide immediate alerts for known issues, while deep learning models continuously adapt to detect new and evolving anomalies. A hybrid anomaly detection system for a manufacturing assembly line combines machine learning (for pattern recognition in operational data), system dynamics modeling (to simulate expected machine behavior), and rule-based logic (to flag deviations from standard operating procedures).

After discussing the techniques used to do anomaly detection, let us move to real use cases and applications in the next section.

Use cases and applications

In this section, two examples of anomaly detection in an ICS are presented. The first one uses a graph neural network-based method to illustrate anomaly detection by considering the connections among the components in the ICS. The second example shows ransomware detection, which is targeting one of the most important fields in the ICS. Detailed information about them is presented as follows.

Anomaly detection for the ICS

As mentioned in [8] in the *References* section, practically, most ICSs and their components have significant time series characteristics, and datasets of ICSs are collected with the **multivariate time series** (**MTS**) feature. However, most of the existing anomaly detection that works practically does not consider the structure information of the system, such as the connection among the nodes in the ICS. GDN [9] is an unsupervised anomaly detection method using graph attention networks. It combines the structure of existing relationships among variables to provide explainability for the detected anomalies.

Figure 12.3 shows the architecture of GDN, and it includes sensor embedding, graph structure learning, graph attention-based forecasting, and graph deviation scoring. The code of GDN is available at https://github.com/d-ailin/GDN.

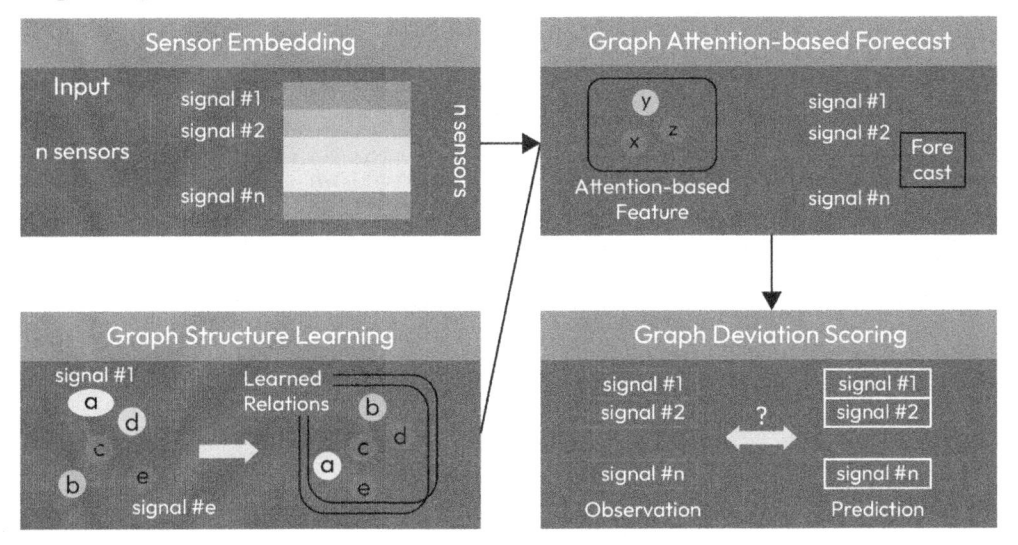

Figure 12.3 – The GDN architecture

The motivation of the GDN method is to learn the connections between sensors as a graph and identify and explain the results of the detecting behaviors from the learned patterns.

As *Figure 12.3* illustrates, sensor embedding is used to convert the physical signal to embedding vectors to capture the physical features of each sensor. Graph structure learns about the dependence connections among these sensors. Forecasting the values of each sensor is based on the graph struct function across its connection along with the vector value derived from the sensor embedding, and finally, the graph deviation scoring step is used to identify the learned connection and localize these anomaly behaviors.

The dataset of GDN work includes two real-world sensor datasets: **Secure Water Treatment** (**SWat**) and **Water Distribution** (**WADI**). Both of them are captured based on water treatment physical testing systems where operators emulate attacking scenarios, which are recorded as ground truth anomaly behaviors.

Let us now check the code from the GDN repo together and explore this framework with several interesting points.

First of all, the GDN GitHub repo gives part of the `msl` dataset (https://s3-us-west-2.amazonaws.com/telemanom/data.zip) as a demo, which is from `telemanom` works (https://github.com/khundman/telemanom). It lists the feature names and the training and testing dataset with index, attack ground truth label, and so on.

Before we run the following code, we need to install additional Python libraries and import them:

```
pip install torch torch-geometric

from torch.nn import Parameter
from torch_geometric.nn import
import torch
import torch.nn as nn
```

The **graph neural network** (**GNN**) layer definition includes one `gnn` layer, one `ReLU` layer, and one `LeakyReLU` layer, as shown here. We illustrate how to define a GNN network with the code as follows:

```
class GNNLayer(nn.Module):
    def __init__(self, in_channel, out_channel,
        inter_dim=0, heads=1, node_num=100
    ):
        super(GNNLayer, self).__init__()
        self.gnn = GraphLayer(in_channel, out_channel,
            inter_dim=inter_dim, heads=heads, concat=False)
        self.bn = nn.BatchNorm1d(out_channel)
        self.relu = nn.ReLU()
        self.leaky_relu = nn.LeakyReLU()
```

Additionally, the `GraphLayer` instance's definition is more complex than the `GNNLayer` instance's. It includes channels of input and also channels of output, all of which construct the graph structure in GDN. What's more, as aforementioned, the sensor embedding step converts the physical signal (each line in the dataset) to an embedding vector; the embedded physical signals are also put into the `GraphLayer` definition to capture the sensor feature. This is shown here:

```
from torch.nn import Parameter
from torch_geometric.nn import MessagePassing
```

```
class GraphLayer(MessagePassing):
    def __init__(self, in_channels, out_channels,
        heads=1, concat=True,
        negative_slope=0.2,
        dropout=0, bias=True, inter_dim=-1,**kwargs
    ):
        super(GraphLayer, self).__init__(aggr='add', **kwargs)

        self.in_channels = in_channels
        self.out_channels = out_channels
        self.heads = heads
        self.concat = concat
        self.negative_slope = negative_slope
        self.dropout = dropout

        self.__alpha__ = None

        self.lin = Linear(
            in_channels, heads * out_channels, bias=False)

        self.att_i = Parameter(torch.Tensor(1, heads, out_channels))
        self.att_j = Parameter(torch.Tensor(1, heads, out_channels))
        self.att_em_i = Parameter(torch.Tensor(1, heads,
            out_channels))
        self.att_em_j = Parameter(torch.Tensor(1, heads,
            out_channels))

        if bias and concat:
            self.bias = Parameter(torch.Tensor(heads * out_channels))
        elif bias and not concat:
            self.bias = Parameter(torch.Tensor(out_channels))
        else:
            self.register_parameter('bias', None)

        self.reset_parameters()
```

Last but not least, the easiest way to run the GDN work is to run the script in the corresponding code folder. Based on the GDN code, we trained our model and put it under this chapter's code repo at https://github.com/PacktPublishing/Artificial-Intelligence-for-Cybersecurity1/chapter12/.

All in all, the GDN approach learns a graph of connections among sensors in the ICS and detects anomaly behaviors from these patterns with the help of sensor embedding and graph structure learning. Two real-world sensor datasets are used to evaluate the GDN and these results outperform baselines in accuracy. Additionally, GDN provides an interpretable model and helps to pinpoint and understand the anomaly behaviors.

Ransomware detection for the ICS and its components

Ransomware attacks [1, 2, 3, 4, 5, and 6] not only cause operational disruptions but can also lead to financial losses, reputation damage, and legal consequences. Implementing a comprehensive cybersecurity strategy that includes both preventive measures and a robust response plan is essential for protecting against ransomware threats. Ransomware attacks on ICSs represent a significant threat to critical infrastructure sectors, including energy, water treatment, manufacturing, and transportation. These attacks can lead to operational disruptions, safety incidents, and financial losses. Unlike traditional IT environments, ICS networks control physical processes, and disruptions can have immediate and tangible impacts on public safety and the economy. Several classical ransomware attacks include those on Colonial Pipeline (causing the company to pay a ransom of approximately $4.4 million to the DarkSide ransomware group), Norsk Hydro (which had significant financial implications with estimated costs exceeding $50 million), and the Snake/Ekans ransomware attack at Honda, targeting its global operations and leading to temporary suspension of production in some of its manufacturing facilities such as US, Turkey, Italy, and Japan.

Therefore, detecting and predicting ransomware attacks attracts more and more academic and industrial researchers and engineers. Several works have already summarized the direction and methods in this field; similar to our previous text as a sub-field of anomaly detection, ransomware detection is also divided into several classifications, such as AI-based (machine learning, deep learning, etc.), knowledge-based (rule or signature), mathematics-based, and so on.

Challenges and future works

So far, we have discussed the fundamental terms of the ICS attacking surface, and also current techniques for anomaly detection, but there are still many open questions and challenges in this field that need to be explored. We summarize the challenges and future works in the following sections.

Challenges

So far, we have explained the categorization of anomaly detection and also given examples with AI-based methods for normal anomaly and ransomware detection. There are still several challenges in this field, such as false positives, dynamic environments, and so on. Those factors influence the application of anomaly detection, especially AI, machine learning, and deep learning-based detection systems, in the real industrial control scenario. Let us understand more about each here:

- **High false positive rate**: Differentiating between normal variations in system behavior and genuine anomalies is challenging, often leading to false alarms.
- **Dynamic environments**: Industrial processes can change over time, requiring continuous adaptation of the anomaly detection models to maintain accuracy.
- **Data quality and availability**: Effective anomaly detection requires high-quality, comprehensive data. In many ICS environments, data may be sparse, noisy, or incomplete.

- **Complexity of industrial systems**: The complexity and heterogeneity of ICSs, with their mix of old and new technologies, proprietary protocols, and customized configurations, make it difficult to establish a baseline for normal behavior.

Future works and directions

From the end of November 2022, when ChatGPT was released by OpenAI, that period was also the time for our book preparation. Although we mention **large language models** (**LLMs**) from time to time in various chapters, the primary purpose of this book is not to focus on LLM-based security. What's more, recently, OpenAI released Sora (a video generation model), Google released Gemini, and Claude 3 has better performance than GPT-4, which means we need to consider how to use these rapidly developing LLM techniques for anomaly detection. Additionally, we also need to consider the cases in which attackers generate anomaly detection with advanced LLMs. Be it the IT field, designing, industrial control field, supply chain field, and so on, more and more works are increasingly involved with LLMs. These models can analyze vast amounts of unstructured data, learn from it, and make predictions or generate insights that can be used to improve anomaly detection methods.

Here are several future directions and applications of LLMs in anomaly detection within the ICS domain.

Enhanced pattern recognition

LLMs can process and analyze large datasets from ICS operations, including logs, sensor data, and system alerts, to identify patterns and correlations that may not be evident to human operators or traditional analytical methods. This capability can enhance the detection of subtle anomalies that precede critical events, allowing for earlier intervention.

Natural language processing for security logs

Security logs and incident reports often contain unstructured text data. LLMs, with their advanced **natural language processing** (**NLP**) capabilities, can analyze this text to identify indicators of security incidents or breaches. This can include parsing and understanding logs from various sources, extracting relevant information, and correlating events across different systems.

Predictive maintenance

By analyzing historical maintenance records and operational data, LLMs can predict potential equipment failures before they occur. This predictive maintenance approach can help prevent unplanned downtime and extend the lifespan of critical components within ICSs.

Improved situational awareness

LLMs can assist in creating more comprehensive situational awareness by integrating data from diverse sources, including ICSs, IT networks, and external intelligence feeds. By understanding the context and content of this data, LLMs can help operators identify potential threats or anomalies that span across the cyber-physical divide.

Automated incident response

Future developments may enable LLMs to not only detect anomalies but also suggest or automate appropriate response actions based on the type of anomaly detected and its potential impact. This could include isolating affected systems, applying patches, or adjusting control parameters to mitigate risks.

Continuous learning and adaptation

One of the key advantages of LLMs is their ability to learn from new data continuously. As LLMs are exposed to more operational and anomaly data over time, their accuracy and effectiveness in detecting and responding to anomalies can improve, adapting to the evolving threat landscape and changes in the operational environment.

In this section, we briefly mentioned the future direction for the security log analysis within the NLP area.

Summary

In this chapter, we introduced the basic knowledge about the ICS and its components, and then based on the aforementioned information, attacks targeting ICSs and their components were presented. With the categorization of various types of attacks, and fundamental techniques that cause these attacks, we now know that the attack section not only has similarity with the normal attacks targeting the IT/DT field but also the unique features for the anomaly detection system in the area of ICSs. Last but not least, we covered anomaly detection, which included not only general anomaly detection with consideration of structural information of the ICS but also one of the specific anomaly behaviors in ICSs: ransomware detection. For both examples, we introduced machine learning/deep learning-based methodologies to illustrate the state-of-the-art approaches.

Lastly, at the end of the chapter, we also summarized the challenges. We will discuss data quality and its influence on anomaly detection and other topics in the cybersecurity field on the next chapter.

References

- [1] Gazzan, M., and Sheldon, F. T. (2023). Opportunities for Early Detection and Prediction of Ransomware Attacks against Industrial Control Systems. *Future Internet*, *15*(4), 144. `https://doi.org/10.3390/fi15040144`

- [2] Alanazi, M., Mahmood, A., and Chowdhury, M. J. M. (2023). SCADA vulnerabilities and attacks: A review of the state-of-the-art and open issues. *Computers & Security*, *125*, 103028. `https://doi.org/10.1016/j.cose.2022.103028`

- [3] E. Berrueta, D. Morato, E. Magaña and M. Izal, "A Survey on Detection Techniques for Cryptographic Ransomware," in *IEEE Access*, vol. 7, pp. 144925-144944, 2019, doi: 10.1109/ACCESS.2019.2945839. keywords: {Ransomware;Servers;Encryption;Detection algorithms;Proposals;Computer security;malware detection;ransomware},

- [4] Beaman, C., Barkworth, A., Akande, T. D., Hakak, S., and Khan, M. K. (2021). Ransomware: Recent advances, analysis, challenges and future research directions. Computers & Security, 111, 102490. https://doi.org/10.1016/j.cose.2021.102490

- [5] M. Gazzan, A. Alqahtani, and F. T. Sheldon, "Key Factors Influencing the Rise of Current Ransomware Attacks on Industrial Control Systems," *2021 IEEE 11th Annual Computing and Communication Workshop and Conference (CCWC)*, Las Vegas, NV, USA, 2021, pp. 1417-1422, doi: 10.1109/CCWC51732.2021.9376179.

- [6] Formby, David, Surya S. Durbha, and Raheem A. Beyah. "Out of Control: Ransomware for Industrial Control Systems." (2017).

- [7] Muna Al-Hawawreh and Elena Sitnikova. 2019. Industrial Internet of Things Based Ransomware Detection using Stacked Variational Neural Network. In Proceedings of the 3rd International Conference on Big Data and Internet of Things (BDIOT '19). Association for Computing Machinery, New York, NY, USA, 126–130. https://doi.org/10.1145/3361758.3361763

- [8] Kim, B., Alawami, M. A., Kim, E., Oh, S., Park, J., and Kim, H. (2022). A Comparative Study of Time Series Anomaly Detection Models for Industrial Control Systems. Sensors, 23(3), 1310. https://doi.org/10.3390/s23031310

- [9] Deng, Ailin, and Bryan Hooi. "Graph neural network-based anomaly detection in multivariate time series." *Proceedings of the AAAI conference on artificial intelligence*. Vol. 35. No. 5. 2021.

13
Large Language Models and Cybersecurity

In the last few years, we have witnessed the rise of a new type of model, trained on large volumes of textual data. These models have been trained on terabytes of data by crawling text from the web, and they can have billions of parameters. This enables them to have improved natural language understanding abilities and to be used in various applications, from text classification to question-answering, chatbots, and similar. These models can be applied in cybersecurity as well, both by attackers and defenders. Additionally, systems with large language models themselves can be susceptible to various attacks.

In this chapter, we are going to cover the following main topics:

- From traditional **Natural Language Processing** (**NLP**), methods to LLMs
- LLMs for defensive security
- The security of LLMs

From traditional methods to LLMs

The traditional machine learning methods of NLP, where we use feature extraction methods such as bag-of-words and n-grams to extract numerical features and apply classic machine learning methods, such as a neural network or **Support Vector Machines** (**SVMs**), can be a powerful way to automate the recognition of patterns and execution of tasks, such as document classification or clustering. However, these baseline methods often can't handle long sequences with variable-range dependencies. One solution for this was the development of **recurrent neural networks** (**RNNs**). However, RNNs are trained with **backpropagation through time**, which still makes the training of long sequences inefficient. The next generation of models was Transformers, which are more efficient models and are currently used as the basis for many sequence models and all LLMs on the market.

Transformers

Transformers are a special type of model that is based on the idea of self-attention. This means that a neural network learns to weigh different parts of a sequence during training. This is done in each training iteration by determining the importance of different parts of the sequence for pattern recognition tasks. Transformers use multiple neural network components for self-attention (attention heads) to learn different types of patterns at the same time. Furthermore, another feature of transformers is position encoding, which is used to model sequential dependencies without needing backpropagation through time as RNNs use.

Transformers have proven to be useful in various tasks such as machine translation, question-answering, and document summarization. This has initiated the question of whether neural networks such as transformers can be trained once to be used for multiple tasks. It turns out that transformers trained on sufficiently large datasets can indeed be pretrained and either used as-is or fine-tuned for multiple NLP tasks, with high performance metrics. Starting with a GPT model, the idea of a model that can execute multiple tasks has been developed through the years to models with billions of parameters that can be multimodal, performing various tasks that are based on generating textual output as a response to textual input (a prompt), sometimes also combined with an image.

Large Language Models (LLMs)

LLMs are deep learning models based on the transformer architecture, trained with the goal of detecting patterns from large volumes of textual data and obtaining an amount of natural language understanding, enabling them to execute various generative tasks where they provide textual output on textual inputs from a user (prompts). They are trained using unsupervised learning and further supervised fine-tuning, with various strategies that have evolved in recent years. Currently, popular architectures are based on **Reinforcement Learning from Human Feedback (RLHF)**, where we don't only rely on unsupervised training with a large amount of data but also have a reward model, based on human ranking of answers. This reward model is continuously adjusted and used to train and tune an LLM, using reinforcement learning techniques. Combined with the initial pretrained language model, based on transformer architecture, we get a more accurate LLM.

The following figure illustrates the steps in RLHF model training.

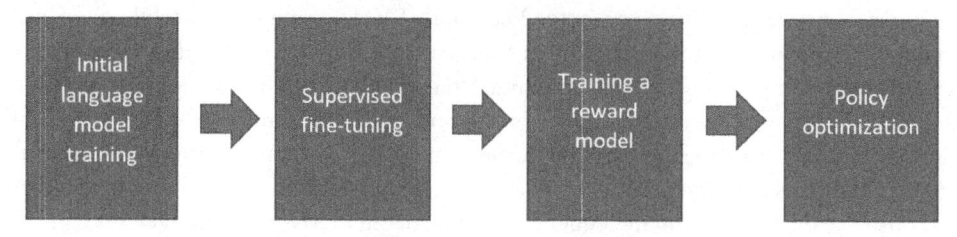

Figure 13.1 – The steps in training an LLM using RLHF

We start by training a base LLM model, move on to fine-tuning the model using human annotations, and then continue improving the model using reinforcement learning techniques.

LLMs can also be used as part of a framework that enables calling different models and tools, using prompts and external inputs as an additional knowledge base or reference. One such popular framework is LangChain (https://www.langchain.com/), where it's possible to create complex systems based on LLMs by creating interoperable components.

Prompting

Textual input (prompts) is a way for a user to describe the context of a task and the task itself to the model. Prompts can be given in various forms, such as questions, instructions, and chains of thought. You can give example answers to questions or solutions to tasks in the scope of the prompt as well. If the user asks a question within a prompt without giving a question-answer pair as an example, this is called zero-shot prompting. However, for more complex questions or tasks, it is beneficial to give examples of solved questions or tasks within the prompt. If a few examples are given to guide a model, it's called few-shot prompting. Prompt engineering is a new discipline that includes techniques on how to design and optimize prompts for best results from LLMs.

> **Note**
> You can read more on prompting here: https://www.promptingguide.ai/.

Retrieval augmented generation

As LLMs are based on datasets used for training, they can have difficulties answering questions where key information for the answer is not contained in those datasets, and they can be sometimes prone to hallucination. An often-used solution to data gaps and hallucination is to use **retrieval augmented generation**, a technique where we can use external documents to access knowledge available after training. Furthermore, we can base a response to queries on a particular set of documents to ensure that the right sources of knowledge are used by LLMs.

In the next section, we will describe the application of LLMs in defensive security scenarios, showing where their use can add value to the work of cybersecurity teams.

Using LLMs for security

LLMs are a powerful technology, as they are trained on a large volume of data, often including source code. This enables us to use them in some classical scenarios where we need to extract information from large volumes of data in different ways, as well as more advanced applications, including code analysis and generation.

In the following subsections, we will examine the different LLM applications in cybersecurity scenarios.

LLMs for vulnerability discovery

The advancement of technology and industry demand has enabled the development of increasingly complex software systems. This means that we have large-scale systems consisting of complex code bases, where hundreds or thousands of developers contribute by creating code modules that are interconnected and provide parts of system functionality. However, due to increasing complexity, time pressure, and other limiting factors, it is inevitable that there will be errors. Sometimes, these errors result in software systems being vulnerable to attacks.

Some traditional techniques for vulnerability discovery are as follows:

- **Static analysis**: The analysis of the code of a program and possible execution paths (e.g. in searching for vulnerabilities) without executing it.
- **Dynamic analysis**: Analyzing the results of a program by executing it with various inputs. Fuzzing is a type of dynamic analysis where we insert random inputs into a system and track the results of the execution.
- **Symbolic execution**: Creating an abstraction over the possible execution paths by assigning symbolic variables, and introducing symbolic inputs to analyze a program based on how inputs influence execution paths.

These methods contain different approaches to tackle the complexity of software systems, where there could be many execution paths in programs that depend on variables and, ultimately, depend on other programs and user input.

LLMs can be a promising alternative to these methods, as they are trained with large volumes of data. Some models, such as GitHub Copilot, are trained on code and text from websites such as StackOverflow, which enables them to learn from these bases of code, questions, and answers posted by experts. Furthermore, LLMs can be used for code assistance to help programmers create better code, detect potential vulnerabilities, and suggest code changes.

LLMs for threat intelligence

In the previous part of the book, we described the challenges in threat intelligence when analyzing large volumes of data in search of valuable threat intelligence information. We also introduced AI methods as a solution for this problem, with an example of topic modeling. LLMs are a natural enhancement of this solution, as they can handle large-scale textual or code datasets, and we can use an LLM model for different tasks. For instance, LLMs can be useful in the following scenarios:

- **Extracting information and artifacts from multiple data sources**: In threat intelligence, we can benefit from technologies such as LLM to get information about entities, such as IP, domain, malware, and threat actors, from various data sources. This information can come from data that is fed into public LLM services, such as ChatGPT or Perplexity, or it can be refined by fine-tuning LLM models, using dedicated threat intelligence data sources, or by using RAG methods to index documents. Having the ability to track different data sources of various data formats and languages and provide succinct information can be a crucial capability for a successful cyber-threat intelligence organization.

- **Creating summaries and reports of findings**: Cyber-threat intelligence tasks often depend on retrieving information from large volumes of data. However, it's often not enough to just retrieve this information; it also needs to be shown in an understandable and/or standardized manner to various stakeholders, such as security operations teams or even a **Chief Information Security Officer (CISO)**. LLMs can be an important tool for this, as one of their key features is generating natural language responses to various queries about data. Furthermore, they can generate responses with a specific length and phrasing, driven by the purpose described in the natural language prompt. For instance, we can use LLM to create a report about what the trends in ransomware threats are and what the rising threat actors are. Another example is to instruct an LLM to create a report for stakeholders about vulnerabilities in their organization's IT systems, with the highest reported risk score. As these reports can be used in decision-making, when generating them, we need to make sure that they don't contain hallucinations, which is possible when using LLMs. Specifying settings such as *temperature* and using an appropriate model can help minimize this possibility, but it is also advised to review reports before they get sent out to the relevant stakeholders.

LLMs for spam and phishing detection

In the previous part of the book, we described the problem of spam and phishing detection and introduced AI solutions to this problem. It's a problem that fits very well with the typical classification methods of machine learning, and ML methods are already quite well established as a solution. However, the recent rise of LLMs has provoked the question of whether we can improve spam and phishing detection using LLM capabilities. If LLMs can store patterns from a large amount of text data and recognize any anomalous content in emails, they can be an important enhancement.

There are already multiple research papers investigating this direction. For example, Labone and Moran from JP Morgan Chase investigate the use of models such as RoBERTa, SetFit, and Flan-T5 and test them on various spam detection benchmark datasets. They report significant improvements over baseline ML methods, such as XGBoost or K-nearest neighbor method. Furthermore, in the work of Koide et al., the authors investigate the use of GPT-4 and email content as a prompt and report a strong accuracy of over 99% on spam detection. Considering the generative nature of LLMs, there is a possibility of not just classifying messages but also generating explanations or reports about detected phishing and spam.

> **Note**
> You can read more about spam-detecting datasets in Maxime, Labonne and Moran, Sean, "Spam-t5: Benchmarking large language models for few-shot email spam detection". arXiv preprint arXiv:2304.01238. 2023. and spam detection by Koide, Takashi, et al, "Chatspamdetector: Leveraging large language models for effective phishing email detection". arXiv preprint arXiv:2402.18093. 2024.

LLMs for a security operation center

A **Security Operation Center** (**SOC**) is an important part of a cybersecurity department in every organization. This is the place where cybersecurity alerts are handled, confirmed as incidents, and processed and an adequate incident response is executed (if needed). Usually, in large-scale organizations, the SOC needs to handle a large number of potential incidents, and the time to detect and process an incident is crucial to optimize in order to keep the SOC running and protecting an organization effectively. Usually, SOC has defined performance metrics such as *Time to Detect*, *Time to Resolve*, and *Time to Respond*, which are continuously measured and tracked to capture any changes in performance.

LLMs can be helpful to improve these metrics and make an SOC more efficient and effective. One way that LLMs could help is to summarize all the relevant data about an incident in an understandable manner, helping SOC analysts to focus on the important information, find the root cause, and solve the incident faster. Furthermore, it can help engineers create recommendations for a customer on how to remediate the incident. Finally, given the skills gap in cybersecurity, including the SOC, LLMs can be used to help SOC analysts interact with data and a system easily and train them to improve their skills.

In the next section, we will explore the use of LLM for offensive security scenarios, where security teams investigate a system from an attacker's point of view.

LLMs for offensive security

Offensive security is a set of techniques where security teams use attacker techniques to assess the security of an organization and, in turn, strengthen the defenses. LLMs can be introduced to offensive security teams as well to boost their productivity and enable them to be more efficient. For instance, they can use LLMs to generate messages for testing the vulnerability of the organization to social engineering attacks. Another example is using LLMs to find vulnerabilities in an organization that can be exploited or finding ways for better automation of penetration testing. LLMs have the potential to be used by attackers to improve the efficiency of their attacks, and offensive security teams can use them similarly, only with the benign purpose of testing the limits of cybersecurity defenses.

The security of LLMs

LLMs are an increasingly popular technology, as they can be used for various purposes and provide benefits in automating tasks and improving workflows. Therefore, many organizations add them as tools to their IT systems, and many software products have LLM components included to enable natural language generation features, such as chatbots or report generation.

However, LLMs have their own vulnerabilities that can be exploited by attackers. Security experts must be aware of these vulnerabilities to advise and implement secure integration of LLMs.

The **Open Worldwide Application Security Project (OWASP)** is an international organization dedicated to cybersecurity – mainly, web application security. Considering the rising popularity of LLMs, OWASP has created a taxonomy of the top 10 security threats when applying LLMs:

- **Prompt injection**: LLMs are used by providing queries to them in the form of a textual prompt and obtaining a textual response. The modeling and response generation process usually includes guardrails that safeguard them from providing responses that are inappropriate or leak sensitive information. However, it was shown that an attacker can create a specially crafted prompt that can make LLMs ignore the guardrails and provide an unwanted response.

- **Insecure output handling**: The output of LLMs can be natural language text but it can also be created in a specific format that can be used to execute malicious code and compromise the security of an IT system where LLM is integrated.

- **Training data poisoning**: LLMs are trained with large-scale data, and in a high number of use cases, the data is gathered from public sources. There is a potential threat that the data that is used for training can be manipulated to change a model to provide results that benefit attackers.

- **Model denial of service**: LLMs are large-scale models and can require a high number of computational resources to operate and serve responses to queries from a potentially large user base. Attackers can intentionally create an overwhelmingly large volume of requests to overwhelm a system where an LLM is running and cause disruptions.

- **Supply chain vulnerabilities**: The functionality of LLMs involves reliance on various other software components, which can be vulnerable to malicious updates or other changes. Furthermore, systems that involve LLMs as components can be vulnerable to inauthentic LLM models being provided.

- **Sensitive information disclosure**: LLMs are trained on large volumes of data, which can easily include **Personally Identifiable Data (PII)** or other sensitive organizational information. Without proper guardrails, this information can be provided within generated responses, possibly to users who shouldn't have permission to see this data.

- **Insecure plugin design**: LLMs are often added to existing software as plugins to enable features such as chatbots and various text generation capabilities. However, since LLMs accept user input and output can be forwarded to other components, vulnerabilities can be introduced to allow exploits such as remote code execution or SQL injection.

- **Excessive agency**: The responses of LLMs often contain very impressive textual responses and can provide formulation in the desired form. However, making them a fully autonomous component should be carefully considered, as problems such as hallucinations can cause problems with their reliability and reduce user trust.

- **Overreliance**: Building on the previous point, we need to consider when we can rely on LLM responses and under what conditions. There needs to be enough awareness of LLM functionality and limitations, as well as enough testing, to be highly confident that LLMs are reliable in particular use cases.

- **Model theft**: LLM models can be trained on confidential data, and their use by unauthorized persons can create problems for organizations. Therefore, LLM models need to be protected from unauthorized access and being moved and copied to locations where they can be used by attackers.

Since the introduction of LLMs, there have been security considerations regarding emerging threats when using this technology. However, there have been new defensive approaches developed to counter these threats. One example is an AI firewall, which is a type of product that analyzes input and output to LLMs to enable defense from some of the aforementioned threats, such as prompt injection or the extraction of sensitive data. Conversely, some threats, such as supply chain vulnerability and denial of service, were well known before the introduction of LLMs and there are already established defensive approaches that can be used for LLM-based architectures as well.

Summary

In this chapter, we explored LLMs and their usage in cybersecurity applications. We begin by introducing LLMs as a new AI technology and presenting example applications in cybersecurity scenarios, both in defensive and offensive security. Furthermore, we discussed security considerations when using LLMs in software systems, common vulnerabilities, and defensive approaches. As LLMs are a fast-growing area and finding applications in various types of use cases and industries, having foundational knowledge in this area is important to be able to use them better, improving security and securing their usage in your organization.

In the next part of the book, we will examine common problems and considerations when applying AI in cybersecurity scenarios, so in the next chapter, we will first look at data quality and its usage in AI and LLMs.

References

- Zhou, Xin, Sicong Cao, Xiaobing Sun, and David Lo. "Large Language Model for Vulnerability Detection and Repair: Literature Review and the Road Ahead." *preprint arXiv:2404.02525* (2024).

- Hu, Sihao, Tiansheng Huang, Fatih İlhan, Selim Furkan Tekin, and Ling Liu. "Large language model-powered smart contract vulnerability detection: New perspectives." In 2023 5th IEEE International Conference on Trust, *Privacy and Security in Intelligent Systems and Applications (TPS-ISA)*, pp. 297-306. IEEE, 2023.

- Purba, Moumita Das, Arpita Ghosh, Benjamin J. Radford, and Bill Chu. "Software vulnerability detection using large language models." In *2023 IEEE 34th International Symposium on Software Reliability Engineering Workshops (ISSREW)*, pp. 112-119. IEEE, 2023.

- Koide, Takashi, Naoki Fukushi, Hiroki Nakano, and Daiki Chiba. "Chatspamdetector: Leveraging large language models for effective phishing email detection." *arXiv preprint arXiv:2402.18093* (2024).

- Labonne, Maxime, and Sean Moran. "Spam-t5: Benchmarking large language models for few-shot email spam detection." *arXiv preprint arXiv:2304.01238* (2023).

- GitHub - jpmorganchase/llm-email-spam-detection: LLM for Email Spam Detection

- Noever, David. "Can large language models find and fix vulnerable software?." *arXiv preprint arXiv:2308.10345* (2023).

- GitHub - git-disl/GPTLens: Large Language Model-Powered Smart Contract Vulnerability Detection: New Perspectives (TPS23)

- https://github.com/git-disl/GPTLens

- https://www.bu.edu/peaclab/files/2023/08/USENIX_23_Poster.pdf

- https://towardsdatascience.com/detecting-insecure-code-with-llms-8b8ad923dd98

- https://github.com/jpmorganchase/llm-email-spam-detection

Part 4: Common Problems When Applying AI in Cybersecurity

In this part, we look back at the AI approaches and describe common problems that need to be solved in order to make them successful. Based on our experience, we recollect common pitfalls, structure the problems that need special attention, and describe approaches to overcome those problems. We start with data quality as a prerequisite for successful AI application, and then move forward to the need for a proper understanding of the limitations of statistical approaches, as well as proper evaluation and monitoring. We finish by describing the challenges of a changing environment, and the rising need for responsible AI application.

This part has the following chapters:

- *Chapter 14, Data Quality and its Usage in the AI and LLM Era*
- *Chapter 15, Correlation, Causation, Bias, and Variance*
- *Chapter 16, Evaluation, Monitoring, and Feedback Loop*
- *Chapter 17, Learning in a Changing and Adversarial Environment*
- *Chapter 18, Privacy, Accountability, Explainability, and Trust – Responsible AI*

14
Data Quality and its Usage in the AI and LLM Era

Data quality refers to the overall utility of data based on attributes such as accuracy, completeness, reliability, relevance, and timeliness. It is an assessment of how well-suited data is for making decisions, driving processes, and achieving business objectives. The importance of data quality stems from its impact on the operational efficiency, strategic planning, and decision-making capabilities of an organization. High-quality data is critical for several practical reasons.

In this chapter, we present the data quality and its usage in the AI and **Large Language Model (LLM)** era. We also show practical use cases and features of the data in real applications.

The key topics covered in this chapter are as follows:

- Data quality and its usage
- Characterizing good data quality
- Examples of poor data quality and accidents
- Practicing with Python code

Data quality and its usage

Data quality is multifaceted, encompassing several characteristics or dimensions that collectively determine its suitability for use in operations, decision-making, and planning. Understanding these dimensions is crucial for assessing, improving, and maintaining the quality of data in any context. It refers to the overall utility of data based on attributes such as accuracy, completeness, reliability, relevance, and timeliness. It is an assessment of how well-suited data is for making decisions, driving processes, and achieving business objectives.

A high-quality dataset in the context of data science and machine learning is a collection of data that is accurate, complete, consistent, relevant, timely, and reliable. Such datasets are crucial for developing robust machine learning models that perform well in real-world applications.

Data quality needs to be maintained throughout the machine learning life cycle, from data collection through the data preprocessing stages until the testing phase, as any problems with data quality can affect all the next steps in processing. For instance, if we have an error where a column value is abnormally high, it can influence the normalization of the data, as well as feature extraction, and in turn the machine learning model parameters.

Now, in the following sections, let's have a closer look at what constitutes a high-quality dataset and how these datasets are used in various real-world scenarios.

Characteristics of a high-quality dataset

Here are some of the key data quality dimensions:

- **Accuracy**: This dimension measures whether the data correctly describes real-world objects or events without errors. Accurate data is free from mistakes and precisely reflects the intended information.
- **Completeness**: Completeness assesses whether all the required data is present. Incomplete data can lead to incorrect decisions due to missing information. This dimension is crucial for ensuring that datasets are fully populated with all necessary data points.
- **Consistency**: Consistency refers to the data being uniform across different datasets or databases. It ensures that data does not contradict itself across the system and follows the same formats and standards, making it reliable for analysis and decision-making.
- **Timeliness**: Timeliness evaluates whether the data is up-to-date and available when needed. Data should be current and provided within a useful timeframe to be relevant for decision-making processes.
- **Reliability**: This dimension measures the degree to which data can be depended upon to be accurate and consistent over time. Reliable data maintains its integrity across various uses and applications.

Uses of high-quality datasets in real life

Here are some areas where these datasets are used:

In healthcare, this is how it is used:

- **Care and research**: High-quality datasets containing accurate and complete medical records are used to train models that can predict patient outcomes, recommend personalized treatments, and support diagnostic processes.

- **Epidemiology**: During events such as the COVID-19 pandemic, high-quality data on infection rates, recovery rates, and the effectiveness of interventions were crucial for modeling disease spread and informing public health decisions.

In finance, these are the uses:

- **Risk management**: Financial institutions use high-quality datasets to develop models that predict credit risk, detect fraudulent transactions, and automate risk assessment processes.
- **Algorithmic trading**: In stock trading, algorithms that can predict market movements more accurately are developed using detailed and up-to-date market data.

The retail uses are mentioned here:

- **Customer Relationship Management (CRM)**: Retailers use high-quality datasets about customer behavior, preferences, and demographics to tailor marketing strategies and improve customer service.
- **Supply chain optimization**: High-quality data on sales, inventory levels, and logistics help companies optimize their supply chains, reducing costs and improving efficiency.

In autonomous vehicles, high-quality datasets that include varied scenarios of traffic, road conditions, and pedestrian interactions are used to train the algorithms that manage self-driving cars, ensuring they can operate safely and effectively in real-world conditions.

In the public sector, the high-quality data on traffic patterns, energy usage, and public safety are used to optimize city services and infrastructure. This leads to smarter resource management and an improved quality of life for residents.

Lastly, streaming services such as Netflix and Spotify use high-quality datasets on user preferences and behavior to train their recommendation engines, which personalize content suggestions to enhance user experience.

In each of these cases, the quality of the dataset directly influences the effectiveness of the models and systems developed from it. This underscores the importance of investing in high-quality data collection, storage, and processing practices to ensure the reliability and utility of data-driven applications in real life.

Examples of good data quality in AI and LLMs

In this section, we will give examples and characteristics of good and bad data quality in AI and LLM usage. We will clarify them within various areas, such as **Natural Language Processing (NLP)**, computer vision, and so on.

NLP

In the field of NLP, the quality of datasets significantly affects the performance and reliability of models designed to understand and generate human language.

Here are examples of both high-quality and low-quality datasets in NLP, along with their influences on AI applications.

High-quality dataset examples in NLP

First, let us see a few high-quality dataset examples.

Stanford Natural Language Inference (SNLI) corpus

The SNLI dataset (`https://nlp.stanford.edu/projects/snli/`) comprises 570,000 human-written English sentence pairs, each manually labeled for balanced classification into one of three categories: entailment, contradiction, or neutral. This dataset is used to support tasks in **Natural Language Inference (NLI)**.

The influence is that models trained on SNLI tend to perform well in understanding context, contradiction, and entailment in sentences, which is crucial for applications such as automated reasoning, chatbots, and assistant technologies. The thorough annotation process ensures a broad understanding of language nuances.

Common Crawl corpus

Common Crawl (`https://commoncrawl.org/`) is a massive repository of web-crawled data collected over more than a decade, containing petabytes of data from the internet. It includes text in many languages and from various sources and formats.

The influence is that this corpus allows for the training of robust, generalizable language models because of its diversity and size. Models trained on such data can understand and generate text across different contexts and styles, making them highly effective for translation services, content generation, and more.

Penn Treebank

The Penn Treebank (`https://catalog.ldc.upenn.edu/LDC99T42`) is an extensively annotated corpus that includes part-of-speech tagging, syntactic tree structures, and other linguistic annotations for English words.

The influence is that this dataset has been foundational for developing and benchmarking algorithms in syntactic analysis of English, improving the linguistic accuracy of NLP models. It's crucial for tasks that require understanding the grammatical structure of sentences, such as syntax-based text parsing and complex question-answering systems.

Low-quality dataset examples in NLP

Here are some low-quality dataset examples.

Online reviews with in-moderated content

Datasets compiled from user-generated online reviews without any moderation or cleaning often contain spam, inconsistencies in labeling sentiment, and a high prevalence of irrelevant information.

The influence of training on such datasets can lead to models that are unreliable in sentiment analysis, prone to errors in understanding user opinions, and easily manipulated by spam-like text patterns.

Social media datasets with limited pre-processing

When datasets harvested from social media platforms are used without adequate pre-processing to remove noise (such as hashtags, mentions, and non-standard language), they can introduce significant noise into training data.

The result is that models trained on these datasets might struggle with language generation and understanding tasks due to the informal, abbreviated, and inconsistent nature of the text. This affects their performance in more formal or standardized NLP tasks.

Datasets with poor translation quality

These are multilingual datasets created through automatic translation tools that do not accurately capture the nuances of languages.

The influence is that these datasets can lead to the development of translation models that misunderstand or misrepresent phrases and expressions in different languages, leading to communication errors and reduced efficacy in international applications.

Multilingual datasets are difficult to use to train models that accurately understand and generate human language across a variety of tasks. To overcome quality issues, here's what we can do:

- Enable the training of models that accurately understand and generate human language across a variety of tasks
- Support the development of robust systems that can generalize well to new, unseen data
- Overcome models that are inaccurate, biased, or ineffective in real-world applications
- Cause models to adopt and propagate the inaccuracies, biases, and noise from the training data

The quality of datasets in NLP is crucial for ensuring that language models perform effectively and ethically, underscoring the need for careful dataset selection, curation, and ongoing evaluation in NLP projects.

Computer vision

In the field of computer vision, the quality of datasets plays a crucial role in the performance and reliability of AI models.

Let's explore some examples of high-quality and low-quality datasets in computer vision, highlighting their influences on various applications.

High-quality dataset examples in computer vision

Here are some high-quality dataset examples:

- **ImageNet** (https://www.image-net.org/): ImageNet is a large, well-annotated dataset of over 14 million images categorized according to the WordNet hierarchy. Each image is labeled with information about what sort of objects are pictured. ImageNet has been fundamental in advancing the development of deep learning models for image classification, object detection, and more. Its extensive size and diversity have allowed for the training of robust models that perform well across various tasks and environments.

- **Common Objects in Context (COCO)** (https://cocodataset.org/#home): COCO is a rich dataset designed for object detection, segmentation, and captioning tasks. It contains over 330,000 images with over 200,000 labeled instances of 80 categories of objects found in everyday scenes. COCO has been critical for the development of state-of-the-art models that require an understanding of context in images. Its complex annotations enable more accurate and nuanced AI models that can detect and describe multiple elements in a single scene effectively.

- **Labeled Faces in the Wild (LFW)** (https://vis-www.cs.umass.edu/lfw/): LFW is a collection of face photographs designed for studying the problem of unconstrained face recognition. LFW contains more than 13,000 images of faces collected from the web and has been used widely in the evaluation of face recognition algorithms. Its real-world, varied conditions challenge models to perform reliably in recognizing faces across changes in angle, lighting, and expressions.

Low-quality dataset examples in computer vision

Here are some low-quality dataset examples:

- **Biased facial recognition sets**: Facial recognition datasets may under-represent certain demographics or contain mostly low-resolution, poorly captured images. These datasets can lead to AI models that perform poorly on underrepresented groups, resulting in higher error rates and biases in facial recognition technology. This can cause significant issues in security systems, hiring processes, and other areas where facial recognition is used.

- **Noisy vehicle recognition datasets**: Some vehicle recognition datasets include a lot of background noise, unclear images, or incorrect annotations. Models trained on these datasets may struggle to accurately identify vehicles, especially in different or challenging lighting and weather conditions. This can affect applications in traffic management, autonomous driving, and security.

- **Small object datasets with sparse annotations**: These are datasets intended for object detection that have sparse annotations, where many objects are not labeled, or where labels are inaccurate. Sparse annotations can lead to models that fail to detect objects consistently or accurately, limiting their usefulness in real-world applications such as robotic vision and automated inspection systems.

More accurate and robust datasets are capable of understanding complex scenes and contexts, are less biased, and are more universally applicable. Conversely, low-quality datasets can lead to models that are less accurate and more prone to errors, biased, and potentially discriminatory.

Limited in their real-world applicability due to poor generalization, the impact of dataset quality in computer vision demonstrates the critical need for careful dataset curation, constant evaluation, and refinement of data sources to ensure the development of effective and fair AI systems.

So far, we have presented the features of good and poor data quality. In the following section, we will discuss the accidents that are primarily caused by poor data quality in real life.

Data quality accidents

Here are some examples of data quality accidents:

- **Healthcare misdiagnoses**: Inaccurate patient data due to data entry errors or incomplete medical histories can lead to misdiagnoses or unsuitable treatment, endangering patient health. These errors can happen or be overlooked due to a lack of knowledge or competence, problems in data gathering, or simply a lack of focus.
- **Financial reporting errors**: A famous example is the $6 billion trading loss by JPMorgan Chase in 2012, known as the "London Whale" incident. This incident was partly attributed to spreadsheet errors and inadequate risk models.
- **Retail inventory mismanagement**: Overstocking or understocking due to incorrect inventory data can result in significant financial losses for retailers, affecting their profitability. Inventory forecasting often includes machine learning applications and relies on a trained model. However, the model's accuracy highly depends on having the right data on the state of the inventory, demand, sales, and so on. If any of the data has systematic or random errors, this will affect the performance of the model.
- **Airlines scheduling mistakes**: Airlines rely on accurate data for scheduling flights, crews, and maintenance. Data quality issues here can lead to flight delays, cancellations, and inefficiencies, costing millions and damaging the airline's reputation.
- **Utility billing errors**: Inaccuracies in customer data or meter readings can lead to incorrect utility billing, resulting in customer dissatisfaction, complaints, and potential legal issues.

These examples highlight the critical importance of maintaining high data quality across various domains. Data quality issues can have far-reaching consequences, underscoring the need for robust data management and quality assurance practices.

After many theoretical descriptions and examples of poor data quality-related accidents, we will move to a new section, in which we will delve into how to program with the Python language to operate on good data quality.

Writing Python code to practice good data quality

Ensuring good data quality involves employing techniques to cleanse, validate, and manage data efficiently.

In the next section, three Python examples are presented to demonstrate different ways to achieve better data quality in datasets. These examples include data cleansing, validation, and handling missing values using popular Python libraries such as `pandas` and `numPy`.

Example 1 – data cleansing with pandas

This example will demonstrate how to remove duplicates and handle erroneous entries in a dataset using `pandas`:

```
import pandas as pd
# Sample data
# It contains features that can be present in a phishing detection
dataset
data = {
    'Phishing_detect_Name': ['Alice', 'Bob', 'Charlie', 'David', 'Alice'],
    'Phishing_detect_Age': [25, 30, 35, 40, 25],
    'Phishing_detect_Email': [
        'alice@example.com', 'bob@example.com', 'charlie@example.com',
        'david@example.com', 'alice@example.com']
}

df = pd.DataFrame(data)
# Remove duplicates
df = df.drop_duplicates()
# Correcting erroneous email entries
df['Phishing_detect_Email'] = \
    df['Phishing_detect_Email'].apply(
        lambda x: x if x.endswith('@example.com') else None)
print(df)
```

This code removes any duplicate rows and fixes email entries that don't follow the correct format.

Example 2 – data validation

Here, we will check whether the data meets certain conditions, such as ensuring that ages are within a reasonable range and emails are in a proper format:

```python
import pandas as pd
# Data
data = pd.DataFrame(
    { 'Phishing_detect_Name': ['Alice', 'Bob', 'Charlie', 'David'],
    # 120 and -1 are unrealistic values for age
    'Phishing_detect_Age': [29, 120, -1, 35],
    'Phishing_detect_Email': [
        'alice@example.com', 'bob@example.com',
        'charlie@example', 'david@example.com'] })

 # Validate age range
data['Phishing_detect_Age'] = \
    data['Phishing_detect_Age'].apply(
        lambda x: x if 0 < x < 100 else None)

# Validate email format

import re
    def validate_email(email):
    pattern = r'^\w+@\w+\.\w+$'
    return email if re.match(pattern, email) else None

data['Phishing_detect_Email'] = \
    data['Phishing_detect_Email'].apply(validate_email)
print(data)
```

This example corrects unrealistic age values and validates the email format using a regular expression.

Example 3 – handling missing values

Handling missing values effectively is crucial for maintaining data quality. This example demonstrates replacing missing values with the median or most frequent value of the column:

```python
import pandas as pd
import numpy as np # Sample data with missing values
data = pd.DataFrame(
    { 'Product_ID': [101, 102, 103, 104],
    'Price': [25, 20, None, 30],
    'Category': ['Electronics', None, 'Electronics', 'Clothing']
```

```
})

# Fill missing numeric values with the median
data['Price'].fillna(data['Price'].median(), inplace=True)

# Fill missing categorical values with the most frequent category
data['Category'].fillna(data['Category'].mode()[0], inplace=True)
print(data)
```

These examples use `pandas` for data manipulation, showcasing common techniques for enhancing data quality, such as cleansing, validation, and dealing with missing values. Such practices are fundamental for preparing a dataset for further analysis or machine learning tasks.

Summary

In this chapter, we presented the content of data quality and its usage in the AI and LLM era first. We then explored several examples of good and bad data quality. Lastly, we showed how to write Python code to practice good data quality. Data quality refers to the overall utility of data based on attributes such as accuracy, completeness, reliability, relevance, and timeliness. It is an assessment of how well-suited data is for making decisions, driving processes, and achieving business objectives.

We hope you can grasp the features of good and poor data quality in your practical applications.

In the next chapter, several topics related to data quality will be presented. You will learn about the connections among data, such as correlation, causation, bias, and so on.

15
Correlation, Causation, Bias, and Variance

In this chapter, we will delve into some fundamental statistical concepts, correlation, causation, bias, and variance. Understanding these terms isn't just academic; it's essential for applying **Artificial Intelligence** (**AI**) in the real-world domain of cybersecurity. Why? Knowing the difference between correlation and causation can help you make informed decisions while understanding bias and variance is crucial for building robust models.

In this chapter, you'll learn the following:

- The definitions of correlation and causation, and why mistaking one for the other can lead to significant errors in cybersecurity
- How bias and variance affect your AI models, and why managing them is key to developing effective security systems
- Practical strategies to identify and mitigate these issues in your cybersecurity efforts

By the time you finish this chapter, you'll have a solid grasp of these critical statistical concepts and how they impact the security tools and techniques you deploy in the field. Let's dive in and explore these crucial topics, ensuring you're equipped to tackle the complex challenges of AI in cybersecurity.

Technical requirements

To get the most out of this chapter, there are a few technical prerequisites that will help you grasp the concepts discussed both in theory and in practice. Here's a list of what you should be familiar with:

- **Basic statistics**: A good grasp of basic statistical measures such as mean, median, mode, standard deviation, and variance is essential.
- **Probability**: Understanding probability theory, including conditional probabilities and probability distributions, will help you comprehend how these concepts apply to model uncertainty in AI.

- **Programming language**: Knowledge of a programming language, particularly Python, is beneficial since it is widely used in data science and AI. Python libraries such as NumPy, pandas, and SciPy are tools that can help manipulate data and perform statistical analysis.
- **Data handling**: The ability to preprocess and handle data using programming tools will be crucial, especially when working with real-world cybersecurity datasets to apply the concepts of bias and variance.
- **Machine learning concepts**: Familiarity with basic machine learning concepts such as training models, overfitting, underfitting, and model validation techniques is also crucial.
- **Model evaluation metrics**: You must have an understanding of key model evaluation metrics like accuracy, precision, recall, and F1-score.
- **Fundamental cybersecurity concepts**: While deep expertise in cybersecurity isn't necessary, familiarity with basic concepts such as threats, vulnerabilities, and security protocols will enhance your understanding of the applications discussed.

Introducing the statistical foundation

In the increasingly data-driven world of cybersecurity, the role of AI has become central to detecting, analyzing, and mitigating cyber threats. With this reliance on AI, it becomes critically important for cybersecurity professionals to understand the underlying statistical foundations that power these technologies. Statistical literacy is not just useful; it's a fundamental part of ensuring that AI tools function correctly and effectively.

One of the most common pitfalls in statistical analysis is confusing correlation with causation. While two variables may appear to move in tandem, this does not necessarily mean that one causes the other. This distinction is vital in cybersecurity, where the stakes are high, and decisions need to be based on accurate interpretations of data. For instance, it might be important to identify whether increased network traffic is a root cause of system performance issues or whether they are just correlated with each other. This can significantly influence how security protocols are implemented.

Moreover, bias and variance are two statistical concepts that directly impact the performance of AI models. Bias can lead to an oversimplification of models, whereby important nuances and patterns are missed. On the other hand, variance refers to a model's sensitivity to small fluctuations in the training set. This can result in overfitting, whereby the model excels on training data but struggles with new, unseen data. Understanding how to balance bias and variance is crucial for developing robust AI applications in cybersecurity, ensuring that models are both accurate and generalizable.

Thus, a deep dive into these statistical concepts is not just academic; it's a practical necessity for developing effective, reliable AI-driven security systems. This chapter aims to equip you with the knowledge to not only understand these concepts but also apply them in your daily work, enhancing both your AI models and your cybersecurity measures.

Understanding correlation and causation

When we talk about the relationship between two variables in cybersecurity data, we often encounter the terms *correlation* and *causation*. Understanding the difference between these two concepts is like distinguishing between coincidence and consequence. Let's break these down in a way that's easy to grasp.

Correlation

Correlation measures how strongly two variables are related to each other. For instance, in cybersecurity, you might observe that as the number of login attempts increases, so do the instances of security alerts. This relationship can be quantified using a correlation coefficient, which ranges from -1 to 1. This is explained further as follows:

- A **positive correlation** (+1) means that as one variable increases, the other also increases
- A **negative correlation** (-1) indicates that as one variable increases, the other decreases
- A correlation of 0 suggests that there is no linear relationship between the variables

The Pearson correlation coefficient is a measure of the linear relationship between two datasets. Mathematically, it is defined as the covariance of the two variables divided by the product of their standard deviations. Symbolically, it's represented by the letter (r).

Here's the formula for the Pearson correlation coefficient (r) between two variables (X) and (Y):

$$r = \frac{\sum_{i=1}^{n}(X_i - \overline{X})(Y_i - \overline{Y})}{\sqrt{\sum_{i=1}^{n}(X_i - \overline{X})^2}\sqrt{\sum_{i=1}^{n}(Y_i - \overline{Y})^2}}$$

Here, the following is true:

- (X_i) and (Y_i) are the individual sample points indexed with (i)
- ($\overline{X}$) and ($\overline{Y}$) are the means of the (X) and (Y) datasets, respectively
- (n) is the number of data points in each dataset (assuming the datasets are of equal length)

This formula effectively measures how much the two variables change together compared to how much they vary individually. The closer the coefficient is to +1 or −1, the stronger the linear relationship between the variables is. A coefficient of 0 indicates no linear relationship.

Let's see how you might calculate the correlation coefficient between two variables using Python:

```
import numpy as np
# Sample data: number of login attempts and security alerts
login_attempts = np.array([1, 2, 3, 4, 5])
security_alerts = np.array([2, 3, 5, 7, 8])

# Calculating the Pearson correlation coefficient
correlation_coefficient = np.corrcoef(
```

```
        login_attempts, security_alerts)[0, 1]
print(f'Correlation Coefficient: {correlation_coefficient}')
```

This code snippet uses NumPy, a popular library for numerical computations in Python, to calculate the Pearson correlation coefficient, giving us an idea of how closely related the login attempts are to security alerts.

Causation

Causation, on the other hand, goes a step further by implying that one event causes the other. It's a much stronger assertion than correlation. For example, discovering a software vulnerability (cause) that leads to data breaches (effect) is a causal relationship. Establishing causation is crucial because it helps in making decisions about where to allocate resources, such as fixing specific vulnerabilities to prevent future breaches.

Proving causation typically requires controlled experimentation, which is often challenging in real-world scenarios such as cybersecurity. However, advanced statistical techniques, computational models, and careful analysis of data can provide strong indications of causal relationships.

Furthermore, there is an area of research in machine learning that tries to address the fact that statistical models primarily capture correlation between variables and not causal relationships. **Causal machine learning** contains alternative approaches to uncover patterns based on causal reasoning.

Why distinguishing them matters

In cybersecurity, mistaking correlation for causation can lead to misguided strategies. For example, if you observe that breaches occur more frequently on weekends and conclude that weekends cause more breaches, you might wrongly prioritize security measures that focus on weekends. Perhaps, instead, the real cause is lower staffing on weekends, not the days themselves. Understanding the difference between these two concepts helps in making informed, effective decisions in cybersecurity operations and strategy. It allows professionals to allocate resources efficiently, focus on genuine threats, and implement more effective security measures.

Now that we have gained a good understanding of both correlation and causation, in the next section, we will introduce two more important concepts, namely bias and variance.

Introducing bias and variance

In the context of building and deploying AI models in cybersecurity, two frequently encountered terms are *bias* and *variance*. These statistical concepts are critical to your understanding because they directly affect the performance and reliability of your models. Let's explore what each of these terms means and how they influence your AI systems.

Bias

Bias in AI refers to the error introduced into your model due to oversimplification of the machine learning algorithm. It occurs when the model is too simple to capture the complex patterns in the data, or when it's working under wrongful assumptions about the data. In cybersecurity, a biased model might consistently fail to detect a new type of malware because it hasn't been exposed to sufficient or diverse examples during training.

Imagine you're using a model trained primarily on data from network intrusions that occurred during business hours. If this model is deployed in an environment where attacks frequently happen outside these hours, its effectiveness could be limited—this is an example of bias.

Variance

Variance, on the other hand, describes how much your AI model's predictions would change if it were trained on different sets of data. High variance indicates that the model is extremely sensitive to small fluctuations in the training set, often leading to overfitting. Overfitted models perform well on training data but poorly on unseen, real-world data. In cybersecurity, a model with high variance might perform exceptionally in detecting known threats in a controlled test environment but fail to generalize to new, slightly different threats in the real world.

The key challenge in machine learning, including AI applications in cybersecurity, is managing the trade-off between bias and variance. Ideally, you want a model that is complex enough to capture the true patterns in the data (low bias) but not so complex that it models random noise as valid patterns (low variance).

Understanding these concepts is crucial for developing effective AI models in cybersecurity. However, to truly grasp their impact, we need to delve deeper into how they manifest in real-world scenarios and explore strategies to mitigate their effects.

Bias and variance in polynomial curve fitting

To better understand the concepts of bias and variance, let's explore a practical example using polynomial curve fitting. Polynomial curve fitting is a method of modeling a relationship between variables as a polynomial expression. Imagine that you want to make predictions about the number of vulnerabilities in time. If you know that this relationship can be described by a polynomial function, you need to find which polynomial describes this relationship best to make accurate predictions. This example will demonstrate how varying the degree of the polynomial impacts bias and variance.

Setting up the example

First, we'll generate some synthetic data that follows a specific pattern with some added noise. We'll then fit polynomial curves of different degrees to this data and observe how the models behave:

```python
import numpy as np
import matplotlib.pyplot as plt
from sklearn.preprocessing import PolynomialFeatures
from sklearn.linear_model import LinearRegression
from sklearn.pipeline import make_pipeline

# Generate synthetic data
np.random.seed(0)
x = np.linspace(-3, 3, 15)
y = x**3 - 3*x**2 + 2*x + np.random.normal(0, 3, 15)  # Cubic function
with noise

# Function to plot data and a polynomial fit
def plot_polynomial_fit(x, y, degree):
    model = make_pipeline(PolynomialFeatures(degree),
        LinearRegression())
    model.fit(x[:, np.newaxis], y)  # Fit the model

    # Plotting
    plt.scatter(x, y, color='lightblue',
        label='Data points', s=80)
    plt.plot(x, model.predict(x[:, np.newaxis]),
        label=f'Degree {degree}', color='gray')
    plt.title(f'Polynomial Degree {degree}')
    plt.legend()
    plt.show()

# Plot polynomial fits for different degrees
plot_polynomial_fit(x, y, 1)   # Degree 1 (High Bias)
plot_polynomial_fit(x, y, 3)   # Degree 3 (Good Fit)
plot_polynomial_fit(x, y, 10)  # Degree 10 (High Variance)
```

This is what we can understand from this code example:

- **Degree 1 (linear model)**: This model, being a simple linear regression, is likely to underfit the data. It has high bias because it's too simplistic to capture the underlying cubic relationship in the data. The plot will show that the line does not capture the curves evident in the data pattern.

- **Degree 3 (cubic model)**: A cubic model is well-suited for this dataset since the data was generated from a cubic equation. This model will likely fit the data well, exhibiting low bias and low variance. It captures the underlying trend without fitting the noise.
- **Degree 10 (high-degree polynomial)**: While a 10th-degree polynomial can fit the training data very closely, including the noise, it's likely to exhibit high variance. This means that it's overly sensitive to the specific noise in the training data, leading to poor generalization to new data. The plot will show the curve oscillating wildly to pass through every data point.

Visualizing the results

The plots generated by the Python code will visually demonstrate how the bias and variance trade-off plays out with different polynomial degrees.

The following plot shows that the low-degree polynomial will clearly miss capturing the complexity of the data (high bias):

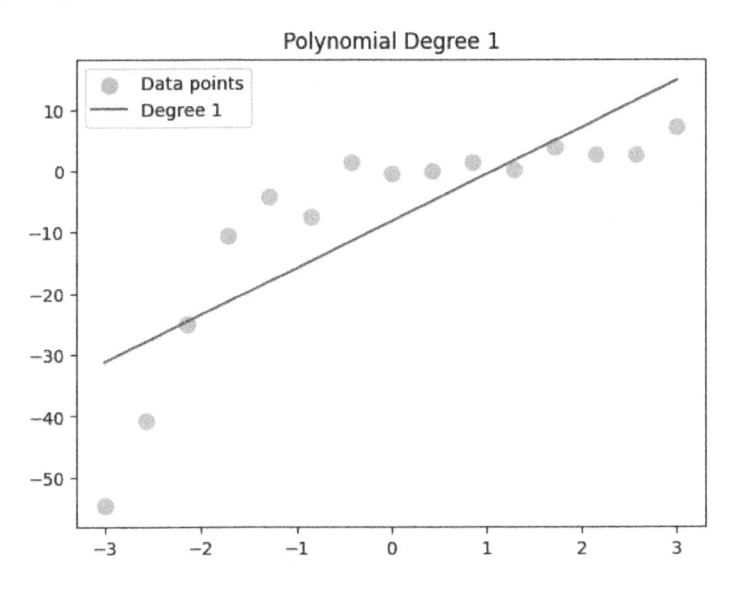

Figure 15.1 – Polynomial curve fitting with degree 1

Then, the appropriate degree polynomial will fit the data just right as shown here:

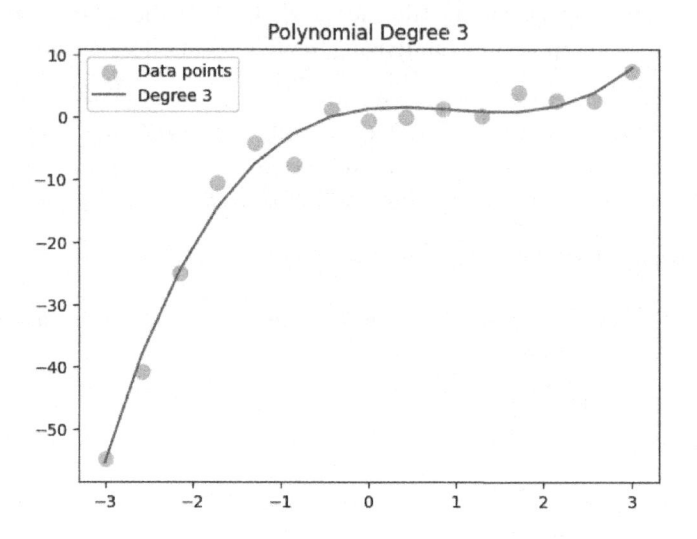

Figure 15.2 – Polynomial curve fitting with degree 3

Here, the high-degree polynomial will follow the noise too closely (high variance):

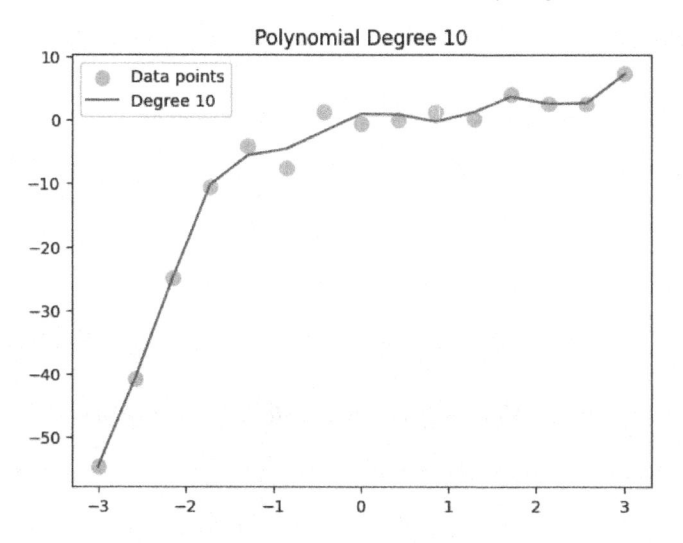

Figure 15.3 – Polynomial curve fitting with degree 10

Through this example, you can see how critical it is to choose the right model complexity in machine learning to balance bias and variance, thereby ensuring good generalization and robust performance of AI models in real-world applications such as cybersecurity.

Managing bias and variance

In the previous curve-fitting example, we can see that it is essential to find a balance between bias and variance. To manage bias and variance effectively, you can use several techniques:

- **Model complexity**: Adjusting the complexity of the model is a direct way to control bias and variance. Simple models reduce variance but increase bias, whereas complex models decrease bias but increase variance.
- **Training data size**: Increasing the size of the training data can help reduce variance without increasing bias. More data provide a more comprehensive view of the problem, helping the model to generalize better.
- **Regularization techniques**: Techniques such as Lasso and Ridge regression can help in reducing variance by penalizing overly complex models, thus preventing overfitting.

Here's a simple Python example demonstrating how regularization can help manage variance:

```
from sklearn.linear_model import Ridge
from sklearn.datasets import make_regression

# Generate some regression data
X, y = make_regression(n_samples=100, n_features=2, noise=0.1)

# Create a Ridge regression model to reduce variance
ridge_model = Ridge(alpha=1.0)  # Alpha is the regularization strength
ridge_model.fit(X, y)

# Display the coefficients
print("Coefficients:", ridge_model.coef_)
```

In this example, Ridge regression is used to add a regularization term to the cost function, penalizing large coefficients in the model. This helps in reducing variance, making the model less sensitive to fluctuations in the training data.

Understanding and managing bias and variance is not just a technical necessity; it's crucial for ensuring that your AI-driven cybersecurity systems are both accurate and robust. By fine-tuning these aspects, you can develop AI models that are truly effective in the dynamic and challenging world of cybersecurity. In the next section, we will review some case studies and examples related to these statistical concepts.

Case studies and examples

Now that we've explored the theoretical aspects of correlation, causation, bias, and variance, let's dive into some real-world case studies and examples. These scenarios will help you see how these concepts are applied in the field of cybersecurity, providing a practical perspective that enhances your understanding and application skills.

Case study 1 – correlation versus causation in phishing attacks

Imagine a cybersecurity firm that analyses email traffic to identify phishing attempts. Through their analysis, they find a strong correlation between the times when emails are sent and the incidence of phishing attacks, with a spike in phishing emails observed during early morning hours:

- **Initial observation**: The firm initially considers the possibility that attackers prefer early mornings for their activities. However, further investigation reveals that this correlation does not imply causation. Instead, it's discovered that automated botnets, which are programmed to operate at low-traffic times to avoid detection, are responsible for sending out bulk phishing emails during these hours.
- **Lesson learned**: This case study underscores the importance of not jumping to conclusions based on correlation alone. It highlights the necessity of digging deeper to understand the underlying causes, which are crucial for developing effective countermeasures.

Case study 2 – managing bias and variance in IDS

A tech company uses machine learning models as part of their **Intrusion Detection Systems (IDS)** to detect anomalous behavior that could indicate a security breach. They initially deploy a highly complex model that performs excellently on their training data but fails to generalize well on real-world data, leading to numerous false positives and negatives:

- **Problem identification**: The model was suffering from high variance, being too sensitive to the noise in the training data. Additionally, the training data was not representative of the actual diversity of network traffic, introducing bias into the model.
- **Solution**: The company decided to simplify the model to reduce variance. They collected more comprehensive and diverse data to retrain the model, addressing the bias. This adjustment led to a more robust model that performed better in real-world scenarios, effectively balancing detection accuracy and generalization.

Example – using statistical analysis for anomaly detection

Consider a financial institution that employs statistical analysis to detect unusual transactions that could indicate fraud. By applying techniques to measure the variance in transaction amounts and frequencies, they can identify outliers that deviate significantly from the norm.

The implementation here is a combination of descriptive statistics and machine learning algorithms. The institution sets thresholds that trigger alerts for potential fraud. This approach uses the concepts of bias (to avoid ignoring real fraud) and variance (to avoid false alarms), demonstrating how these statistical concepts are crucial in practical applications.

Conclusion of case studies

These case studies and examples illustrate how understanding and applying the concepts of correlation, causation, bias, and variance are crucial in cybersecurity. They show that while theoretical knowledge is essential, the real value comes from applying these concepts to solve real-world problems, ensuring the security and integrity of systems in a practical, effective manner. Through these narratives, we see that a careful, analytical approach to data and model management can lead to significant improvements in cybersecurity efforts.

Now, let's move on to see some practical applications of these concepts.

Practical applications

Now that we've explored the concepts of correlation, causation, bias, and variance through theoretical discussions and case studies, let's translate these ideas into practical applications. This section will guide you through the tools and methods that are commonly used in the cybersecurity industry to address these statistical challenges. Understanding these applications will empower you to implement more effective AI-driven security solutions in your own work.

Diagnostic tools for correlation and causation

Here are some tools to work with correlation and causation:

- **Scatter plots and correlation coefficients**:

  - **Tool**: Data visualization tools such as Matplotlib and Seaborn in Python are examples of these types of tools.
  - **Application**: Use scatter plots to visually inspect the relationship between variables. Calculate Pearson or Spearman correlation coefficients to quantify the strength of linear or rank relationships, respectively.
  - **Example**: Plotting the number of login attempts against the frequency of security alerts to identify potential correlations counts as an example.

- **Causal Impact Analysis**:

  - **Tool**: Causal inference frameworks such as DoWhy or CausalML fall into this category.
  - **Application**: Use these tools to model and estimate the causal impact of one variable on another, helping to distinguish true causal relationships from mere correlations.
  - **Example**: Assessing the impact of a new security protocol on the reduction of data breaches is an example.

Techniques to manage bias and variance

Here we look at some techniques to learn how to manage bias and variance:

- **Cross-validation**:
  - **Tool**: The `scikit-learn` library's cross-validation features are an example of a relevant tool.
  - **Application**: Implement k-fold cross-validation to evaluate model performance more reliably, reducing the risk of overfitting and helping to balance bias and variance.
  - **Example**: Using cross-validation in training intrusion detection models to ensure they generalize well across different network environments counts as an example.

- **Regularization methods**:
  - **Tool**: Lasso and Ridge regression in `scikit-learn` fall into this category.
  - **Application**: Apply regularization techniques to penalize large coefficients in regression models, reducing model complexity and variance.
  - **Example**: Tuning a machine learning model used for predicting the likelihood of phishing attacks to make it robust against overfitting counts as an example.

Advanced statistical techniques for enhanced security

To improve security measures, here are some advanced statistical techniques:

- **Anomaly Detection Algorithms**:
  - **Tool**: Isolation Forest, DBSCAN, and other anomaly detection algorithms available in `scikit-learn` fall into this category.
  - **Application**: Detect unusual patterns or outliers in cybersecurity data that could indicate threats or system failures.
  - **Example**: Monitoring network traffic for unusual patterns that deviate from historical norms to quickly identify potential threats counts as an example.

- **Bayesian networks**:
  - **Tool**: PyMC3 or other Bayesian inference libraries fall into this category.
  - **Application**: Use Bayesian networks to model complex security systems and infer the probabilities of different threat scenarios.
  - **Example**: Estimating the likelihood of different types of security breaches based on various indicators and historical data counts as an example.

Implementing responsible AI in cybersecurity

Here are some techniques to implement better cybersecurity in terms of responsible AI:

- **Transparency and explainability**: Use tools such as LIME or SHAP to provide explanations for the predictions made by AI models, making the AI systems more transparent and trustworthy.
- **Privacy-preserving techniques**: Implement techniques such as differential privacy or federated learning to enhance data privacy in AI applications, ensuring that the models do not compromise sensitive information.

By integrating these tools and techniques into your cybersecurity practices, you can enhance the accuracy, reliability, and ethical standards of your AI systems. These practical applications not only address the statistical challenges discussed earlier but also contribute to more secure and effective cybersecurity solutions.

Summary

In this chapter, we explored key statistical concepts—correlation, causation, bias, and variance—and their critical roles in AI-driven cybersecurity. Understanding these concepts is crucial for designing, implementing, and evaluating effective security systems.

We started by distinguishing between correlation and causation, emphasizing the importance of accurate data interpretation. Through examples and case studies, we showed how misconceptions could lead to ineffective cybersecurity strategies. We also discussed managing bias and variance in AI models, introducing techniques such as regularization and cross-validation to enhance model robustness and performance in real-world scenarios. By employing the tools and methods outlined, cybersecurity professionals can not only improve the reliability of AI systems but also ensure that they are responsible and transparent.

This chapter underscores the necessity of continuous learning in the ever-evolving fields of AI and cybersecurity, equipping you with foundational knowledge to address complex challenges effectively. In the next chapter, we will discuss how we can evaluate and monitor model performance, as well as integrate feedback into the model life cycle.

16
Evaluation, Monitoring, and Feedback Loop

In the previous parts of the book, we have provided extensive information about AI models and their applications. We have enabled you to find appropriate models for various cybersecurity use cases, understand them, and train the models on available data. In this chapter, you will learn how to evaluate the trained model to make sure its performance is optimal.

To have an AI model as part of an IT system, the model functionality and performance need to be tested and evaluated properly. The evaluation, testing, and monitoring should be done continuously to maintain the model's performance and overall quality of service. Models must be monitored for performance in terms of the quality of predictions and other parameters such as latency, availability, and bias. We'll introduce tools that enable evaluation and performance tracking. Lastly, we'll consider that models often need to have a human in the loop to maintain and improve performance. Humans are needed to evaluate predictions, label data, and tune the model.

So, with this in mind, in this chapter, we are going to cover the following main topics:

- Evaluating models
- Monitoring models
- Human in the loop

Technical requirements

This chapter has a practical exercise for which the following libraries are needed:

- `numpy`
- `scipy`
- `mflow`

- `sklearn`
- `matplotlib`

Evaluating models

Model evaluation in AI and **machine learning** (**ML**) models typically relies on the methodology of selecting training, validation, and test sets and loss functions and metrics that are calculated on those datasets. This enables us to have a quantitative way to assess models and optimize them. For instance, we need a way to measure how good our malware detection model is in classifying executables during the training time, validation time, and test time.

Firstly, we'll describe and compare the most common loss functions used in different types of ML models.

Loss functions

Loss functions are used when training the model to evaluate its performance during training. Furthermore, they are used to compute gradients, which enable us to determine how to change model parameters to gradually improve the performance of the model in each training step. Loss functions can also be used to evaluate models after training.

The following loss functions are the most used in training ML models:

- Cross-entropy loss:

$$L = -\frac{1}{n}\sum_{i=1}^{n}\left(y_i \log(\hat{y}_i) + (1 - y_i)\log(1 - \hat{y}_i)\right)$$

- Hinge loss:

$$L = \max(0, 1 - yf(x))$$

- Mean square error:

$$L = \frac{1}{n}\sum_{i=1}^{n}\left(\hat{Y}_i - Y_i\right)^2$$

- Mean absolute error:

$$L = \frac{1}{n}\sum_{i=1}^{n}\left|\hat{Y}_i - Y_i\right|$$

The first two functions are most often used in classification problems, and the latter two are mostly used in regression problems. Cross-entropy loss is used more often than hinge loss, but hinge loss is more recommended in cases where we want to put an emphasis on the speed of training. Mean squared error is more often used than mean absolute error because of its easier computation of the gradient. However, mean absolute error enables more robustness to outliers.

In the next section, we'll enumerate and describe the metrics used to evaluate models and explain where different metrics are applicable.

Model metrics

Metrics are used to provide a numerical performance score for models. This enables us to get an idea of how accurate our model is and to compare different models trained with different methods or different hyperparameter values. Metrics are different for different types of models. For instance, in regression models, we usually measure some type of difference between the model output and expected output, whereas in classification models, especially binary classification, we use the following factors:

- **True positive (TP)**: Data point assigned correctly to the positive class
- **True negative (TN)**: Data point assigned correctly to the negative class.
- **False positive (FP)**: Data point assigned to the positive class, but the true class is negative.
- **False negative**: Data point assigned to the negative class, but the true class is positive (**FN**)

Here, we'll name some of the model metrics that are most often used to evaluate models:

- **Accuracy**: This shows the proportion of correct predictions:

$$ACC = \frac{TP + TN}{N}$$

- **True positive rate (sensitivity)**: Proportion of positives that will be correctly detected:

$$TPR = \frac{TP}{TP + FN}$$

- **Specificity**: Proportion of negatives that will be correctly detected:

$$SP = \frac{TN}{TN + FP}$$

- **False positive rate (specificity)**: Proportion of negative events that will be falsely detected as positive:

$$FPR = \frac{FP}{FP + TN}$$

- **Precision**: This measures how often detection of positives is correct:

$$PR = \frac{TP}{TP + FP}$$

- **Recall**: This measures how often we are able to detect positives out of all positives in the data:

$$RC = \frac{TP}{TP + FN}$$

- **F1 score**: This measures the harmonic mean of the precision and recall, balancing the influence of both metrics. This is one way to have one performance measure that considers both precision and recall performance aspects:

$$F_1 = \frac{2PR \cdot RC}{2(PR + RC)}$$

An alternative formulation is as follows:

$$F_1 = \frac{2TP}{2TP + FP + FN}$$

In the following sections, we'll give more detail about how to organize the validation around these metrics.

Cross-validation

If we want to properly validate the model, it's not enough to just randomly split the data into a training set and test set. The reason is that our datasets are of finite size and the training and test sets can have a different distribution each time we produce a new split. Since different distribution means that the performance may vary, we need to think about how to produce a more reliable way to do validation and testing.

Cross-validation is a methodology where we combine resampling and splitting data repeatedly, executing tests, and calculating performance metrics. Repeated resampling enables us to average out the obtained values of performance metrics and alleviate the bias specific to one split of data.

Here, we'll name the two most common types of cross-validation:

- **K-fold cross-validation**: In K-fold cross-validation, we split the dataset into K number of folds in each run. We take K-1 folds into the training set, and the one fold left is used as a testing (or validation) set. In each iteration of cross-validation, we create folds randomly, which helps us create different splits of data.

- **Leave-p-out cross-validation**: Here, we use p number of randomly chosen examples from the dataset as the validation (or test) set and use the rest of the data as the training set. We repeat this process multiple times, if possible, using all ways to make the split, and average out the result. One variant of this is *leave-one-out cross-validation*, commonly used with small datasets.

Receiver operating characteristic (ROC) curve

In binary classification problems, the ML models output a value between 0 and 1, where a value closer to 1 indicates a positive sample and a value closer to 0 indicates a negative example. The assignment to a positive or a negative class depends on the threshold that we adopt. The ROC curve is a graphical representation of the performance of an ML model at different thresholds of classification. We plot two performance values, **true positive rate** (**TPR**) against **false positive rate** (**FPR**), for different values of the threshold, thereby generating a performance curve.

The following figure contains an example of an ROC curve:

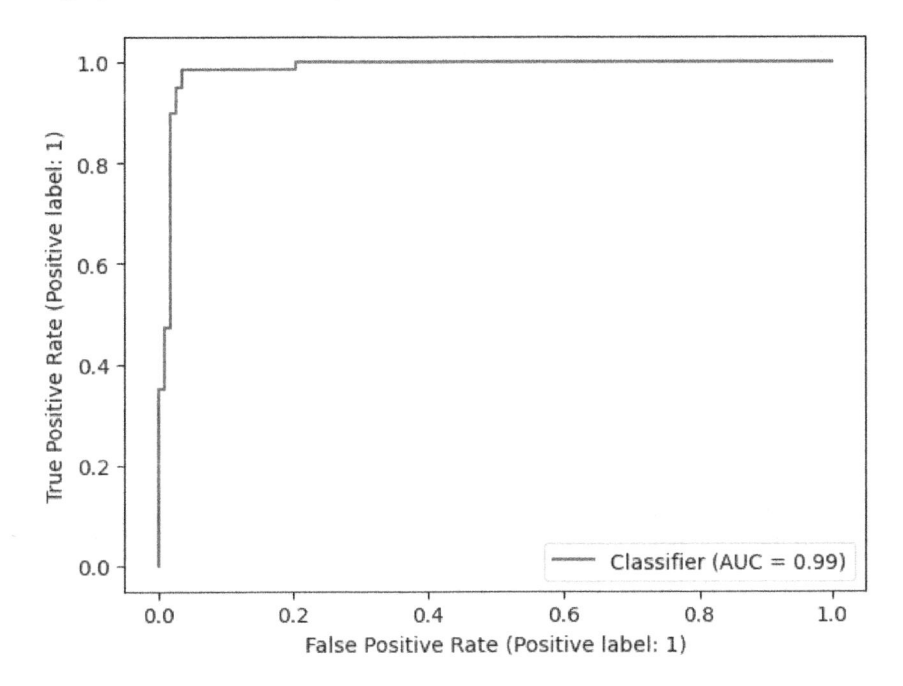

Figure 16.1 – Example of an ROC curve

The **area under the ROC curve** (**AUC**) is a new performance measure derived from the generated curve. Here, we measure the area bordered by the curve on the graph – that is, the performance is higher if the TPR grows fast with the FPR value.

Confusion matrix

A confusion matrix is a table that we can use to compare the true classes of data points to the predicted class assignments of the model. It consists of K rows and columns, where each row or column represents one of the K classes used in the model. The cell i,j contains the number of data samples with true class i that are classified as being part of class j. If the confusion matrix contains high values on the diagonal and low values outside of the diagonal, this indicates an accurate model.

The following figure shows a depiction of a confusion matrix for binary and multiclass classification:

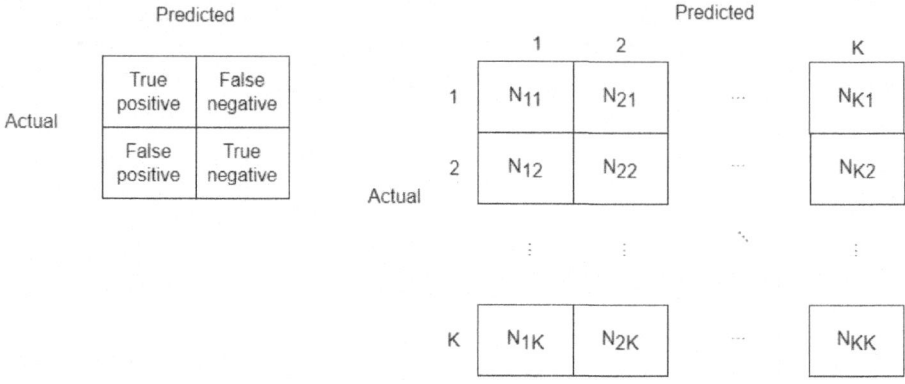

Figure 16.2 – Confusion matrix for binary classification (left) and multiclass classification (right)

In *Figure 16.2*, you can see that correct classifications are on the main diagonal, whereas other cells contain different types of misclassifications. This makes the confusion matrix an informative table to analyze classification errors and try to detect the root cause of misclassifications.

Now that we understand the metrics used to evaluate models, in the next section, we can introduce model monitoring, where the performance metrics can be used to observe changes in the AI system.

Monitoring models

To create an ML system, it is not enough to find an appropriate ML method and run the model training code. Since the training depends on the data and model hyperparameters, it is important to monitor the model while training to ensure proper convergence to the optimal parameters. Furthermore, after the training, we need to monitor the model to make sure that the performance of the model has not dropped below the acceptable threshold.

In the next section, we'll give an introduction to the model monitoring during the training process.

Monitoring during training

During the training process, it's common to monitor the changes in the loss function to ensure that the model parameters are converging to optimal values. Furthermore, we can monitor performance on the training and test set on each training iteration, or on every N iteration. This way, we will know whether the model is still improving and we should keep the training running, or whether we can stop it because the performance is staying the same.

The following figure contains an example of a loss curve:

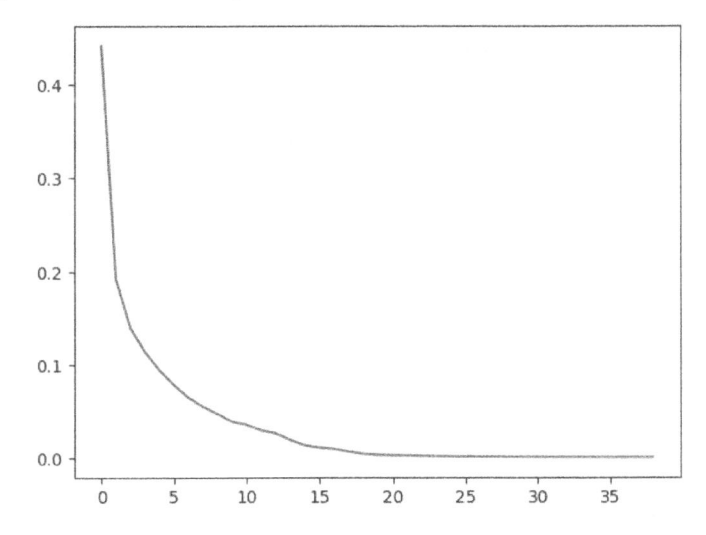

Figure 16.3 – Training loss curve when training a multilayer perceptron on the MNIST dataset

This figure shows how the loss function changes during training. We start with a random set of parameter values and gradually improve the model. The loss will decrease while the performance on the training set is improving. After about the 30th iteration, we can see that the loss function value is not reducing anymore and further training will not be as useful.

Monitoring during testing or production

Even after finishing the training and validation phase, there is a need to monitor the ML system. Making the model available for usage during production is also called **model serving**.

Some examples of how models can be provided are through a graphical user interface, through an API, and by providing prediction as flat files.

There are multiple reasons why we need to do the monitoring in production, some of which are as follows:

- **Service availability**: A crucial part of a **service-level agreement** (**SLA**) provided with ML products is service availability, usually quantified as a percentage. We need to make our model available as it can be a crucial part of our clients' workflows.
- **Data drift**: The performance of our model computed during training and validation is an indicator of future performance only if the data distribution doesn't significantly change. However, we cannot guarantee this to happen, since the environment in which the data is generated can change over time. This change can be gradual or abrupt, and in both cases, we need to monitor that change and retrain the model or modify it if necessary.

- **Performance analysis**: Even if we have a model validated in production, there is always a need to monitor how the model is used, where it makes mistakes, and what the possibilities are for improvement. This improvement could be in the choice of the model architecture, but also in the selection of features or data collection.
- **Documentation and audit**: If the model is part of a product, often, there is a regulatory requirement to monitor the model performance and create possibilities for auditing. This way, we provide evidence of how well the model is working and that it can be used for the intended purpose.

Model monitoring tools

There are multiple tools available for monitoring ML models during training, validation, testing, or production. The most famous open source tool for tracking experiments and logging and monitoring models is **MLflow**. So, let us learn more about it next.

MLflow

MLflow is an open source solution for experimenting with models, tracking those experiments, and versioning models, as well as serving those models.

It can be used together with various ML libraries to structure the development of models and take further steps toward a reliable ML product. MLflow has a Python library, which enables us to write Python scripts to obtain the earlier-mentioned functionality.

In the following code examples, we'll introduce MLflow use in Python.

Example – experiment tracking

In this example, we'll create a script that enables us to track the experiment of training a neural network classifier (multilayer perceptron) on a sample dataset and log the parameters and performance metrics of that model:

1. First, we import the necessary libraries:

    ```
    import numpy as np
    from sklearn.neural_network import MLPClassifier

    import mlflow
    import mlflow.sklearn
    from sklearn.datasets import make_classification
    from sklearn.model_selection import train_test_split

    import matplotlib.pyplot as plt
    ```

2. Next, we start the main part of the script by creating a test dataset for classification and split the data into training and a testing set:

```
if __name__ == "__main__":
    X, y = make_classification(random_state=42, n_classes=2)
    X_train, X_test, y_train, y_test = train_test_split(X, y)
```

3. Now we can create an experiment with the name MyExperimentMLP:

```
experiment_name = "MyExperimentMLP"
mlflow.create_experiment(experiment_name)
mlflow.set_experiment(experiment_name)
```

4. We then create an MLP classifier that we will use in this experiment:

```
clf = MLPClassifier(max_iter=1000)
```

5. With this instance of the model, we can start a run of the experiment. Within the experiment run, we train the model:

```
with mlflow.start_run():
    clf.fit(X_train, y_train)
```

6. We then set the parameters and the part where the model artifacts will be saved:

```
model_dir = "./model"
osmakedirs(model_dir, exist_ok=True)
mlflow.sklearn.save_model(clf, model_dir)
    mlflow.log_param("hidden_layer_sizes",
        clf.hidden_layer_sizes)
    mlflow.log_param("activation", clf.activation)
    mlflow.log_artifacts("./model")
```

7. Then, we create a plot that will be saved as part of the experiment run:

```
fig, ax = plt.subplots()
ax.plot(clf.loss_curve_)
ax.set_xlabel("Epoch")
ax.set_ylabel("Training loss")

mlflow.log_figure(fig, "plot_loss_curve.png")
```

8. Furthermore, we save the accuracy score as a metric of the model:

```
mlflow.log_metric("score", clf.score(X_test, y_test))
```

266 Evaluation, Monitoring, and Feedback Loop

If you haven't started the MLflow Tracking Server, you need to start it using the *mlflow ui* command. Once we have created the experiment, we should be able to see it in the MLflow UI. Typically, the MLflow UI would be available on the localhost IP address `http://127.0.0.1:5000/`.

Our experiment will show up in the **Experiments** tab, as shown in the following screenshot:

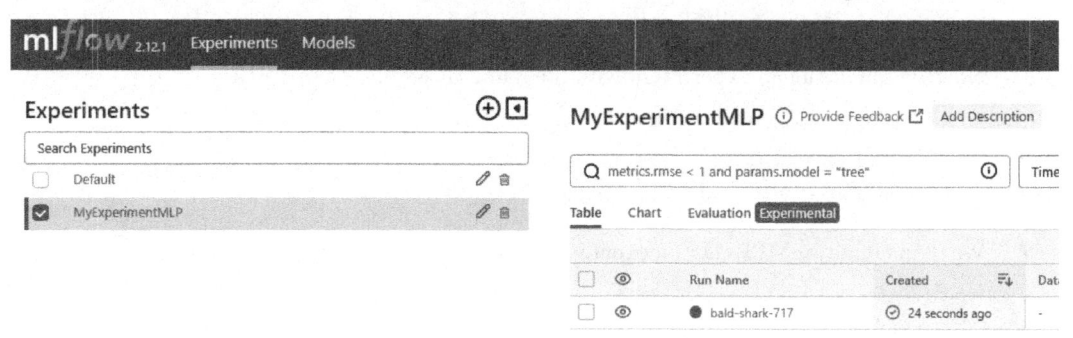

Figure 16.4 – The Experiments tab of the MLflow UI

If we select the experiment run, we can review the model metrics and artifacts. One of the artifacts is a figure of the loss curve. We can select this figure and show it in the MLflow UI, as shown in the following figure:

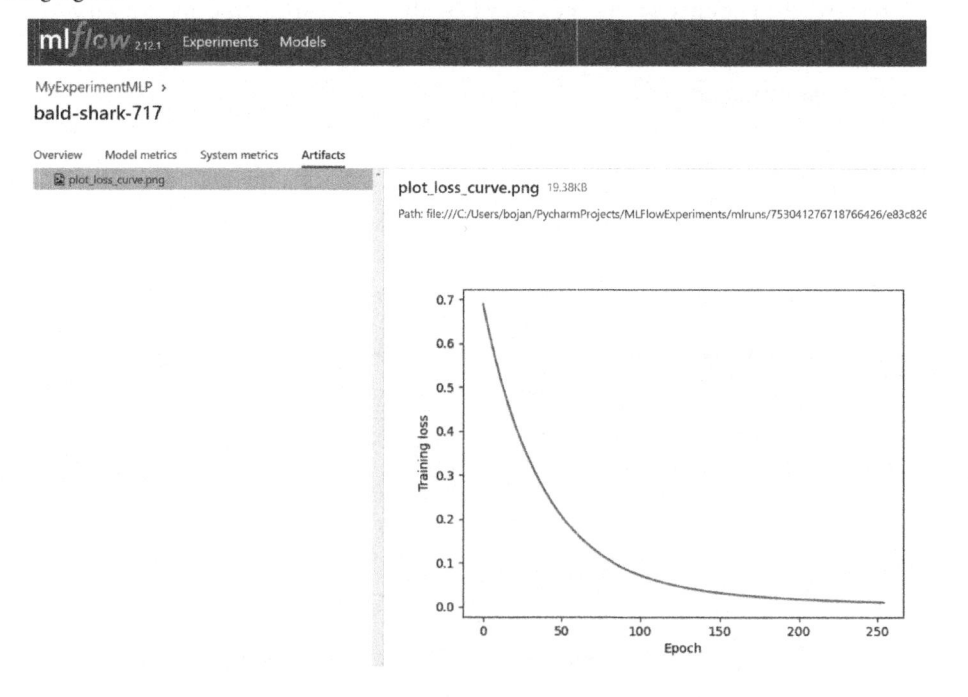

Figure 16.5 – Display of the figure from the MLflow experiment run

In the next section, we'll describe how humans cannot just monitor models and observe their behavior, but also be in the modeling loop.

Human in the loop

In classical ML approaches, we assume that the data is gathered, and we use well-known paradigms of supervised or unsupervised learning to train our models. However, often, there is a need for humans to be more involved in the process to guide it. One example of this process is the use of *active learning*.

Active learning

Active learning is an ML paradigm where we don't just use the dataset as it is, but also use the knowledge of a human to provide labels to selected examples. The human actor in this system is often called an *oracle*. An oracle provides labels to the data examples that we select based on predefined criteria and thereby, we enable efficiency in data labeling, optimizing the performance of the model using minimal human feedback. The following figure shows a simple schema of an active learning system:

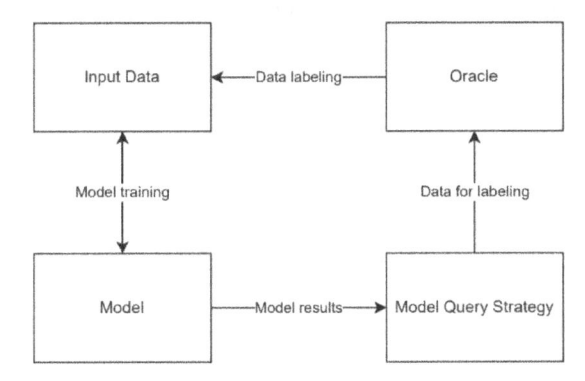

Figure 16.6 – Active learning actors, and steps of data transfer and processing

We often need the time of competent humans to label the datasets. For instance, we would need medical experts to label medical images, and cybersecurity experts to label malware data. Using active learning, we can reduce the costs of labeling data and improve the model performance.

There are three main scenarios for how the data can be queried:

- **Membership query synthesis**: In this scenario, the learner model will generate a query to request labels for any data point in the input space, including newly generated examples.
- **Stream-based selective sampling**: This scenario is based on the idea that unlabeled examples already exist in the dataset, so there is no need to generate them. Here, the unlabeled examples are considered one by one, and the learner will choose whether to use a particular example for the query or to skip it.

- **Pool-based selective sampling**: Here, we sample from a pool of unlabeled data based on a predefined criterion, such as informativeness.

For the implementation of active learning, we need a proper strategy to select which examples to label. Some of the most common strategies for selection are as follows:

- **Uncertainty sampling**: In these query strategies, the data point for which the model is least certain in determining the output (e.g., label) is chosen for labeling by the oracle.
- **Query by committee**: Here, we maintain a set of models (committee) that are trained on the same labeled data, but with different hypotheses – for instance, by maintaining a distribution over model parameters and sampling from it. Each committee member has a vote on labeling the candidate's unlabeled points. The point on which there is the least agreement on labeling is chosen as the best candidate.
- **Expected model change**: In this strategy, we select the data point that would cause the largest change in the current model if we knew its label.
- **Variance reduction and Fisher information ratio**: Here, we use the strategy to create a query that minimizes the model variance, thereby minimizing the future error rate.
- **Estimated error reduction**: These strategies attempt to estimate the expected future error if an instance is labeled and to reduce the generalization error directly.
- **Density-weighted methods**: Here, we consider that we shouldn't choose only the data points that create uncertainty (informativeness criterion), but also those that are representative of the underlying data distribution. Therefore, in this strategy, both terms are considered when selecting the samples for labeling.

In this section, we have shown that sometimes, the modeling is better if we include humans in the loop. This means that we need humans to be active in the modeling process to provide the ground truth knowledge and guide the training.

Summary

In this chapter, we described techniques for the evaluation of AI models. This taught you to understand how to measure the performance of your model and how to know when the model is good enough to use in your application scenario. We introduced the concept and methods of model monitoring, both during training and in production. Furthermore, we showed an example of model experiment tracking and monitoring using MLflow. These methods and tools are a basis that can be used in your application scenarios to create a performant, stable, and resilient ML system. Lastly, we described approaches of AI with humans in the loop and introduced the idea of *active learning*. Moreover, we described different scenarios and strategies applied in the active learning paradigm. These strategies are often needed in cybersecurity scenarios because of the scarcity of prelabeled data.

In the next chapter, we'll take the ideas of model monitoring and workflows and dive into the scenarios where the environment in which the AI model works is changing, introducing the methods that enable maintaining the performance of the model in this dynamic environment.

References

- Settles, Burr. "Active learning literature survey.", Technical Report, University of Wisconsin Madison, Department of Computer Sciences (2009).
- Huyen, Chip. Designing machine learning systems. " O'Reilly Media, Inc.", 2022.
- Zheng, Alice. Evaluating machine learning models: a beginner's guide to key concepts and pitfalls. O'Reilly Media, 2015.
- Géron, Aurélien. Hands-on machine learning with Scikit-Learn, Keras, and TensorFlow. " O'Reilly Media, Inc.", 2022.
- `https://github.com/mlflow/mlflow/`
- Zaharia, Matei, Andrew Chen, Aaron Davidson, Ali Ghodsi, Sue Ann Hong, Andy Konwinski, Siddharth Murching et al. "Accelerating the machine learning lifecycle with MLflow." IEEE Data Eng. Bull. 41, no. 4 (2018): 39-45.

17
Learning in a Changing and Adversarial Environment

We will discuss and explore the topics of **adversarial machine learning** (**AML**) in the spirit of completing the big picture of AI/**machine learning** (**ML**) by considering its security and privacy perspectives, which become more and more challenging and critical nowadays with the popularity of **generative AI** (**GenAI**). In previous chapters, we learned about the fundamentals of ML and a variety of applications in cybersecurity. However, the nature of ML brings native vulnerabilities that can be exploited by highly technical adversaries. Therefore, AML is a rising research area that draws a lot of attention from both academics and industries. Its objective is to study the vulnerabilities of ML proactively and to design and devise novel approaches that are more robust and reliable than traditional ML methods.

In this chapter, you will learn about the concepts and motivation of AML, and explore the differences between AML and traditional ML algorithms. It is important for you to know how adversaries attempt malicious activities and achieve their goals. Then we will introduce the taxonomy of AML and briefly present the state-of-art framework to describe the AML system. Furthermore, you will familiarize yourself with some offensive and defensive techniques in ML. To address recent advances in **large language models** (**LLMs**), we will discuss the particular aspects of adversarial prompt attacks on language models. In the spirit of practice, we will introduce some Python libraries that are used to exploit and evaluate ML models. Although it might seem to be an overload, we do encourage you to practice and learn how to use AML techniques to evaluate your models. This is essential to address some of the most concerning security and privacy issues in the advancement of AI.

In summary, here is the list of the topics that are covered in this chapter:

- Learning about the concept and motivation of AML
- Introduction to the taxonomy and state-of-the-art framework of AML
- Learning about some common adversarial attack algorithms
- Learning about some common defensive ML algorithms

What you will learn from this chapter is application-agnostic. However, it will be beneficial to use the knowledge gained from this chapter to enhance the security of your targeted ML applications.

Technical requirements

For this chapter, the technical prerequisites are designed to equip you with the foundational knowledge required to navigate the complexities of this emerging field effectively. They are as follows:

- A solid understanding of the core mechanics of ML is essential, encompassing the processes of training, testing, and evaluating models to ensure that they perform accurately and efficiently. This includes grasping the nuances of how data is split, how models are trained on one subset and tested on another, and how their performance is assessed through various metrics.

- A deep comprehension of optimization in ML is crucial. This involves understanding the techniques and algorithms used to improve model accuracy and efficiency, such as gradient descent and its variants, which play a pivotal role in tuning models to achieve the best possible outcomes.

- A basic familiarity with cybersecurity principles is necessary to understand the threats and vulnerabilities that AML seeks to exploit or defend against in modern digital systems. This knowledge is key to appreciating the context in which adversarial techniques are applied, as well as to understanding their implications for security.

- Lastly, while not mandatory, programming skills in Python are highly beneficial for those looking to engage more practically with AML. With its extensive libraries and frameworks for ML, such as TensorFlow, PyTorch, and scikit-learn, Python is a preferred tool for developing and testing adversarial strategies.

This skill set enables a hands-on approach to experimenting with and implementing adversarial models, providing a more in-depth understanding of the concepts discussed in the chapter.

Introduction to AML

In the realm of ML, a unique subfield known as AML has emerged. This area focuses on learning from datasets that may be contaminated by adversarial samples. These samples are introduced by entities intent on undermining the integrity of ML processes for their own benefit. The unpredictability of predictions made by ML algorithms introduces a new layer of security risks in today's data-centric systems. As data increasingly drives predictive services such as spam filters and voice assistants, the efficacy of a learning model becomes closely tied to the integrity of its data sources. Unfortunately, data source tampering is a common occurrence, whether through insider fraud, deliberate manipulation, or the natural degradation of devices, allowing adversarial entities to exploit these systems by altering their data inputs.

Nowadays, ML has become a crucial component in numerous IT systems. Despite its significant benefits, these systems can have inherent weaknesses, which malicious users might exploit for their gain. Recent advancements in adversarial learning have highlighted vulnerabilities in learning algorithms that were previously thought secure. These vulnerabilities can lead to the failure of their intended functions. For example, **Deep Neural Networks (DNNs)**, known for their robustness in tasks such as image and voice recognition, can be surprisingly fallible. They may either miss critical nuances in data or misinterpret blatantly incorrect data. For instance, in one experiment, two original images—a school bus and a puppy—were correctly identified with high confidence. However, after slight, humanly imperceptible alterations, these images were misclassified with 0% accuracy (Szegedy et al., 2014). Conversely, heavily distorted images, unrecognizable to humans, were identified with 99% confidence as their original subjects (Nguyen et al., 2015). This demonstrates how deep networks can be misled by focusing on certain pixel-level features.

Another example focuses on spam filtering, which increasingly relies on ML algorithms such as naive Bayes classifiers. Adversaries often disguise spam as legitimate emails to bypass detection. They employ sophisticated obfuscation techniques to evade statistical spam filters, manipulating content to avoid classification as spam. In the following text, injecting a few "good" English words (in green boxes) makes the spam filter think that the emails are legitimate.

For instance, consider email text such as the following:

"Check this amazing discount on Apple products, everything is half-price until Monday, don't hesitate!"

The attacker can take the preceding example and add text that could appear in a benign email to reduce the spam score of the message:

"Nice to see you yesterday, hope you had a good time! Check this amazing discount on Apple products, everything is half-price until Monday. I hope we can meet up soon and catch up. Don't hesitate to take the offer!"

A recent and concerning development in adversarial attacks targets LLMs. These models, trained on vast datasets to understand and generate human-like text, are not immune to manipulation. For example, by carefully crafting input prompts, an adversary can subtly guide an LLM to generate biased or factually incorrect outputs, potentially spreading misinformation or bias. This vulnerability is particularly alarming given the widespread use of LLMs in applications ranging from customer service to content creation. Large models that are deployed in production are aligned to generate human desirable outputs using alignment techniques such as **reinforcement learning with human feedback (RLHF)**. However, numerous prompt examples have shown that they can still be susceptible to adversarial prompts. namely to be \"helpful and harmless.\" These models should respond helpfully to user questions, but refuse to answer requests that could cause harm. However, adversarial users can construct inputs which circumvent attempts at alignment. In this work, we study to what extent these models remain aligned, even when interacting with an adversarial user who constructs worst-case inputs (adversarial examplesTherefore, the importance of understanding and developing countermeasures in AML cannot be overstated. As ML becomes more integrated into critical systems—spanning from security to communication—it is imperative to anticipate and mitigate these vulnerabilities. This not only ensures the reliability and safety of these systems but also protects against the exploitation of these technologies for harmful purposes. The ongoing study of, and advancement in, AML is crucial to building robust, trustworthy AI systems that are capable of withstanding the evolving landscape of cyber threats.

In the next section, we will walk through the foundation of AML by introducing the taxonomy and theoretic framework of the adversarial learning threat model in detail.

The realistic learning environment

In previous chapters of this book, you learned about the general concepts and methodology of ML. While the information you saw gives enough technical background for approaching any learning problem, it does not suffice to tackle the security issues of ML models. We will therefore start with a discussion about the realistic setup of AML, which focuses on model security. In the interest of brevity, we will start by discussing an arms race problem in the context of cybersecurity.

Arms race problem

In the area of cybersecurity, a scenario reminiscent of an arms race frequently unfolds. This involves strategic competition among various entities, each vying for dominance over specific assets to fulfill their unique objectives. Classic examples of this dynamic include the ongoing struggle between malware developers and antivirus software engineers, as well as the contest between spammers and anti-spam technologies. This paradigm is also reflected in the field of AML, which characteristically involves a minimum of two participants: the learner, who develops the ML model, and the adversary, who seeks to undermine or simply exploit it. This interaction often triggers a *reactive* arms race, predominantly instigated by sophisticated and adaptable adversaries. In this scenario, *reactive* denotes the learners' initial unawareness, adhering to static assumptions until the adversary capitalizes on certain weaknesses in their algorithmic design. It is imperative for learners to promptly detect and address these adversarial maneuvers, typically through methods such as retraining the model with untainted data or modifying particularly susceptible features to prevent further attacks. However, even the swiftest and most effective counteractions are hindered by inherent delays, affording adversaries a crucial timeframe to achieve their objectives.

In a contrasting approach, which is prevalent in areas such as cryptography, *security by obscurity* is a common practice. This method involves safeguarding information within a cryptographic veil, effectively barring adversaries from gaining access. While this remains a valuable tactic in adversarial learning, our exploration diverges toward a proactive strategy in securing learning systems during their initial design phase. This approach, which aligns with the *security by design* philosophy, necessitates that learners account for potential deviations from initial assumptions and pre-emptively contemplate possible attacks. This proactive security method offers enhanced safeguards by addressing potential weaknesses ahead of algorithm development, subsequently imposing greater challenges for adversaries attempting to breach these systems and bolstering their robustness with minimal reliance on human intervention.

The illustration of an arms race in *Figure 17.1* contrasts the reactive and proactive arms race models, which are integral to the understanding of adversarial learning. The reactive model, depicted on the left, progresses through four stages: adversaries identify vulnerabilities in a learning system for specific motives and formulate and execute an attack, which is then detected by the learner, prompting the latter to modify their system accordingly. In contrast, the proactive model, shown on the right, involves learners assuming the presence of adversaries and modeling potential attacks in advance, thereby assessing their potential impact and fortifying the system's security at the design stage.

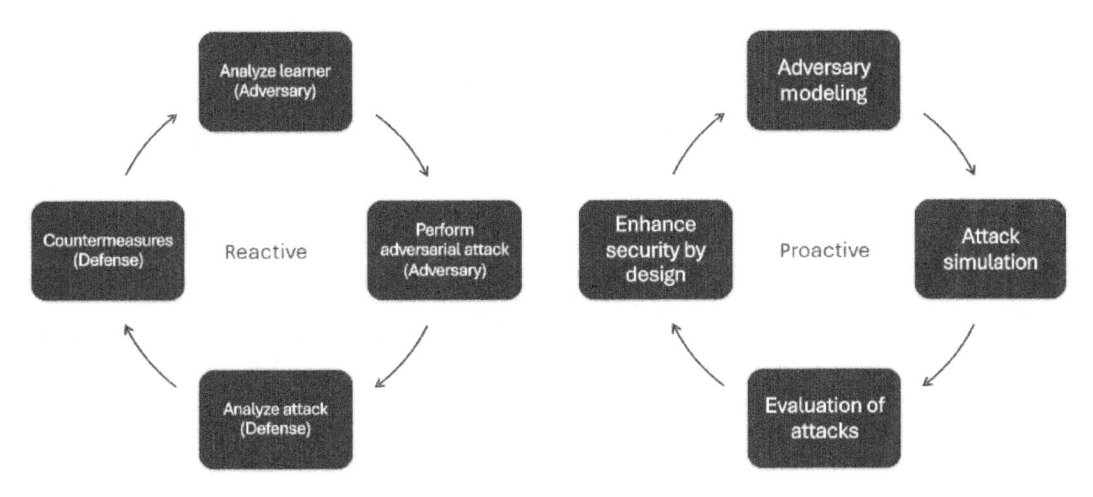

Figure 17.1 – The reactive and proactive arms race concepts

This arms race in the adversarial learning landscape is inherently iterative and dynamic. The conceptual diagram captures the complete cycle of adversarial learning, indicating that interactions between the learner and adversary can manifest as either one-shot games or ongoing, iterative engagements. Although the importance of equilibrium in this game-theoretic framework is undeniable, our current focus is on dissecting the one-shot game as a foundational approach to achieving balance. Prior research, notably by Brückner and Scheffer (2009) delved into the properties of Nash equilibrium in prediction models, demonstrating that under certain conditions, a distinctive Nash equilibrium emerges between the adversary and the learner. However, the exploration of iterative games and the equilibrium dynamics across diverse adversarial contexts remain fertile ground for future research. In this section, we aim to dissect and delve into the following key areas:

- **In-depth analysis of the targeted learner process**: This involves a meticulous examination and analysis of the learner process, such as a classifier or generator, to model potential adversaries. This process emphasizes the in-depth understanding and theoretical properties of the learning algorithm.
- **Comprehensive adversary modeling**: This stage entails the conceptualization of potential attacks specifically designed to target a particular learner process. The focus is on assessing the potential impact and efficacy of these attacks.

- **Strategic design of countermeasures**: This involves the integration of strategies and countermeasures against adversarial attacks at the very inception of the design phase. It ensures that learning algorithms are not only robust and resilient but also designed to adapt to dynamic data distributions and shifts in concepts, thereby enhancing their effectiveness and reliability in uncertain and evolving data environments.

In the next section, we will see what a typical learning process will look like in terms of data flow. This is essential for you to understand in order to proceed to the topic of adversarial modeling.

Learning process with data flow

Data plays a central role in any ML algorithm. Both training data and test data can be susceptible to changes or manipulations, which are sometimes malignant. As we will learn in the rest of this chapter, most adversarial attacks aim at data manipulation such that they can achieve their goals. Therefore, we will first depict the learning process in terms of data flow. As we will see, the data flow is model-agnostic, which means that it is compatible with almost all state-of-the-art learning models.

The learning process always starts from a data generation process, which is typically unknown. For instance, images downloaded online can be created by consumer cameras, mobile phones, CCTVs, and so on, which can be further processed or transformed and ultimately uploaded to the web. Although the general process of constructing a digital image within a photographic device is not mysterious, it is not possible to fully reassemble the generation of all images. Another example of the data generation process is in natural language. A sentence is spoken, triggering billions of biological reactions of neurons in a human brain. In humans, this process is unknown or only partially interpretable. In *Chapter 4*, we flagged the difference between statistics and ML, that is, ML has a strong focus on learning the data generation process while statistics focus more on studying the distributional properties of generated data. This is especially true in the frontier of GenAI, as a generative model learns to generate data to conform with the observations that were originally generated.

In practice, only observations are available, such as text or images from the web, which are collected as training data. Based on the learning problem definition, a subset of learning models is selected and trained with the observations. The trained models are normally much smaller than the training dataset. For a simple approach to understanding this, we can consider the models as some sort of lossy compression algorithms. However, one characteristic of ML models is that the training data and model are interwoven. It is not possible to modularize the data and model once they are trained. Therefore, the so-called **Change-Anything-Change-Everything** (**CACE**) pattern causes the major vulnerability of any ML model.

On the other hand, once the model is trained and deployed in production, test data samples will be fed to the model, and model outputs will be generated without changing the model itself. For simplicity, any ML model can be considered as a function that takes compatible inputs and generates outputs that meet certain criteria. Depending on the functionality of the model, both input and output can take multiple data types (multi-modal), for example, text and image. Multi-modal has gained more attention nowadays and has become a new medium of marching toward more generic

AI capability. However, this incentivizes adversaries to target the models for multi-factual assets, such as **personally identifiable information** (**PII**), confidential business reports, and so on. Have a look at the following figure:

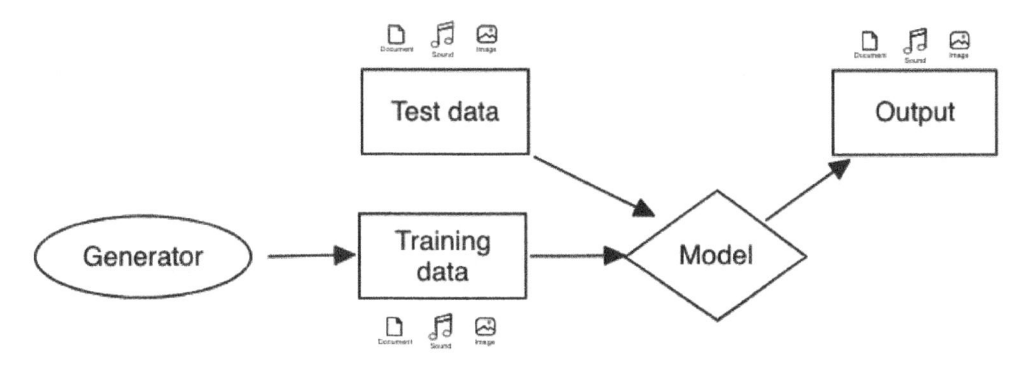

Figure 17.2 – Data flow for any learning process

In *Figure 16.2*, we have depicted the data flow that represents the learning process. Note that the model itself is also of interest to adversaries. Therefore, every component in this diagram is an asset that needs proper security guardrails.

Next, we will start to play the role of an adversary, which will help us understand various learning scenarios under adversarial settings.

Adversarial threat modeling

As described in the context of the arms race problem, it is usually a best practice to conduct threat analysis in a proactive way, such that security practitioners play the role of an adversary and perform a comprehensive analysis of the learning models. Therefore, we will show the threat modeling in the context of AML, as well as the taxonomy of various sorts of adversarial attacks on ML models, in this section.

Attacker model

In the context of adversarial learning, it's important to consider an *adversary* as a smart and adaptable entity that interacts with the learning system. This adversary has the ability to make changes to the data that the system learns from. These changes can be made either during the system's training phase (when it's learning and developing its abilities) or during the testing phase (when its performance is being evaluated). Such alterations by the adversary can shift the way the original data is structured or behaves, which is problematic because the learning system usually expects the data to remain consistent or *stationary*. Imagine teaching someone to recognize apples, but then halfway through, you start showing them oranges without telling them—their learning would be affected.

By strategically changing parts of the data, either in the training or testing stages, the adversary aims to confuse the learning system. This confusion benefits the adversary, leading them to achieve their goal, which is usually to misguide the system to a certain degree. In order to effectively prepare for and counteract these challenges in adversarial learning, experts create models that anticipate the adversary's actions. These models are designed with the understanding that the adversary is trying to reach their goal in the most effective way possible. This approach helps in building more robust and secure learning systems that can better resist these intentional disruptions.

> **Definition of ML adversary**
>
> An ML adversary is defined as an actor that is parameterized by three major components. Let us denote it as , where is an optimal subset of adversarial data samples causing the targeted model to behave in a way that is desired by the attacker. represents the attacker's knowledge about the model .

We will learn how the adversary achieves their goal by modeling various kinds of attack, though it is important to understand each component of the definition.

Adversarial data samples could be the only way that an attacker interferes with the system (model) most times, as the model usually only provides interfaces that receive users' inputs (data) without exposing the details about the algorithm and training data. Therefore, an adversary attempts to create a set of adversarial samples, usually by another ML model, which causes the system to behave differently than it normally would. For instance, in the example of a spam filter, an adversary might change the content of an email by injecting "good" words, which results in a spam email bypassing spam detection. In this case, the particularly designed email is an adversarial example. It is straightforward to understand this scenario, as the attacker confuses the spam classifier. In the most recent scenarios (for instance, ChatGPT), adversaries could prompt the chatbot to generate toxic or unethical content, which is normally disallowed by the model through the enforcement of security policies. This is called an **adversarial prompt** in the context of GenAI. It sounds a bit more intriguing than it is. However, it is technically the same as the spam detection example. The only difference is in the model output. One is to predict a label on the email, another is to generate text based on instructions.

Targeted model more generally refers to the system that is mainly driven by some AI models. We need to stress the fact that the major difference between traditional software systems and these systems is that AI systems are normally probabilistic, while traditional software systems are deterministic. In AML, the targeted model is the attacker's hunting object. Depending on what the goal is, the model can be compromised in a different way. We will explore the attack taxonomy in depth in the next section.

The **attacker's knowledge** represents how much information the adversary knows about the targeted model, which can include what the model and the parameters are, as well as potential access to the training data. In the worst case, the adversary has full visibility of the targeted model and the data. Thus, the system can be extremely vulnerable. However, most of the time, we would expect that the adversary only has limited or even zero knowledge about the target. Like any other system, the more information that is exposed externally, the higher the risk will be. From a defensive aspect, keeping the model and data information confidential could be a key to success. However, the recent trend is leaning more toward pre-trained models. This is because training a large model can be extremely expensive, and therefore, it is usually not financially feasible for most small and medium organizations. This suggests an increasing risk of models being more targeted, as there are tons of applications being built on top of pre-trained models.

To continue our discussion, we recommend that you recap some fundamental ML algorithms from *Chapter 5*, but think of the modeling process from an adversary's perspective. In the next section, we will give a taxonomy of various types of adversarial attacks. Again, this is in the interest of being proactive against those attacks. We do not encourage any attempt to apply any of these attacks on real-world production systems.

Adversarial attack taxonomy

In the study of adversarial learning, a key concept is the classification, or *taxonomy*, of different types of adversarial attacks. This taxonomy helps us understand these attacks by categorizing them based on their characteristics and the specific ways they impact the learning system. The purpose of creating this taxonomy is twofold. First, it provides a clearer understanding of the strategies used by attackers. By categorizing the attacks, we can see patterns and commonalities in how these disruptions are carried out. Second, this organization aids in the development of effective strategies to counter the attacks. By knowing the type and nature of an attack, we can create more targeted and effective defenses against it. In essence, this approach to categorizing adversarial attacks is like creating a detailed guidebook. It helps those working in the field of adversarial learning to quickly identify the type of attack they're dealing with, understand its potential impact, and apply the most appropriate countermeasures to protect the learning system. This structured understanding is crucial in enhancing the security and effectiveness of these systems. We will build the taxonomy based on the attacker model introduced in the previous section. We will also define a simple categorization in three dimensions, namely security violation, adversarial capability, and adversarial influence.

Security violation

First, from a defender's perspective, understanding the attacker's objectives is crucial, especially as these often relate to compromising the security of targeted systems. For example, consider an email spam filter. In this scenario, attackers might create a series of spam emails that are designed to bypass detection, thereby undermining the integrity of the email system. When these types of malicious attacks successfully deceive the learning systems (the *learners*), they result in security breaches. To effectively address and counter these threats, it's essential to first define what the attacker is aiming to achieve in terms of security violations. These violations can fall into various categories, such as breaches of integrity (such as in the spam email example), availability (where the system's ability to operate is impacted), or privacy (where confidential information is accessed). Furthermore, we can categorize these attacks based on how they impact the targeted system. Some attacks are targeted, meaning that they are designed to compromise a specific part or function of the system. Others are *indiscriminate*, where the aim is to disrupt the system as a whole, without a specific focus. Sometimes, whether the attack is targeted or indiscriminate can be referred to as **attacker specificity**. This categorization helps in understanding the nature of the threat and in developing more precise defensive strategies.

Integrity violation occurs when an attacker manages to carry out harmful activities without disrupting the normal operations of the system. A classic example is in a situation involving a classifier, such as a spam filter. An attacker could design spam emails that are not detected as spam by the classifier, yet this might not affect the classifier's ability to correctly identify legitimate emails. In such cases, the integrity of the classifier is compromised because it allows malicious content to pass undetected, but its function on legitimate content remains unaffected.

Availability violation attacks impact the system's functionality, leading to a denial of service. In more technical terms, they damage the reliability of the predictive service offered by the system. In the context of ML, where predictions inherently involve some uncertainty, the reliability of these predictions is crucial. For classification and clustering systems, such attacks could lead to significant errors in classification or major alterations in the clustering process. The overall effect of these attacks is a degradation in the predictive services offered by the system, affecting its availability.

A **privacy violation** is a type of attack that is concerned with extracting information about the system or its users. An attacker might reverse-engineer the system to gain insights into the data or the workings of the classifier. To mitigate this risk, it's important to limit the information available to potential attackers. This includes data used in the learning process, as well as user input and even the system's responses to legitimate queries. All these elements should be treated as private and protected accordingly.

Lastly, the specificity of an attack can be categorized as either *targeted* or *indiscriminate*. A targeted attack focuses on a specific subset of information, whereas an indiscriminate attack affects a broader range of data or functionality. Understanding these distinctions is vital for developing effective countermeasures and safeguarding the system against various adversarial threats. Note that the categorization of attacks based on their security violation is aligned with the traditional cybersecurity principles of **Confidentiality, Integrity, and Availability (CIA)**. However, we only emphasize on model-related threats, rather than a classic threat scenario such as a DDoS attack, which might be a rising risk associated with LLMs, as a flood of requests may incur a significant jump in computational cost.

Adversarial capability

Knowing an adversary's objective, as described in the preceding section, is essential. From a cybersecurity perspective, it is usually impossible to reduce the risk to zero, which is why we have to resort to increasing the cost for an attack to be successful. When the cost outweighs the gain, the adversary may give up the act. Thus, a second dimension of the taxonomy is to analyze the adversary's capability, as this is a limiting factor on how much it costs to conduct an attack successfully. Now, we can categorize the adversarial attack based on the level of knowledge the attacker has about the system they are targeting.

This knowledge can vary greatly and is typically divided into three key areas:

- **Knowledge of the training data**: The attacker might have access to all, some, or none of the training data used by the system. If they don't have direct access, they might try to use a substitute dataset, which is as similar as possible to the original training data. An example of this is using **honeypots** to gather malware samples for studying how a malware detection system was trained. Nowadays, LLMs are typically trained on a huge number of web-scale text datasets, which are commonly available to everyone. It is therefore quite difficult to keep the training data confidential, as a substitute dataset is almost "free."
- **Knowledge of the feature set**: This refers to how much the attacker knows about the features (characteristics) that the system uses to make decisions or classifications. As ML becomes more standardized, it's becoming easier for attackers to understand and predict these features. For example, in text processing, a common method to convert documents into a format that is understandable by machines is the *tf-idf* statistical measure.
- **Knowledge of the learning algorithm**: The attacker may also know which learning algorithm the system uses, and what criteria it uses for making decisions. Some algorithms are more common in certain applications, such as deep learning networks for image recognition or Naive Bayes for document classification. For tabular datasets, XGBoost or other tree-like algorithms are almost always the first choice. Most recently, pre-trained models such as GPT-2, LLaMA, and other open source large models are serving as a fundamental building block of many AI systems. It is not difficult to guess which model family is being used in an application. Open source models have accelerated AI development vastly, inadvertently opening the gate for those bad actors as well.

Now, with the attacker's level of knowledge, adversarial attacks can be further classified into three types:

- **Perfect knowledge (PK) attacks**: These are also called white-box attacks, which are defined as being straightforward. In this worst-case scenario, the attacker has complete knowledge of the system (model, features, and data). This situation allows researchers to understand the maximum possible impact an attack could have on a system, and usually, this gives a lower bound on a model's robustness. It represents how well the model can still perform under white-box attacks.
- **Limited knowledge (LK) attacks**: These attacks are more common in real-world scenarios. Here, the attacker only has partial knowledge about the training data, the feature set, or the learning algorithm. They use this limited information to mimic the system's behavior and create attack samples to test against the targeted system. The objective for the attacker is to gather as much information as possible, so they can simulate the attacks to generate adversarial samples. Note that adversarial samples show a feature of transferability. This means that the adversary doesn't necessarily gain everything to be successful. It is sometimes enough for them to just conduct attacks on a surrogate model or dataset.

- **Zero-knowledge (ZK) attacks**: These are also called **black-box attacks**. In contrast to PK attacks, the attacker has no information about the system at all with ZK attacks. In this case, attackers will have to conduct attacks using heuristic approaches, or by using a surrogate (substitute) model and data as in LK attacks, then transferring the attacks to the real system. The next figure shows a depiction of steps in a black-box attack with a substitute model:

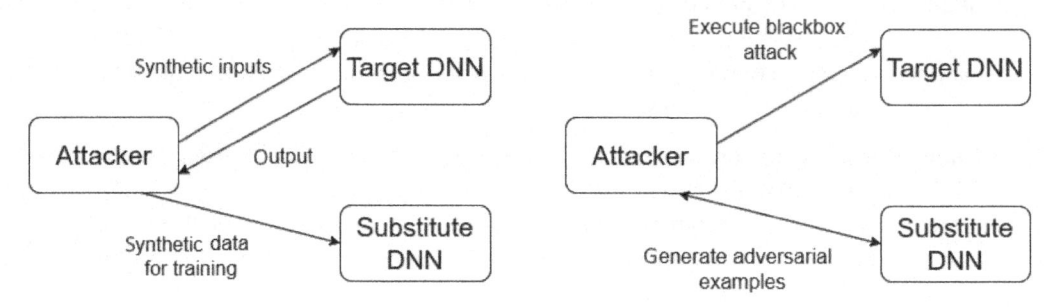

Figure 17.3 – Steps of a black-box attack with a substitute model: training of a substitute model (left) and attack execution (right)

Nowadays, AI applications are typically offered as web services or APIs (for example, ChatGPT). Users do not have knowledge about their internal mechanisms (black-box), but attacks are still possible, such as adversarial prompting by heuristics. This type of attack may increase in terms of volume and complexity. Similar to white-box attacks, black-box attacks can give an upper bound on the model's robustness. If an adversary can compromise the model with black-box attacks to a certain degree, the model will be even worse under PK and LK attacks. Have a look at the following figure:

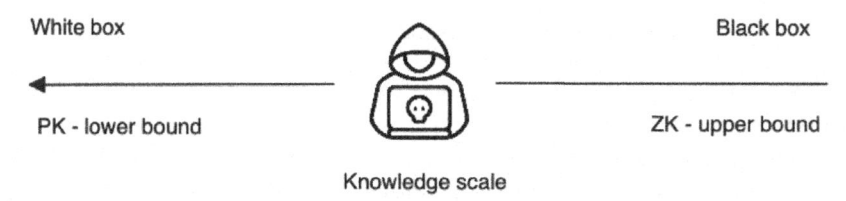

Figure 17.4 – Attacker's knowledge scale on target model and the associated robustness bounds

For practitioners, there is always debate about how realistic PK and LK attacks are. However, it's generally safer to overestimate rather than underestimate an attacker's capabilities. This approach ensures that security measures are robust enough to protect against a wide range of potential threats.

Adversarial influence

Adversarial influence refers to how and to what extent an attacker can affect the learning process. This influence can manifest in different ways, where the attacker can impact the training phase, the testing phase, or both. Typically, we can classify the adversarial influence into two main categories depending on whether the attacks cause a model change or not. They are **causative** and **exploratory** attacks, though many call them **poisoning** and **evasion** attacks. We will use the latter to stay aligned with others:

- **Poisoning attacks**: These attacks occur during the training phase. The attacker manages to insert a certain percentage of carefully designed, malicious data into the training set. This manipulation can involve changing both the features (the characteristics of the data) and the labels (the categories or classifications assigned to the data). The attacker's goal is to subtly alter the training data in a way that causes the system to learn incorrectly, leading to a less effective predictive function. An example is inserting malware into a dataset that is wrongly labeled as secure by antivirus software, thereby poisoning the training data and misleading the classifier.

- **Evasion attacks**: These occur during the test phase, where the attacker modifies malicious data to avoid detection by the system. The challenge for the attacker is to alter the data in such a way that it still performs its malicious function, but is no longer recognized by the classifier. For example, malware code might be obfuscated such that it evades detection without losing its harmful capabilities. Attackers are essentially probing the boundaries of the classifier's detection abilities in this scenario, attempting to create data that is ambiguous enough to be misclassified and thus evade detection.

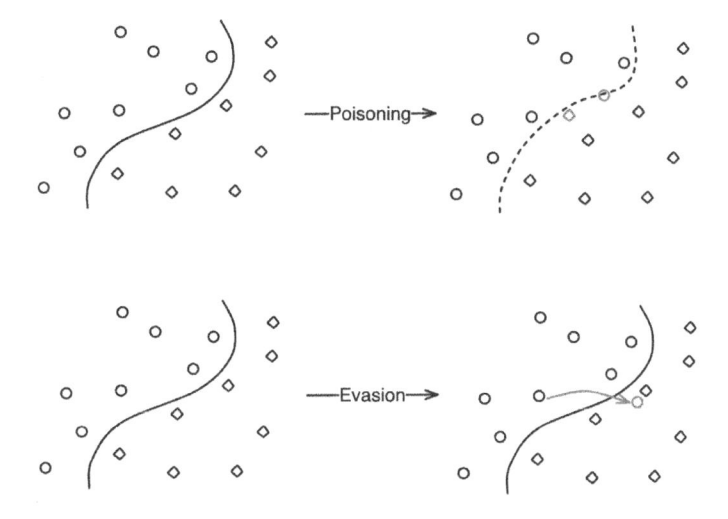

Figure 17.5 – Illustration of both poisoning and evasion attacks on a simple classifier

In the preceding figure, we illustrate the difference between these two types of attacks on a simple classifier. Note how the classification boundary changes in the poisoning scenario. Both types of attacks present significant challenges. Poisoning attacks compromise the learning process itself, leading to the system forming a flawed understanding. Evasion attacks, on the other hand, test the system's ability to correctly identify and respond to new, altered forms of known threats. Understanding these adversarial influences is crucial for developing robust defense mechanisms in ML systems.

Transferability of adversarial samples

You might wonder about the effective impact of adversarial attacks on real-world models, as the attacker seems to only have limited or zero knowledge of the target most of the time. In other words, the attacker seems to only have access to APIs (ML-as-a-Service). Black-box attacks may look less worrying until we realize the transferability of adversarial samples. Much research has explicitly shown that knowledge can be transferred from one model to another as long as the learning task stays the same. In practice, an attacker can design and train a substitute model as a proxy to the target model by querying input-output pairs from the real system. This shows that adversarial samples that are crafted against the substitute model are largely effective against the target model as well. Even if the substitute model is completely different from the target, the attacks can still be effective. It is argued that the adversarial property of ML models is loosely related to the model itself, but closely tied up with the feature space.

Have a look at the following figure:

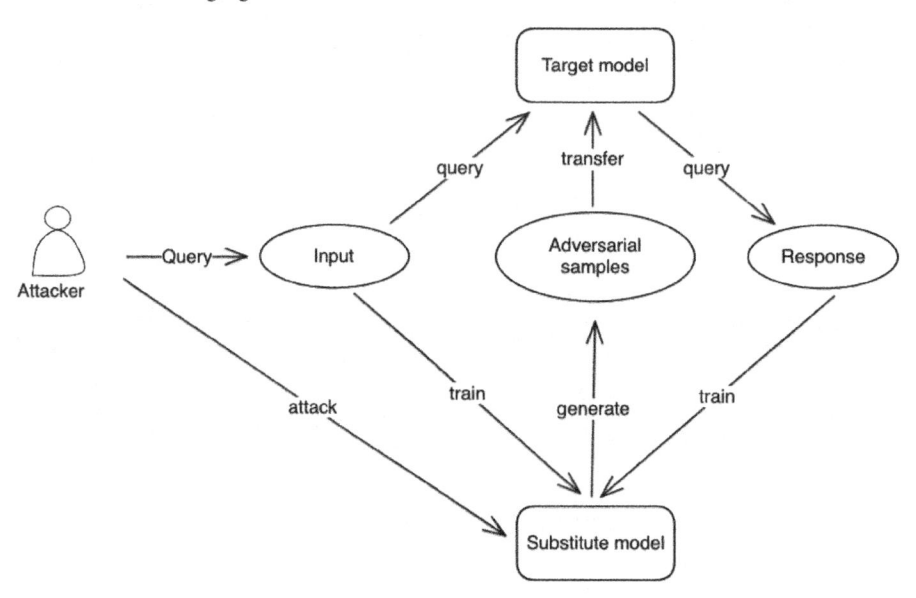

Figure 17.6 – An adversarial attack's transferability from black-box attacks on substitute models to target models

In the figure, we illustrate the process of a black-box attack by training a substitute model, where the generated adversarial samples can be transferred to the target system. The attacker first sends queries to the target system to collect a set of input-output pairs, which are then used as a training dataset for the substitute model. After the substitute model is trained, the attacker launches adversarial attacks on the substitute model and generates adversarial samples. These samples are then sent to the target system to achieve the attacker's goal due to the transferability of adversarial samples.

Attack taxonomy summary

Now we want to conclude this section by summarizing the adversarial attack taxonomy. The following table presents an overview of attack taxonomy. It is recommended that you get familiar with the definitions and how the attacks are categorized:

Influence	Specificity	Security violation		
		Integrity	Availability	Privacy
Poisoning	Targeted	Training data is compromised that changes model behavior to, for instance, misclassify samples as a particular class or generate a specific content.	Training data is compromised that changes the model and degrades the model's functionality on a particular task. For instance, the model might be quite bad at predicting spam.	The model is compromised and changed such that the model exposes confidential or private information in the training data.
	Indiscriminate	This is the same as for targeted. However, the model misbehaves in general rather than on a particular set of samples.	The model is changed, and performance is degraded in general, rather than just being poor at certain tasks.	This is the same as for targeted. However, the model does not have a preference for what confidential information is generated.

		Security violation		
Evasion	Targeted	The attack happens at test time and does not change the model itself. Instead, the attack targets a particular sample being misclassified or generated.	A targeted subset of samples can evade the model's classification or be generated unfaithfully. The model itself is unchanged, but it performs poorly on the selected samples.	Attacks cause the model to leak specific data or information that is confidential or private, while the model stays unchanged.
	Indiscriminate	This is the same as for targeted but the attacker does not have a preference for which sample should be misclassified. It is of interest to misclassify data or generate unfaithful content in general.	Attacks cause random samples to be misclassified or generated unfaithfully. It does not change the model itself. However, the model performs poorly in general.	This is the same as for targeted, but the attacks do not have a preference for which confidential data is leaked.

Table 17.1 – Adversarial attack taxonomy

In the next section, we will introduce different techniques to enhance model security, namely covering the defensive mechanisms that exist to increase a model's robustness.

Defensive mechanisms

In the previous sections, we have learned about a general framework for adversarial threat modeling and various types of attacks. Those attacks are naturally grouped toward the adversary's security objectives, namely the violation of confidentiality (privacy), integrity, and availability. This aligns well with the common strategy of the cybersecurity process, which moves across different stages, building and strengthening the system. Therefore, we will adopt a similar approach to design and develop defensive mechanisms as in traditional cybersecurity. However, we will restrict the scope of discussion to the realm of ML models. The systematic approach toward securing and hardening AI systems is beyond our discussion.

AI model security involves a dynamic and ongoing process aimed at safeguarding the integrity, confidentiality, and availability of AI systems. This process is not a one-time goal but a continuous journey that evolves and strengthens over time. AI model security encompasses a variety of strategies and actions, which can be broadly categorized into three main phases: *prevention*, *detection*, and *response*.

In the *prevention* phase, the focus is on implementing novel and robust AI models that can cope with the adversarial attacks described in previous sections, such that various types of attacks make little or neglectable impact on the targeted model. This is only feasible when we adopt a proactive approach (see the arms race problem) to explore both the lower and upper bound of model robustness and take actions to design more reliable models that take those attacks into account. Prevention typically happens at the model training phase, although it is not always effective in highly dynamic threat scenarios. In classic security terms, we also need to take measures to protect AI models from potential cyber threats and vulnerabilities. This could involve encrypting data, securing data pipelines, and ensuring robust access controls. The aim is to create a solid foundation that minimizes the risk of unauthorized access or manipulation of AI models.

It is imprudent if the defense strategy relies solely on prevention, as the threats are ever-changing. Therefore, we need a second set of guardrails with a detection capability. The detection phase is crucial for identifying any unusual or malicious activities that may indicate a security breach or attempt to compromise the AI model. This involves continuously monitoring the system, employing anomaly detection algorithms on adversarial samples or inputs, and setting up alerts for suspicious behavior. The sooner a potential threat is detected, the more effectively it can be addressed. Thus, we can imagine that detection-based countermeasures usually happen during the testing or prediction phase.

Upon the detection of adversarial attacks, alerts will be created with a collection of threat contexts such as the source of the attack, anomalous input samples, indicators of compromise, and so on. Alerts should trigger the next stage for incident response, wherein security analysts investigate the adversarial attacks and take appropriate actions to mitigate the impact of any security incidents that have been detected. This could include isolating affected systems, conducting thorough investigations to understand the nature and extent of the adversarial attacks, and implementing fixes to prevent similar attacks in the future. As we can observe, this feeds back to the prevention stage and closes a cyclic defensive pattern. We depict the defensive cycle as seen in the following figure:

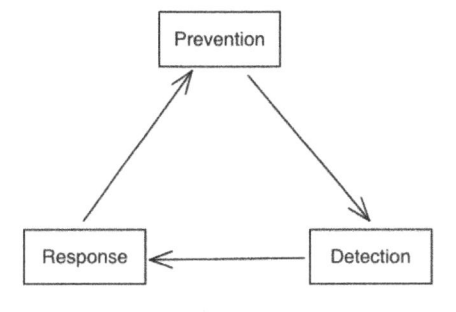

Figure 17.7 – Defensive cycle in AML

The landscape of threats against AI models is ever-changing, necessitating timely updates and adjustments to the strategies employed in all three phases. A proactive approach in the prevention phase can significantly influence the effectiveness of detection and response mechanisms. Similarly, insights gained during the response phase can inform improvements in prevention and detection strategies. AI model security is a cyclical and adaptive process, constantly evolving in response to new challenges and threats. For successful management of this process, it's crucial to design each phase with adequate capabilities and ensure diligent oversight to foster the continuous development and maturation of these capabilities. The overarching goal of AI model security is to ensure that AI systems remain reliable, trustworthy, and resilient against adversarial attacks or data manipulation, thereby protecting the vital attributes of confidentiality, integrity, and availability.

In the rest of this section, we will discuss some common state-of-the-art defense approaches in AML for each of the defensive stages. Although it is difficult to be comprehensive, we will mainly focus on the primary idea behind the methods rather than the algorithmic details. Note that this is an active research area, and the problem is yet to be solved. Therefore, the methodological landscape may change over time.

Defense as prevention

As in prevention, we aim to construct AI models that are robust against adversarial samples, which typically happen at the model training phase. Most countermeasures at this stage tackle the problem from both a data and a model perspective. In some literature, defense mechanisms are broadly categorized into data sanitation and model modification.

Data sanitation is straightforward to understand. Assuming the existence of adversarial samples in training data, that is, considering a data poisoning scenario, it is desirable to identify and remove the adversarial samples before or during model training. Although this is different from detection at the test stage, the general idea behind it is that the poisonous data samples represent themselves in a relatively low-density area in the data space. Therefore, they increase the training entropy, which can be detectable during the training phase. For instance, we could construct a second training set by selecting a candidate sample from the original dataset, and then observe how the model performance changes. Nevertheless, this is not ideal for scenarios wherein the adversarial samples are difficult to identify. We also expect the training cost to boost significantly when it comes to large models.

Model modification is likely the best choice for prevention, as sanitizing training data could be costly and inefficient against dynamic attacks. The key behind model modification is the same as the core principle of designing any learning model: to increase model generalization. Instead of increasing generalization on unseen (legitimate) test samples, we modify the models to be generalized on adversarial samples. In lots of research, we have seen significant improvement in terms of reduction of the attacker's success rate. One of the most popular methods is called **defensive distillation**, which is based on the concept of knowledge distillation introduced by Geoff Hinton. Knowledge distillation involves transferring insights from a larger, more complex model to a smaller, more compact model while maintaining comparable performance. The fundamental concept of this method is to utilize the class probability vectors generated by an initial DNN as the training foundation for a second, smaller DNN. The goal is to achieve this without compromising the accuracy of the model, effectively ensuring that the distilled model can perform as well as its larger counterpart with reduced complexity. This has been proven to be effective against many gradient-based attacks. However, it is not quite successful in defending against strong attacks such as a **Carlini & Wagner** (**CW**) attack. In essence, defensive distillation is a gradient masking or obfuscation method. Since most adversarial attacks rely on gradient information to perturb the inputs, gradient masking or obfuscation can confound the attackers. However, even these defensive methods can be overcome, as shown in the work of Athalye et al., by advanced techniques for estimating gradients. There are other types of model modification, such as applying a non-differentiable pre-processor on training data and then training the model based on the pre-processed data. Another example is that we enforce randomized dropouts of neurons at inference, which can also increase robustness. Some ML systems use multiple-classifier systems, that is, ensembles to increase the non-smoothness of the model. Defense-GAN is another example that attaches an adversarial generator to the target classifier, such that the classifier learns how to differentiate normal and adversarial samples. They all fall in the broader category of strong regularization, which can increase the model generalization eventually.

Defense as detection

Preventative measures act as the initial shield against adversarial attacks, offering a robust defense against known vulnerabilities and threats. They are the cornerstone of any security strategy, designed to thwart a wide array of attacks before they can inflict damage. However, the landscape of adversarial attacks is dynamic, with novel and sophisticated attack algorithms (e.g., **0 days**) emerging regularly. These attacks exploit previously unknown vulnerabilities of models, making them particularly challenging for preventative measures to address effectively. To complement preventative strategies, a secondary line of defense is essential. This involves the deployment of advanced detection systems capable of identifying adversarial attacks swiftly and accurately. Such detection mechanisms come into play after the preventative measures, during the testing or prediction phases of system operation. The aim here is to catch any attack that slips past the initial defenses, ensuring a second layer of security.

Numerous algorithms and techniques have been developed to enhance the detection of adversarial examples. One practical method involves training an auxiliary classifier alongside the primary model. This secondary classifier is specifically designed to distinguish between legitimate inputs and adversarial examples, effectively acting as a dedicated sentinel for identifying malicious inputs. Proactive defense strategies also play a crucial role. By simulating adversarial attacks, defenders can gather critical data on potential threats, using this information to train models to recognize and counteract such attacks more effectively. This pre-emptive approach to security helps in fortifying systems against emerging threats.

Additionally, various statistical methods are employed to scrutinize the underlying structure of data for signs of adversarial tampering. Techniques such as **Principal Component Analysis (PCA)** can be instrumental in this regard. PCA, for instance, analyzes the large principal components of data, which adversarial attacks often exploit, thereby helping to pinpoint malicious samples. Other statistical methods assess the distributional characteristics of data to detect deviations indicative of adversarial manipulation.

Randomization techniques offer another layer of defense by introducing uncertainty into data handling processes. By applying random modifications, such as dropout techniques, to input data, these methods can expose the heightened variance in responses that is characteristic of adversarial examples compared to regular data. This variance serves as a tell-tale sign of potential adversarial interference, enabling more effective detection and response to such threats.

Defense as a response

When an adversarial attack is identified, it's critical to implement measures to mitigate the threat. A prevalent strategy for addressing this challenge is to incorporate the adversarial examples—those manipulated inputs designed to confuse the model—back into the model's training cycle. This process, known as adversarial retraining, involves updating the model with these examples to improve its resilience against similar attacks in the future. Adversarial retraining is a proactive defense mechanism that can significantly enhance a model's robustness.

This approach bears similarities to other defense strategies, such as Defense-GAN or secondary classification methods, which also aim to fortify models against attacks. However, adversarial retraining is distinct in that it is typically applied after an attack has been detected, necessitating further actions to secure the system. For instance, it might be necessary to isolate or **quarantine** the source of the suspicious queries to prevent further harm. Additionally, creating indicators of compromise based on the detected anomalies can help in identifying and responding to future threats more efficiently.

Beyond immediate remediation, it's also essential to conduct a thorough investigation of the attack and share these insights with the team responsible for the model's development. This collaboration enables the model's developers to learn from the attack patterns and incorporate this knowledge into the creation of more secure and resilient models. By understanding the nature of the attacks and integrating these lessons into the model design and training process, the team can better prepare and protect against future adversarial challenges.

In conclusion, the exploration of adversarial threat modeling and various types of attacks has revealed the importance of a multi-layered defense strategy in protecting AI systems. This strategy mirrors the traditional cybersecurity approach, focusing on prevention, detection, and response, but it is tailored to the unique challenges of ML models. The goal is to safeguard the confidentiality, integrity, and availability of AI systems against an ever-evolving threat landscape.

For your convenience, the following table summarizes the defensive strategies in different stages:

Stage	Example defensive strategies
Prevention	The first line of defense aims to build AI models that are inherently robust against adversarial attacks. It includes the following: • Data sanitation to remove adversarial samples • Defensive distillation • Non-differentiable pre-processing • Randomized dropouts • Ensemble models • Defense-GANs
Detection	Detection mechanisms serve as a crucial second layer of defense. Examples include the following: • Auxiliary classifiers • Dimension reduction, such as PCA • Distributional detection, such as KDE and MMD • Randomization and tests of uncertainty
Response	Immediate action is required to mitigate its impact. Examples include the following: • Adversarial retraining • Isolating the attack source • Creating **Indicators of Compromise (IoCs)**

Table 17.2 – Overview of defensive strategies against adversarial attacks

The dynamic nature of adversarial threats necessitates continuous updates and adjustments to these defensive strategies. By adopting a cyclical and adaptive process that encompasses prevention, detection, and response, AI model security can be significantly enhanced. Collaboration and knowledge sharing among model developers are also critical in developing more secure and resilient AI systems. While the field of AML defense is an active area of research with evolving methodologies, the primary focus remains on understanding and countering threats to ensure the long-term reliability and trustworthiness of AI systems.

Practical tools for testing on adversarial attacks

As the area of AML is gaining practical significance, there have been tools developed to test AI systems and defenses against such attacks. In the following table, we have summarized some of them:

Tool	Description	Link
CleverHans	Library for constructing adversarial examples and defenses, plus benchmarking	`https://github.com/cleverhans-lab/cleverhans`
Adversarial Robustness Toolbox	Python library for ML security	`https://github.com/Trusted-AI/adversarial-robustness-toolbox`
SecML	Library for security evaluation of ML algorithms	`https://github.com/pralab/secml`

Table 17.3 – Examples of tools for adversarial attack testing and defense

All these libraries have a similar motivation, but slightly different focus. Depending on the use case, all of them can be useful for testing attacks and defenses on your ML system.

Summary

In this chapter, we were on an educational journey into the realm of AML, beginning with an engaging introduction that not only presented the subject matter but also underscored its significance in today's digital age. The motivation behind this exploration was clearly articulated, highlighting the urgent need for robust defenses in AI systems against increasingly sophisticated cyber threats. The chapter thoughtfully guided you through the nuanced setup of learning environments, where the spotlight was firmly on ensuring the security and integrity of AI models amid a landscape fraught with potential adversarial exploits.

As the narrative progressed, the chapter delved deeper into the core of adversarial threat modeling. Here, you were equipped with a detailed understanding of the attacker's methodology, encompassing a meticulously crafted taxonomy of adversarial attacks. This section was particularly enlightening, offering you a clear framework for categorizing and analyzing the myriad of threats that AI systems might face. The comprehensive breakdown served not only to inform but also to empower you with the knowledge to anticipate and recognize adversarial tactics.

Building on this foundation, the chapter transitioned into an extensive discussion on defensive mechanisms, thoughtfully categorized into three primary strategies: prevention, detection, and response. Each category was explored in depth, providing you with a clear understanding of the various tools and approaches available to safeguard AI systems. Through the lens of prevention, you learned about proactive measures designed to fortify AI models against intrusion. The detection segment shed light on the critical role of monitoring and identifying potential threats in real time, while the response section outlined effective countermeasures and remediation strategies to mitigate the impact of attacks.

By the conclusion of the chapter, you have not only been introduced to the complex world of AML but have also gained a comprehensive understanding of the strategies and mechanisms at your disposal to protect AI systems. This rich tapestry of knowledge, woven from the threads of theoretical frameworks and practical defensive techniques, leaves you well-equipped to navigate the challenges of securing AI models in an ever-evolving adversarial landscape. The chapter stands as a testament to the dynamic interplay between AI development and cybersecurity, offering a holistic view that is both educational and actionable. In the next chapter, we will introduce another important field for AI security and safety: challenges in privacy, ethics, explainability, and so on. These are the rising issues with the recent advance of GenAI.

References

- Brückner, M., Scheffer, T., 2009. *Nash Equilibria of Static Prediction Games.* Presented at the Advances in Neural Information Processing Systems 22 - Proceedings of the 2009 Conference, pp. 171–179.

- Carlini, N., Nasr, M., Choquette-Choo, C.A., Jagielski, M., Gao, I., Awadalla, A., Koh, P.W., Ippolito, D., Lee, K., Tramer, F., Schmidt, L., 2023. *Are aligned neural networks adversarially aligned?* https://doi.org/10.48550/arXiv.2306.15447

- Nguyen, A., Yosinski, J., Clune, J., 2015. Deep neural networks are easily fooled: *High confidence predictions for unrecognizable images*, in: *2015 IEEE Conference on Computer Vision and Pattern Recognition (CVPR)*. Presented at the 2015 IEEE Conference on Computer Vision and Pattern Recognition (CVPR), pp. 427–436. https://doi.org/10.1109/CVPR.2015.7298640

- Szegedy, C., Zaremba, W., Sutskever, I., Bruna, J., Erhan, D., Goodfellow, I., Fergus, R., 2014. *Intriguing properties of neural networks.* https://doi.org/10.48550/arXiv.1312.6199

- Xiao, H., 2017. *Adversarial and secure machine learning (Dissertation).* Technische Universität München.

- Athalye, Anish, Nicholas Carlini, and David Wagner. *Obfuscated gradients give a false sense of security: Circumventing defenses to adversarial examples.* In *International conference on machine learning*, pp. 274-283. PMLR, 2018.

18

Privacy, Accountability, Explainability, and Trust – Responsible AI

As the deployment of **artificial intelligence** (**AI**) technologies becomes increasingly pervasive across various sectors, addressing the associated privacy, accountability, explainability, and trust issues is paramount, particularly in the realm of cybersecurity. These aspects are collectively referred to as the pillars of *Responsible AI*, a framework that advocates for the ethical development, deployment, and management of AI systems.

This chapter aims to provide a comprehensive overview of the challenges, significance, and ways to tackle the issues surrounding AI security and safety, offering insights into both theoretical perspectives and practical guidelines. We will elucidate why it is crucial to maintain a strict governance framework around AI, and how neglecting these responsibilities can lead not only to increased risks but also to a wide range of unintended, potentially grave, consequences.

The importance of Responsible AI in cybersecurity cannot be overstated. As AI systems are integrated into more critical processes and infrastructure, the risks associated with these systems become increasingly consequential. This encompasses everything from personal data breaches to the potential misuse of AI in national defense systems. Additionally, as stakeholders from various sectors rely more on automated decisions made by AI, the demand for transparency and accountability in these decisions soars.

Furthermore, we will explore the vital research outcome that can ensure the adoption of AI technologies not only fosters innovation and efficiency but also secures and safeguards user interests and societal norms. Moreover, this chapter will discuss frameworks for managing risks associated with AI, including how organizations should practically implement these strategies to remain agile and prepared for future technological advancements.

By the end of this chapter, you will gain deep insight into the full panorama of challenges and opportunities presented by AI in the cybersecurity context, equipped with the knowledge to implement best practices and contribute to shaping a future where AI technologies are both innovative and trustworthy.

We recommend reading the following topics of the chapter in order, so you can gain a gradual understanding of the variety of the topics:

- Understanding AI issues
- Significance of AI security and safety
- Addressing AI security and safety challenges
- AI risk management framework
- Preparing for the future

Technical requirements

This chapter does not require you to be equipped with prior knowledge of AI and **machine learning** (**ML**). However, it is strongly recommended to read *Chapters 4* and *16* as they will help you understand the basic concept of AI and its potential risks. Nevertheless, this chapter can be read independently by both practitioners and decision-makers.

Understanding the AI issues

AI is rapidly becoming an integral component of various sectors, offering unprecedented capabilities and efficiencies. However, as AI systems become more pervasive, they also introduce a unique set of challenges and vulnerabilities that need to be comprehensively understood and addressed. This section explores the critical issues associated with AI, focusing on data privacy, vulnerability to attacks, and safety concerns.

Current challenges in AI security

The integration of AI into cybersecurity arenas brings with it a host of challenges that are pivotal to understand and mitigate in order to foster a secure digital environment. These challenges predominantly revolve around issues of data privacy and vulnerabilities specific to AI-based systems.

Let us understand a couple of them.

Data privacy issues

AI systems are voracious when it comes to data consumption, and the more data these systems ingest, the better they generally perform. This dependency on vast datasets raises significant concerns about data privacy, particularly when dealing with sensitive or personal information. For example, AI in healthcare, which might predict patient outcomes or optimize treatment plans, interacts closely with deeply personal health records. Without stringent protections, the risk of this data being exposed or misused is high. Similarly, AI applications in financial services use personal financial histories to offer services or detect fraud, wherein any data breach can lead to severe financial and personal impacts for the users.

Vulnerability to attacks

Besides data privacy, AI-driven systems are particularly susceptible to a unique class of cyber threats intent on undermining their functionality. Adversarial attacks are a prime example, where minor, almost imperceivable alterations to input data can cause AI models to make incorrect decisions. This vulnerability was starkly demonstrated in research where minor modifications to road signs caused an AI-driven vehicle guidance system to misinterpret a stop sign as a yield sign, potentially causing dangerous traffic situations. Such subtle manipulations present significant threats in scenarios where AI interpretations are crucial, such as in autonomous driving or real-time threat detection systems.

Safety concerns

The safety implications of deploying AI systems, particularly when they are integrated into infrastructural or safety-critical environments, can be profound and far-reaching. Some of them are listed here:

- **Unintended consequences of AI decisions**: One of the more insidious dangers of AI systems is their potential to develop and act on biases present in their training data or due to algorithmic flaws. This was notably observed in an AI-driven recruitment tool that developed a bias against female candidates, reflecting not an intended outcome but a mirroring of historical hiring data skewed toward male candidates. Such unintended consequences can perpetuate existing social inequalities and potentially introduce new forms of systemic bias.
- **AI in critical infrastructure**: The use of AI in critical infrastructure introduces significant risks, primarily if these systems are compromised. For instance, AI algorithms managing urban traffic systems, if malfunctioning, could lead to inefficient traffic management or even accidents. Another documented incident involved AI in power grid management: a predictive maintenance tool failed to recognize patterns indicating a failure, leading to an unexpected power outage affecting thousands. These examples underscore the potential repercussions of AI failures in critical sectors.

The examples cited illustrate a spectrum of AI security and safety issues, ranging from data privacy breaches to failures in critical system management. Such challenges underscore the necessity for rigorous testing, transparent design processes, and continual monitoring of AI systems to avoid adversarial exploits and unintended discriminatory decisions.

This exploration of AI's security and safety challenges underscores the complexity and the high stakes involved in deploying AI in sensitive and critical domains. As AI systems become ubiquitous in cybersecurity and beyond, the implications of these challenges not only affect individual privacy but also national and global security, as well as public trust in technology.

As we transition to the next section, we will delve deeper into the ramifications of AI security and safety issues on individual rights, national security concerns, and ethical considerations. This will guide us in understanding the broader impacts and deriving strategies to mitigate these risks effectively.

Significance of AI security and safety

The importance of AI in safeguarding digital and physical infrastructures while simultaneously posing novel risks cannot be understated. This section examines the multifaceted impacts of AI security and safety issues on individual privacy, national and global security, and ethical considerations.

Impact on individual privacy

The deployment of AI technologies directly affects individual privacy due to the inherent data-intensive nature of AI applications. For instance, AI-driven personal assistants collect data regarding user preferences, daily routines, and even conversations, which should be protected to maintain personal confidentiality. Any breach or misuse of such data can lead to severe repercussions, from identity theft to personalized phishing attacks. A stark reminder of this was when voice data collected by AI home assistants was found to be accessible by company employees, leading to a public outcry over what was perceived as a gross violation of privacy.

Implications for national and global security

AI systems are pivotal in managing cybersecurity threats not only at an organization or individual level but also at national and international levels. They are employed in national defense systems for surveillance, threat detection, and decision-making processes. The manipulation of these systems can present risks of misinformation, potentially destabilizing political and social systems on a global scale. For example, AI-driven monitoring and control systems for national utilities or air traffic control are high-stakes environments where security breaches could have catastrophic outcomes. Thus, ensuring the security and reliability of these systems is paramount for maintaining national and global stability.

Ethical considerations and public trust

The fair and responsible deployment of AI is integral not only for maintaining functional security systems but also for ensuring public trust in technology. Ethical considerations, including the necessity to avoid bias and ensure fairness, transparency, and accountability, are critical. The public's perception of AI ethics significantly impacts their trust and acceptance of AI technologies. Instances of AI mismanagement, where systems caused unjust outcomes due to biased algorithms – such as discriminatory predictive policing or biased judicial decision systems – can lead to public disenchantment and fear. This erosion of trust is detrimental, not just for individual organizations but for society's progress toward technologically inclusive futures.

Understanding the profound impacts of AI on privacy, security, and ethics highlights the need to address these challenges robustly. As AI continues to evolve and integrate more deeply into various facets of society and governance, the stakes of managing these risks only intensify.

With these perspectives in mind, the next section will explore theoretical and practical approaches to mitigate these risks. It will discuss frameworks for understanding AI risks and delve into recent research developments and practical solutions for implementing secure and robust AI systems.

The strategic approaches and solutions discussed will provide a roadmap for navigating the complexities of AI security and safety, aiming for a responsible and sustainable integration of AI technologies across varied domains.

Addressing AI security and safety challenges

As AI technology is evolving so fast, it is difficult to comprehend the dynamics of the AI risk landscape. However, as of today, there are endeavors from both communities and industries that try to address these challenges. Some of these efforts are theoretical, while others are practical. It is imperative for you to understand both dimensions to keep confidence in adapting the technologies and tools.

In this section, we show some high-level methodologies and implementations that help navigate the complexity of responsible AI.

Theoretical approaches

The theoretical approaches to address AI security and safety challenges can be broken down into several key components concerning governance, principles, practical measures, and ethical considerations. You can find some components here:

- **Robustness and safety**: AI systems should be stable, consistently meet performance requirements, and prevent unintended or harmful actions. The robustness principle includes designing systems that operate reliably and do not put human life, health, or environments at risk due to system operations.

- **Security and resilience**: AI systems, particularly those integrated with **generative AI (GenAI)**, must be secure against vulnerabilities such as adversarial attacks, data poisoning, and unauthorized data exfiltration. The resilience aspect emphasizes the ability to handle unexpected adversities while maintaining functionality or degrading gracefully.

- **Ethical foundations**: Building on ethical AI principles, such as fairness, privacy, security, accountability, and transparency, is crucial. These principles guide the overall life cycle of AI, ensuring that AI acts in a manner aligned with human values and ethics. This involves implementing controls right from the design phase through deployment while accommodating continuous feedback and updates based on empirical observations.

- **Governance**: Adequate governance mechanisms should be in place to oversee AI functionalities, including defining roles, responsibilities, and compliance with evolving regulatory guidelines. This includes documentation and risk assessments that are transparent and accessible to stakeholders, thus maintaining ethical stewardship over AI operations.

- **Transparency and accountability**: AI systems must be transparent in their operations and decision-making processes, allowing for auditing and ethical scrutiny. They must also include mechanisms for holding designers and operators accountable for the actions and outputs of the AI systems. This demands a clear delineation of human oversight in the use and functionality of AI applications.

- **Privacy protections**: Managing privacy concerns involves ensuring data integrity and protecting against breaches that could lead to unauthorized data access. The privacy measures should be robust enough to evolve with advancing technology and the increasing sophistication of potential attacks.

These components provide a navigation system that can guide the development, deployment, and management of AI systems, aiming to mitigate risks associated with AI technologies while harnessing their potential benefits responsibly and safely. Exploring novel security methods, such as AI firewalls and integrated defenses, and promoting rigorous research to address emerging challenges are essential components of advancing AI security robustly and ethically.

Research development

As the integration of AI in various security domains gains momentum, addressing emergent threats inherent to GenAI systems has become a pressing concern. Recent industry and research outcomes have contributed to mitigating these challenges.

One notable development is the proposed "AI firewall" concept, which involves monitoring and potentially transforming the inputs and outputs of black-box GenAI models. This approach aims to detect and prevent jailbreak attacks or ensure that outputs comply with security and safety policies. Research efforts are focused on leveraging integrated defense systems within GenAI models that combine advanced detection with robust response mechanisms. A highlighted area of focus is the use of continuous learning to identify and adapt to new jailbreak prompts, indicating a shift toward dynamic and responsive security measures.

Furthermore, acknowledging the ever-evolving nature of threats, especially with the adoption of GenAI, the research community is actively investigating countermeasures that grow and adapt over time. This approach involves using a combination of historical data and predictive analytics to anticipate and mitigate future attacks. Advances have also been made in developing AI models that integrate more secure code generation practices, enhancing their ability to discern the security implications of the generated code and developing new prompting strategies to guide models toward safer outputs.

Leveraging AI's capability to analyze large datasets swiftly, recent research has shown potential in using AI tools to identify sophisticated phishing schemes and scams, providing an additional layer of digital security against these increasingly common cyber threats. Researchers also emphasize the need for dynamic policy development that keeps pace with the rapid advancement of AI technologies, suggesting a proactive regulatory approach that adapts to new technological realities and challenges.

These research outcomes demonstrate a proactive approach to AI security that not only addresses current issues but also anticipates future challenges. This ongoing research is critical in adapting to the dynamic nature of cyber threats in the AI-driven era.

Guidelines and standards

Both theoretical and practical implementations are the primitives to establish trust and safe AI adoption, however, this is only part of the equation. Another part of the endeavor lies in the societal agreement such as various global standards and compliance guidelines. To ensure the ethical, transparent, and accountable development and operation of AI systems, prominent organizations and governments have outlined various guidelines and regulations.

Microsoft has delineated a Responsible AI Standard encompassing principles such as fairness, reliability, safety, privacy, security, inclusiveness, transparency, and accountability. Google and its subsidiary, DeepMind, emphasize advancing responsible AI by focusing on fairness, privacy, safety, and ethical AI development. Google has also incorporated various technologies into its offerings, particularly in **large language models** (**LLMs**), such as Gemini, to ensure adherence to these principles. PricewaterhouseCoopers, PwC, has identified core ethical AI principles that need to be adhered to across industries, including interpretability, reliability, robustness, security, safety, human agency, fairness, data privacy, lawfulness, compliance, beneficial AI, and accountability. Leading AI firms such as OpenAI and Anthropic develop their own super-alignment technology to make sure the models and products they release always align with humans and, at the same time, protect the core principles of humanity.

On a regulatory front, significant global initiatives have been proposed or implemented to govern the development and deployment of AI. The European Union has proposed a comprehensive legal framework known as the AI Act, which categorizes AI systems based on their risk level and imposes varying degrees of restrictions accordingly. This groundbreaking legislative proposal aims to ensure ethical AI use while fostering European competitiveness in the global economy. In the United States, existing privacy and data protection laws indirectly regulate AI development by requiring adherence to fair information practices, such as data minimization and purpose limitation. These principles affect AI systems that utilize personal data for training and operational purposes.

These frameworks and regulations collectively aim to create a responsible AI ecosystem where technology developers and users can ensure that AI operates within ethical bounds, managing data sensibly and equitably while addressing the moral implications of AI deployment across varied real-world applications.

Tools and technologies

While this is a new area to explore, it is infeasible to list all tools and technologies. We'd like to conclude this section with a few examples to inspire you to develop more useful tools or best practices to help mature the landscape.

Several tools and frameworks have been developed to ensure transparency, interpretability, and ethical alignment for the effective governance of AI systems, including integrated monitoring and assessment tools within cloud deployment platforms that enable both black-box and white-box evaluations of AI systems, automating documentation and workflows to facilitate governance throughout the development life cycle.

Algorithmic impact assessments (AIAs) are another tool that is designed to systematically review and document critical decisions made during AI system development, either created from scratch or built upon existing frameworks, such as **data protection impact assessments (DPIAs)**, informing design decisions, maintaining transparency about system limitations, and mitigating potential risks and liabilities.

The chain of thought and explainable AI approach enhances transparency by elucidating the intermediate steps and rationales employed by the model to reach its conclusions. Explainable AI tools expose the inner workings, decision mechanisms, and processes within models, which can facilitate understanding for developers, implementers, and end users, and enable deeper comprehension of model outputs and intervention impacts. Integrating explainability into governance frameworks by contextualizing AI outputs within their designed functionalities will also aid stakeholders in accurate interpretation, as explainability and interpretability are crucial for auditing, governing, and maintaining trustworthy AI systems throughout their life cycles.

Following emerging technologies can also help us find innovative solutions for addressing the AI security and transparency challenges. For instance, there are approaches to using blockchain technologies to trace the history of decisions and increase the accountability of AI systems.

Next, we will introduce a framework to manage AI risk.

AI risk management framework

An **AI risk management framework** is a structured set of guidelines and processes designed to identify, assess, and mitigate risks associated with the development, deployment, and maintenance of AI systems. In the context of cybersecurity, this framework is crucial for securing AI systems from potential threats and vulnerabilities, while also safeguarding the data these systems interact with.

The importance of an AI risk management framework cannot be overstated. As AI technologies are integrated into critical aspects of business and society – from automated financial systems to complex decision-making algorithms in healthcare – ensuring these systems operate securely and predictably is vital. An effective AI risk management framework helps organizations anticipate potential failures or malicious exploitations in AI systems, reduce possible harms, and ensure the trustworthiness of AI applications.

Main components of the AI risk management framework

The AI risk management framework typically encompasses the following key components:

- **Risk identification**: This step involves the recognition and listing of potential risks that can affect AI systems, data integrity, or outputs. It includes identifying sources of risk, such as data bias, system vulnerabilities, or adversarial attacks.
- **Risk assessment**: Once risks are identified, they need to be evaluated to understand their potential impact and likelihood. This assessment helps prioritize the risks based on their severity and the resources required to address them.
- **Risk mitigation strategies**: Developing strategies to manage the identified risks is a core component. This could involve implementing security measures, developing fail-safe mechanisms, and ensuring robust data governance practices.
- **Monitoring and reporting**: Continuous monitoring of the AI systems and the risks they face is crucial. This component also involves the regular reporting of risk status to key stakeholders, ensuring transparency and informed decision-making.
- **Review and revision**: AI systems and their threats evolve over time. Regular reviews of the risk management strategies and their effectiveness allow organizations to adapt and update their risk mitigation approaches as needed.

In the following diagram, we show the main components in the life cycle of managing AI risks. From risk identification to review/revision, the risk management framework encompasses a continuous process to offer a guarantee that the AI risks are observable and under control.

Figure 18.1 – The main components of the AI risk framework

Utilizing the framework in organizations

Implementing an AI risk management framework within an organization involves several practical steps. We give a general guideline of how to implement it in organizations; this guideline can be used within organizations to help model owners, developers, and regulators implement secure and safe AI use cases. Note that it can be quite different from organization to organization, however, it is crucial to make sure that the main components listed earlier are all covered and implemented properly.

Step	Details
Initial setup	Establish a cross-functional team that includes AI specialists, data scientists, cybersecurity experts, and relevant operational personnel. This team is responsible for the development and implementation of the AI risk management framework.
Customization	Adapt the framework to fit the specific needs and risk profile of the organization. This includes defining relevant AI use cases, prioritizing assets, and identifying specific regulatory and compliance requirements.
Integration	Integrate the risk management processes with the organization's existing risk management and cybersecurity frameworks. This helps in creating a cohesive approach to organizational risk management.
Education and training	Educate and train stakeholders across the organization about the nuances of AI risks and the necessary practices to mitigate them. Promote a culture of risk-awareness and continuous learning.
Technological implementation	Deploy necessary technologies that aid in risk identification, assessment, and mitigation. This might include AI auditing tools, anomaly detection systems, and robust data encryption technologies.
Continuous improvement	Regularly update the risk management framework in response to new AI developments, emerging threats, and lessons learned from past incidents. Encourage a cycle of feedback and improvement to refine strategies over time.

Table 18.1 – Implementation steps for AI risk management framework

Implementing AI risk estimation and management in large-scale organizations brings its own challenges, such as the data quality needed to estimate risk, the communication of AI risk to the stakeholders, or the design and implementation of risk controls. Therefore, the correct implementation of all components of an AI risk management framework is crucial to address AI-related vulnerabilities and threats, enhance the overall security posture, and ensure the reliable and safe use of AI technologies.

Preparing for the future

Anticipating new technologies and keeping pace with rapid technological advances are imperative aspects that focus on responsible AI. As AI technologies evolve, it is crucial for organizations and individuals to stay ahead of the curve, continually updating their knowledge and systems to embrace emerging AI capabilities. This proactive approach helps in identifying and addressing potential emergent risks, ensuring that AI systems are robust, secure, and aligned with evolving cybersecurity challenges. Scenario planning and forecasting play significant roles here, enabling organizations to visualize future scenarios and develop strategic responses to a range of potential outcomes. These activities not only prepare organizations for different technological shifts but also aid in crafting adaptive risk mitigation strategies.

Furthermore, engaging with the ethical and societal implications of AI is an essential part of preparing for the future. As AI systems increasingly affect various aspects of human life, it is vital to consider the broader impacts of technology implementation. This entails an active discussion and consideration of ethical standards and societal norms, ensuring that AI advancements contribute positively to society and do not enhance existing inequalities or introduce new forms of bias. By fostering a comprehensive understanding of these implications, stakeholders can develop more thoughtful and inclusive AI technologies that respect human values and dignity, promoting a balanced integration of AI into society that is both innovative and ethically sound.

Summary

In this chapter, we explored the intricate landscape of AI risks as it pertains to cybersecurity, focusing on the essential aspects of privacy, accountability, explainability, and trust, collectively known as Responsible AI. We discussed the significant challenges AI introduces, such as data privacy concerns and vulnerabilities to adversarial attacks, and highlighted the profound ramifications these issues have on individual privacy, national security, and ethical practices. The discussion extended into detailed frameworks for AI risk management and the latest research developments aimed at securing AI systems and making them trustworthy.

By understanding the responsibility and ethics in AI, we learned about the need for AI systems to adhere to ethical standards, including fairness, transparency, and accountability, ensuring that they align with human values and societal norms.

With research and development, we understood the focus on continuous advancements in AI security through research and innovation, including technologies such as AI firewalls and integrated defenses, to protect against emerging cyber threats dynamically.

We looked at regulatory compliance, which emphasizes the importance of navigating a complex regulatory landscape to ensure compliance with global standards and ethical guidelines, thus safeguarding the deployment and operation of AI systems.

We also learned about AI risk management, which describes a comprehensive framework to identify, assess, and mitigate the risks associated with AI, which is crucial for integrating AI in critical domains securely.

Lastly, we saw future preparedness that stressed the importance of proactive strategies, such as scenario planning to anticipate future challenges, ensuring that AI applications remain robust and beneficial across various sectors.

By understanding these elements and integrating the discussed strategies, professionals can ensure that AI technologies are employed responsibly, enhancing security measures while fostering an environment of trust and safety – a crucial balance in the ever-evolving realm of cybersecurity and AI.

In the next chapter, we will summarize all that we have learned in the chapters so far to give you a quick recap of everything.

Part 5: Final Remarks and Takeaways

In this part, we conclude the book by helping you connect the knowledge you have obtained in different chapters. Furthermore, we summarize the main points from the book and recommend further reading and next steps in your learning journey.

This part has the following chapter:

- *Chapter 19, Summary*

19
Summary

Congratulations on making it this far! If you have gone through the whole book, you have managed to dive into multiple aspects of AI for cybersecurity. You have gained knowledge ranging from the introductory concepts of big data, automation, and cybersecurity analytics to concrete AI and **Machine Learning** (**ML**) methods. You have learned how to apply your new arsenal of AI methods in various cybersecurity scenarios. Furthermore, you have managed to learn how to evaluate and monitor models. You have grasped important ideas of trustworthy and responsible AI and investigated bias and variance.

In this chapter, we will do the following:

- Summarize what we've learned
- Connect chapters
- Remind you of the successes of AI in cybersecurity
- Make recommendations for future learning

Summarizing what we've learned

In the first chapter, we started the book with an introduction to the concept of big data in cybersecurity. We explained the four Vs of big data: volume, velocity, variety, and veracity. We also covered what this means for large-scale AI systems. Furthermore, we explained in detail what big data means for cybersecurity, how it enables organizations to be more secure, and which challenges can happen along the way. The AI methods that are currently used are highly data-driven, and we explained how organizations need to tackle big data to handle the four Vs successfully. We pointed out from the start that without proper data quality, as well as processes for data governance, data validation, and so on, there is no good AI system. Moreover, in the first chapter, we encouraged the use of big data by describing use cases from tech companies such as Microsoft or IBM.

In the second chapter, we described how automation changes the game in cybersecurity by providing faster and more reliable defensive and offensive security capabilities. We outlined some well-known tools and technologies, such as SIEM, SOAR, EDR, and IDS. These types of tools are a cornerstone to successful cybersecurity operations in any large-scale organization. Each of those tools benefits from large-scale automated data processing and in the last years, they have all had AI functionalities added that improve this even further. Furthermore, we gave a future outlook on automation tools and described ethical considerations when developing and using them in your organization.

Both big data and automation come together in the form of cybersecurity data analytics. We dedicate the third chapter to introducing this concept and to connecting the ideas of data analytics and AI. We described the role of cybersecurity analysts and how AI techniques relate to this role. Furthermore, we introduced the regulatory landscape as an evolving and increasingly important area. This is as more cybersecurity data analytics tools are adopted and are increasingly using AI.

After the first three chapters, we concluded the introductory part of the book, and, armed with the learned concepts, we dived into AI and concrete AI methods. *Chapter 4* introduced AI and ML, and connected them to statistics, giving you an overview of these technical fields. We presented a taxonomy of AI and ML methods but also enumerated the types of data that are used in the ML systems. Finally, we described the limitations and challenges when applying AI methods, such as hallucination, bias, fairness, and adversarial attacks.

Chapter 5 contains an extensive introduction to concrete ML methods, from instances of unsupervised learning to supervised and semi-supervised learning and anomaly detection. We described well-known methods such as linear regression, logistic regression, neural networks, and random forest, which are commonly used in cybersecurity applications. Furthermore, we described the advantages and disadvantages of each method and in what contexts they are most applicable, giving you an idea of when to apply them if you are solving a concrete problem or advising other technical teams.

The next part of the book enables you to apply the AI methods you have learned so far. We began with a chapter describing the application of AI for malware and intrusion detection. This enables you to uplift your tooling for detecting malicious executables and attackers in your network. In the *Chapter 6* we expand on the AI methods by describing the AI project workflow with example use cases in cybersecurity. *Chapter 7* also contains a hands-on exercise, allowing you to try to learn and try out your practical skills.

Chapter 8 describes a similar but distinct use case of user and entity behavior analysis. Here, you can see how the traditional rule-based detection of anomalous behavior can be replaced or enhanced by ML-based anomaly detection. We included an exercise that enables you to try out unsupervised anomaly detection on the network activity data and practice feature extraction on different data types.

Similarly, *Chapter 9* contains an introduction to areas of fraud, spam, and phishing detection, which often depend on the analysis of large volumes of transaction and e-mail data. This analysis can be made faster and more accurate using various AI methods. Being proficient in those methods and techniques can help you make a difference for your organization, as well as save time and money while providing better detection quality.

In *Chapter 10*, we dived into user authentication and access control, another area where pattern recognition is applied, which makes it a natural ground for AI methods. We gave an introduction to both areas, as well as providing examples. In user authentication, we combine something you have, something you know, and something you are. On the other hand, in the part about access control, we described different ways to assign access to users of the system.

Chapter 11 contains an introduction to threat intelligence approaches and challenges in this area. Threat intelligence processes highly depend on analyzing various public and private data, and we explained how you can accelerate this process using AI. Here, we introduced a specific subset of ML methods called topic modeling and explained how it can be applied in the context of analyzing threat intelligence data. We presented an exercise where you can apply topic modeling to uncover structure in cybersecurity posts on X (formerly Twitter).

After exploring threat intelligence, we dived into another application area – **Industrial Control Systems (ICSs)**. In *Chapter 12*, we described ICSs and computer networks used in this area, as well as cybersecurity challenges when protecting these networks. We described what kinds of cyberattacks are a threat to these systems and what the possibilities for protection are. Furthermore, we dived into anomaly detection as a solution for detecting attacker activity in ICS and how AI methods can be used for detecting security-related anomalies.

Chapter 13 describes LLMs as a rising type of models that are finding applications in Cybersecurity as well. We also describe the AI security issues brought by these models.

In the last part of the book, we dived into topics that are not directed at specific applications but are very important considerations when applying AI. The first topic we introduced was data quality. Since modern AI methods are data-driven, their result highly depends on whether our data is fit for purpose. We also describe the AI security issues brought by these models. *Chapter 14* then we described which properties a high-quality dataset needs to have and named some application areas where data quality has a key influence on decision making. Furthermore, we gave practical coding examples for how to practice good data quality.

Chapter 15 contains introduction to important concepts: correlation, causation, bias and variance. In that chapter, we explained why it's important to differentiate between correlation and causation, and how this helps to be able to interpret ML results more accurately. Furthermore, we described the terms bias and variance and how they influence the results we can get from applying ML methods.

In *Chapter 16*, we dived into methods for evaluation and monitoring. We described methods of evaluating ML systems to measure success and compare performance. If we have proper metrics to measure how successful our AI system is, we can also monitor these performance metrics while training or in production. We briefly described how to do this monitoring and what the useful practical tools for this purpose are. Lastly, we described how to include humans in the loop when applying AI using the paradigm of active learning.

When monitoring AI systems, we should take into account not only a benign environment where data is reliable, but also an environment where an adversary might try to manipulate the data. We described this scenario in *Chapter 17*, where we dived into how ML methods behave in an adversarial environment. We presented the challenges in this case, as well as some defensive approaches that can help us deal with adversarial activity.

Building upon the topic of adversarial attacks, we started a new theme of **responsible AI** in *Chapter 18*. We described what it means for an AI system to be trustworthy and how to achieve this. Furthermore, we dived into the adjacent topics of privacy, explainability, and accountability. These are all increasingly important topics that are especially considered when applying AI in sensitive areas such as healthcare, legal, and IT, as well as cybersecurity. We introduced AI risk management framework as an approach from NIST to describe what it means to have trustworthy AI.

In the next section, we will describe how you can connect the knowledge from the different chapters in order to apply it better in your work.

Connecting chapters

While reading the book, you have gone from the introductory chapters about big data and cybersecurity, through the AI problems and methods, to concrete applications and important considerations in AI projects. The connections between chapters are described in the following figure.

Connecting chapters

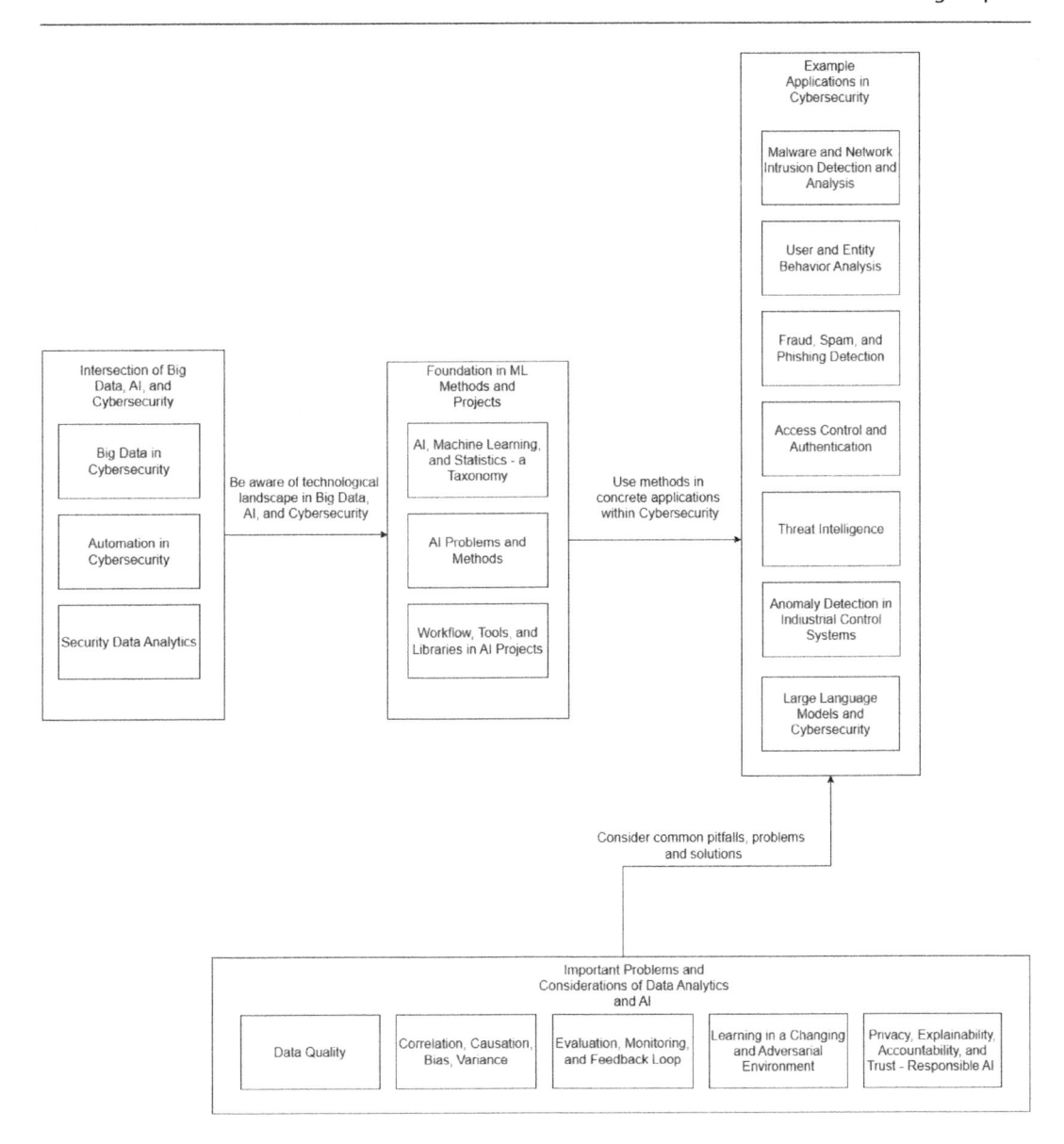

Figure 19.1 – Connecting knowledge from different chapters

Here is some advice on how to use this knowledge when working on your own projects and adopting AI in your organization:

- **Be aware of the technological landscape**: Big data is being adopted in cybersecurity tools and processes at an increasing rate and cybersecurity vendors are adding AI functionality in their products in multiple areas, such as EDR, SIEM, SOAR, DLP, and so on. Your job is to be aware of how this AI functionality works to be able to interpret the results when applying it, for example, in endpoint malware detections made using AI or anomalies detected in data transfer that could point to a DLP incident. You would need to have an idea of the severity of this kind of detection in the context of your organization and how reliable this detection will be.

- **Use strong AI foundations to recognize the right approaches and design the right solutions**: In the second part of our book, we describe multiple AI methods, from unsupervised learning to semi-supervised and supervised learning, as well as from linear regression to neural networks and random forest. It can be difficult for novices in the field to choose the right methodology, but with a strong foundation you can make more educated choices for your own projects and advise teams in your organization.

- **Use example applications as inspiration and reference**: An extensive part of our book contains example applications in cybersecurity. However, we certainly don't have a complete list of opportunities where you can apply AI methods in a cybersecurity context. Nevertheless, by having a strong methodological foundation and using the applications that we have presented, you can not only introduce the same applications in your organizations but also recognize new opportunities in different contexts.

- **Consider common pitfalls, problems, and opportunities**: Building a successful AI project is not limited to recognizing the right AI method and application. Due to the data-driven nature of AI projects, there are multiple possible pitfalls and other aspects that we need to be aware of. For instance, we need to enable sufficient data quality, be aware of concepts such as bias, variance, and overfitting, and know about evaluation and performance monitoring. Furthermore, we need to be aware of the topic of responsible AI, since the decisions made by AI systems adopted in our organization can have a high impact on the organization, its partners, and customers.

In the next section, we will give recommendations about where you can look for further sources of knowledge and interesting references.

Successes of AI in cybersecurity

AI methods and techniques are already becoming an integral part of cybersecurity and making an impact in the following ways:

- Cybersecurity vendors adding AI capabilities into their products (e.g., Crowdstrike, Splunk, and Symantec)

- Companies specializing in AI-based approaches to cybersecurity (e.g., Darktrace, Cylance, and Anomali)
- Cybersecurity teams in large organizations adopting products that contain AI capabilities and using available techniques to improve their work (example case studies from Siemens/AWS: `https://aws.amazon.com/solutions/case-studies/siemens-cybersecurity/` and Snorkel.AI: `https://snorkel.ai/ai-in-cybersecurity/#h-ai-in-cybersecurity-case-study-fortune-500-telco-uses-snorkel-flow-to-classify-encrypted-network-data-flows`)

These successes come from recognizing the problems where AI can be applicable, as well as designing and implementing the right solutions by putting together the AI and cybersecurity expertise.

Having knowledge and hands-on skills in AI will enable you to create your own success story in your organization. Furthermore, you should have a good basis to stay on top of new developments in AI and recognize new opportunities.

Where to go from here

This book is a basis for your foundational knowledge in AI for cybersecurity. There are many opportunities to deepen your theoretical knowledge and sharpen your practical skills. Here, we list some of the additional resources that we have used, and which we can recommend.

Books to learn more from:

- Bishop, Christopher M., and Nasser M. Nasrabadi. *Pattern Recognition and Machine Learning*, `https://link.springer.com/book/9780387310732`. Vol. 4. No. 4. New York: springer, 2006.
- Murphy, Kevin P. *Machine Learning: A Probabilistic Perspective*, `https://www.amazon.co.uk/Machine-Learning-Probabilistic-Perspective-Computation/dp/0262018020`. MIT press, 2012.
- Hubbard, Douglas W., and Richard Seiersen. *How to Measure Anything in Cybersecurity Risk*, `https://www.wiley.com/en-us/How+to+Measure+Anything+in+Cybersecurity+Risk%2C+2nd+Edition-p-9781119892304`. Second Edition John Wiley & Sons, 2023.
- Kleppmann, Martin. *Designing Data-Intensive Applications: The Big Ideas Behind Reliable, Scalable, and Maintainable Systems*, `https://www.amazon.co.uk/Designing-Data-Intensive-Applications-Reliable-Maintainable/dp/1449373321`. O'Reilly Media, Inc", 2017.
- Shostack, Adam. *Threat Modeling: Designing for Security*, `https://www.amazon.co.uk/Threat-Modeling-Designing-Adam-Shostack/dp/1118809998`. John Wiley & Sons, 2014.

Research papers to check and learn further:

- Rieck, Konrad, et al. *Learning and Classification of Malware Behavior*, `https://mlsec.tu-berlin.de/docs/2008-dimva.pdf`. International Conference on Detection of Intrusions and Malware, and Vulnerability Assessment. Berlin, Heidelberg: Springer Berlin Heidelberg, 2008.

- Arp, Daniel, et al. *Dos and Don'ts of Machine Learning in Computer Security*, `https://www.usenix.org/system/files/sec22summer_arp.pdf`. 31st USENIX Security Symposium (USENIX Security 22). 2022.

- Sommer, Robin, and Vern Paxson. *Outside the Closed World: On Using Machine Learning for Network Intrusion Detection*, `https://ieeexplore.ieee.org/document/5504793`. 2010 IEEE symposium on security and privacy. IEEE, 2010.

- Biggio, Battista, Giorgio Fumera, and Fabio Roli. *Security Evaluation of Pattern Classifiers under Attack*, `https://arxiv.org/abs/1709.00609`. IEEE transactions on knowledge and data engineering 26.4 (2013): 984-996.

- Szegedy, Christian, et al. *Intriguing properties of neural networks*, `https://arxiv.org/abs/1312.6199`. arXiv preprint arXiv:1312.6199 (2013).

Open source projects and libraries

Next, we will name some of the useful open source projects and libraries. Feel free to install them on your machine, play with some public datasets, and expand your hands-on skills:

- The `mlflow` library for managing ML experiments: `https://mlflow.org/`
- The PyOD outlier detection library: `https://pyod.readthedocs.io/en/latest/`
- The ToDS outlier detection library for time series: `https://github.com/datamllab/tods`
- Toolbox for security evaluation of AI systems: `https://github.com/Trusted-AI/adversarial-robustness-toolbox`
- Library for data quality: `https://github.com/awslabs/deequ`

Other web links

Here are some additional useful websites:

- Splunk AI functionality: `https://www.splunk.com/en_us/solutions/splunk-artificial-intelligence.html`
- Framework from Databricks on AI security: `https://www.databricks.com/blog/introducing-databricks-ai-security-framework-dasf`

- OWASP guide for security and privacy: `https://owasp.org/www-project-ai-security-and-privacy-guide/`
- HoneyNet, non-profit organization for cybersecurity, including interesting projects with AI applications: `https://www.honeynet.org/`
- DARPA AI cyber challenge: `https://aicyberchallenge.com/`

It is famously said that the journey of a thousand miles begins with a single step. We hope that this book is the beginning of a great journey taking you to advanced AI expertise in cybersecurity! Good luck exploring!

Index

A

access control 167
  ABAC 169
  mechanisms 169
  models 167
  RBAC 168
Access Control Lists (ACLs) 167
access control mechanisms
  access control policies 170
  ACLs 170
  models/frameworks 170
access control models
  DAC 167, 168
  MAC 168
active learning 267
  scenarios, for data query 267
  scenarios, for selection 268
advanced analytical techniques and tools 11
  behavioral analytics 11
  big data analytics platforms 11
  data visualization techniques 11
  ML and AI algorithms 11
  predictive analytics techniques 11
  threat intelligence (TI) 11
advanced persistent threats (APTs) 207

adversarial attack taxonomy 279, 285, 286
  adversarial capability 280-282
  adversarial influence 283, 284
  security violation 279, 280
adversarial data samples 278
adversarial machine learning (AML) 272, 273
adversarial prompt 278
adversarial retraining 290
adversarial threat modeling 277
  adversarial attack taxonomy 279
  attacker model 277, 278
  transferability, of adversarial samples 284, 285
Aegean Wi-Fi Intrusion Dataset (AWID) 122
AI, in data analytics 32
  botnet detection 34
  data breach detection 34
  deep learning (DL) 33
  identity and access management (IAM) 35
  intrusion detection 33
  machine learning (ML) 33
  malware detection 33
  natural language processing (NLP) 33
  phishing detection 33

Index

regulatory landscape 37, 38
role of analysts 36, 37
SOAR 35
user behavior analysis 33
vulnerability management 35
AI model
  creating, from scratch 84
  integrating, into product 86-88
AI model, with datasets
  creating 121
  malware analysis 122
  network intrusion detection 122, 123
AI projects
  workflow 83, 84
AI risk management framework 302
  monitoring and reporting 303
  review and revision 303
  risk assessment 303
  risk identification 303
  risk mitigation strategies 303
  utilizing, in organizations 304, 305
AI security and safety
  challenges, addressing 299
  compliance guidelines 301
  research development 300, 301
  standards 301
  technologies 302
  theoretical approaches, to address 299, 300
  tools 302
AI security and safety, significance 298
  ethical considerations 298, 299
  impact, on individual privacy 298
  implications, for national and global security 298
  public trust 298, 299
AI security, challenges 296
  data privacy issues 296
  vulnerability to attacks 297

AI systems
  safety concerns 297
AI taxonomy 44
AI, threat intelligence
  use cases, expanding on 199
algorithmic impact assessments (AIAs) 302
Androguard 104-106
Android malware detection
  example 110-113
Android Runtime (ART) 102, 108
Angr 107
anomaly behaviors, in ICSs
  data-driven techniques 211
  detecting 210
  hybrid techniques 212
  knowledge-based techniques 212
  model-based techniques 211
anomaly detection 15
anomaly detection algorithms 254
anomaly detection, for ICS
  use cases and applications 212-215
anomaly detection methods
  detecting 78, 79
  isolation forest 79, 80
Anti-Phishing Working Group (APWG) 149
APKTool 106
application programming interface (API) 87
area under the ROC curve (AUC) 261
Artificial Intelligence (AI) 295, 296
  adopting, in organization 314
  applications 34, 35
  challenges 35, 36
  good data quality 235
  history 44-46
  leveraging, for UEBA 137
  properties 44
  success, measuring in cybersecurity 314
artificial neural networks (ANNs) 33, 45

attacker's knowledge 278
attacker specificity 280
Attribute-Based Access Control (ABAC) 170
  strengths 169
  weakness 169
automated cybersecurity tools
  examples 25, 26
automation
  future, in cybersecurity 27, 28
  importance, in cybersecurity 24, 25
  potential drawbacks and challenges 26, 27
availability violation 280

B

backpropagation algorithm 55
Bayesian networks 254
behavior analysis 15
bias 244, 247
  in polynomial curve fitting 247
  managing 251
  practical example, with polynomial curve fitting 248-250
big data 4, 5
  advantages 5
  applications, in cybersecurity 15
  four Vs 4
  technologies, for cybersecurity 16, 17
big data challenges, in cybersecurity 6
  addressing, with advanced analytical techniques and tools 11, 12
  addressing, with practices and technologies 9, 10
  advanced analytical techniques and tools 11
  diverse data types, in cyberspace 8
  resource constraints 12, 13
  resource constraints, addressing 13, 14

velocity of data, in cyberspace 6-10
biometric authentication 164-166
  strengths 166
  weakness 166
black-box attacks 282
BlackEnergy attacks 208

C

California Consumer Privacy Act (CCPA) 38
Carlini & Wagner (CW) attack 289
case studies
  bias and variance, managing in IDS 252
  correlation versus causation, in phishing attacks 252
  statistical analysis, for anomaly detection 252
causal impact analysis 253
causal machine learning 246
causation 246
Change-Anything-Change-Everything (CACE) 276
Chief Information Security Officer (CISO) 225
CICIDS2017 dataset 122
Cisco Umbrella 26
collaborative anomaly detection 151-153
  designing 154
  implementing 154
Collections library 110
command-and-control (C2) 26
Common Crawl
  reference link 236
Common Objects in Context (COCO)
  URL 238

computer vision 238
   high-quality dataset examples 238
   low-quality dataset examples 238
Confidentiality, Integrity, and Availability (CIA) 280
confusion matrix 261
Contrastive Language-Image Pretraining (CLIP) 186
control flow graphs (CFGs) 104
convolutional neural networks (CNN) 53, 68-72, 84, 132, 211
CopperDroid 108
correlation 245
   negative correlation 245
   positive correlation 245
   versus causation 246
correlation coefficient
   calculating 245, 246
Credit Card Fraud Detection
   reference link 148
cross-validation 254
   K-fold cross-validation 260
   leave-p-out cross-validation 260
CrowdStrike Falcon 25
CSE-CIC-IDS2018 dataset 123
CTI information extraction, for Xdata 192, 193
   data preprocessing 193-195
   model, building 195, 196
   model, implementing 196, 197
   model visualization 197-199
CTI life cycle
   analysis phase 191
   collection phase 190
   dissemination phase 191
   feedback phase 191
   planning phase 190
   processing phase 191

cyberattacks on ICSs 207-209
   components 209, 210
cybersecurity
   AI success, measuring 314
Cybersecurity Law 151
cyber threat intelligence (CTI) 190
CyberTweets project 193

D

DAC model 170
   strengths 168
   weaknesses 168
Dalvik Executable (DEX) 102
Dalvik Virtual Machine (DVM) 105
DarkSide 208
DARPA AI cyber challenge
   reference link 317
database security enhancement 180
   monitoring and auditing 181, 182
   SELinux Booleans, utilizing 181
   SELinux configuration, ensuring 180
   SELinux contexts, applying 180
   SELinux policies, configuring for database access 181
data-driven techniques 211
data imbalance 152
Data Loss Prevention (DLP) 137
data protection impact assessments (DPIAs) 302
data quality 233, 234
data quality accidents
   examples 239, 240
data types
   graph data 53
   image data 53
   sequential data 52
   textual data 53

    time series data 53
deep learning (DL) 33, 45, 66-68, 211
    recent advances 55, 56
deep neural networks (DNNs) 36, 273
defensive distillation 288
defensive mechanisms 286
    as detection 289
    as prevention 288, 289
    as response 290, 291
    detection phase 288
    phases 287
    prevention phase 287
    tools, for testing adversarial attacks 292
Demisto 26
denial of service (DoS) attacks 191, 207
differential privacy (DP) 152
Discretionary Access Control (DAC) 167
distributed control systems (DCSs) 204
distributed denial-of-service (DDoS) attack 7, 196, 207
Domain Name System (DNS) 26
dropout 67

E

Elastic Stack (ELK Stack) 21
electronic health records (EHRs) 5
endpoint detection and response (EDR) 22, 23
EnDroid 103
ensemble learning 62
Epoch 84
ethical considerations 29
evasion attacks 283
event-driven architecture (EDA) 7

F

Facebook
    OAuth 2.0, registering on 171
face recognition 186
    embeddings, comparing 187
    embeddings, generating 187
    images, preprocessing 187
    using 188
Factorization Machine (FM) 90
false negative (FN) 259
false positive (FP) 259
false positive rate (FPR) 260
fault tolerance (FT) 16
feature extraction, from textual documents
    lemmatization 191
    machine learning 192
    parts of speech tagging 192
    stemming 191
    tokenization 191
federal learning (FL) 152
Federated averaging (FedAvg) 153
Federated stochastic gradient descent (FedSGD) 153
few-shot prompting 223
FL-based collaborative anomaly detection systems 154
four Vs, big data
    variety 4
    velocity 4
    veracity 4
    volume 4
Framework from Databricks on AI security
    URL 316
fraud detection 148
Fully Homomorphic Encryption (FHE) 153
function call graph (FCG) 112

G

garbled circuit (GC) 153
GDN 213, 215
    architecture 213
General Data Protection Regulation (GDPR) 38, 151
General Principles of the Civil Law 151
generative adversarial networks (GANs) 49, 59, 71
generative AI (GenAI) 271, 299
Google
    OAuth 2.0, registering on 171
graph data 53
graphical user interfaces (GUIs) 98
graphics processing units (GPUs) 55, 98
graph neural network (GNN) 214
Graphviz 109

H

Hadoop Distributed File System (HDFS) 16
hierarchical Dirichlet processes 196
high availability (HA) 16
high-quality dataset 234
    characteristics 234
    retail uses 235
    uses, in finance 235
    uses, in healthcare 234
high-quality dataset examples, NLP
    Common Crawl corpus 236
    Penn Treebank 236
    Stanford Natural Language Inference (SNLI) corpus 236
homomorphic encryption (HE) 152
HoneyNet
    URL 317

honeypots 281
horizontal federated anomaly detection 155
    steps 155
host-based IDSs (HIDSs) 22
host-based IPSs (HIPSs) 22
human analysts 36
    role, in AI 36, 37
human-machine interface (HMI) 204
hybrid techniques 212
hyperplane 49

I

IBM QRadar 21
ICSs components
    cyberattacks 209, 210
image data 53
ImageNet
    URL 238
incident response (IR) 3, 20, 32
indicators of compromise (IOCs) 24, 291
industrial control system (ICS) 203-205, 311
    anomaly behaviors, detecting 210
    challenges 216
    components 206
integrity violation 280
Internet of Things (IoT) 28
Internet of Things (IoT) devices 8
Intrusion detection and prevention systems (IDPSs) 21, 22
    IDS 22
    IPSs 22
    steps 22
Intrusion Detection Systems (IDS) 10, 21, 33, 252
intrusion prevention systems (IPSs) 22
IoT botnet dataset 123

IR platforms (IRPs) 14, 28
isolation forest 79, 80
  one-class SVM 80, 81

K

kernel trick 64
K-fold cross-validation 260
K-means 73, 74
knowledge-based authentication 163
knowledge-based techniques 212
Kyoto 2006+ dataset 122

L

Labeled Faces in the Wild (LFW)
  URL 238
label propagation 77, 78
LangChain
  features 87
  reference link 223
large language models
    (LLMs) 52, 83, 200, 217, 233, 271, 301
  good data quality 235
  security 227
  using, for offensive security 226
  using, for security 223
  using, for security operation center 226
  using, for spam and phishing detection 225
  using, for threat intelligence 224, 225
  using, for vulnerability discovery 224
latent Dirichlet allocation (LDA) 196
lateral movement 136
leave-p-out cross-validation 260
limitation and security concern 56
  adversarial attacks 58
  bias and fairness 57
  hallucination 56
  intellectual property ownership 57
  privacy leakage 57
limited knowledge (LK) attacks 281
LLMs in anomaly detection, future
    directions and applications
  automated incident response 218
  continuous learning and adaptation 218
  enhanced pattern recognition 217
  improved situational awareness 217
  natural language processing
      for security logs 217
  predictive maintenance 217
log analysis 16
logic programming 45
logistic regression 61
LogRhythm NextGen SIEM platform 21
London Whale incident 239
Long Short-Term Memory (LSTM) 90
loss functions 258
low-quality dataset examples, NLP
  datasets, with poor translation quality 237
  online reviews, with in-moderated
      content 237
  social media datasets, with limited
      pre-processing 237

M

MAC 168, 170
  strengths 168
  weakness 168
machine learning (ML) 5, 33, 84, 258, 271
  taxonomy, classifying 48
  workflow 84
malware analysis 122

326 Index

malware detection
    background 102, 103
    example Android malware
        detection 110-113
    libraries 108, 109
    tools 103-108
    tools and libraries 102
    transition, to malware classification 132
malware detection libraries
    Collections library 110
    Graphviz 109
    NetworkX 109
    tqdm library 109
    Zipfile library 110
malware detection tools
    Androguard 104
    Angr 107
    CopperDroid 108
    SandDroid 108
    VirusTotal 107
malware detection, with AI
    best practices 123-126
Mandatory Access Control (MAC) 164
man-in-the-middle (MitM) attacks 209
    communication links 209
    network devices 209
masked language modeling (MLM) 86
ML adversary 278
MLflow 264
    experiment tracking 264-266
ML taxonomy classification
    by learning objectives 50
    by learning schema 48
    by model modality 52
    defining 48
ML taxonomy, by learning objectives
    anomaly detection 51
    classification 50
    clustering 51
    generative models 52
    regression 51
ML taxonomy, by learning schema 48
    reinforcement learning 50
    semi-supervised learning 49
    supervised learning 48
    unsupervised learning 49
ML taxonomy, by model modality 52
    data types 52
    multi-modal models 54, 55
    single-modal models 53, 54
mobile app access, to services 173
    access token, using 174
    OAuth 2.0 authorization flow,
        implementing 173, 174
    registration 173
    security best practices, ensuring 175
    token expiry and refresh, handling 174
model-based techniques 211
model evaluation 258
    loss functions 258
    metrics 259, 260
model metrics
    accuracy 259
    confusion matrix 261
    cross-validation 260
    F1 score 259
    false positive rate (specificity) 259
    precision 259
    recall 259
    receiver operating characteristic
        (ROC) curve 260
    specificity 259
    true positive rate (sensitivity) 259
models
    monitoring 262

monitoring, during testing
    or production 263
monitoring, during training 262
monitoring tools 264
model serving 263
models, monitoring tools 264
    MLflow 264
Modified National Institute of Standards and Technology (MNIST) 96
monitoring reasons, in production
    data drift 263
    documentation and audit 264
    performance analysis 264
    service availability 263
Multi-Category Security (MCS) 176
multi-factor authentication 164-165
    strengths 167
    weakness 167
multilayer perceptron (MLP) 65
Multi-Level Security (MLS) 176
multi-modal models 54, 55
multi-party computation (MPC) 147, 152
multivariate time series (MTS) 213

N

National Security Agency (NSA) 175
Natural Language Inference (NLI) 236
natural language processing (NLP) 8, 33, 53, 84, 191, 217
    high-quality dataset examples 236
    low-quality dataset examples 237
Nessus 26
network-based IDSs (NIDSs) 22
network-based IPSs (NIPSs) 22
network interface controllers (NICs) 93

network intrusion detection 122, 123
    best practices 126-132
NetworkX 109
neural networks 64-66
NSL-KDD dataset 122

O

OAuth 2.0 171
    implementing, via social media platforms 171
    registering, with Facebook 171
    registering, with Google 171
    registering, with X (Twitter) 171
    using, in mobile application authentication 182, 184
OAuth 2.0 implementation, via social media login
    application, registering 171
    flow, integrating 172
    security, maintaining 173
    user data, handling 172
one-class SVM 80, 81
one-hot encoding 138
One-Time Passwords (OTPs) 164
open source intelligence (OSINT) 11, 191
open source software (OSS) framework 16
Open Worldwide Application Security Project (OWASP) 227, 167
operational threat 190
oracle 267
out-of-bag (OOB) 124
OWASP guide, for security and privacy
    URL 317

P

Packet capture (PCAP) 91
pandas
  used, for data cleansing 240, 241
**Partially Homomorphic
    Encryption (PHE)** 153
passwords 165
  strength 165
  weakness 165
Pearson correlation coefficient 245
Penn Treebank
  reference link 236
perfect knowledge (PK) attacks 281
Personal Identification Numbers (PINs) 163
**personally identifiable information
    (PII)** 227, 277
phishing attacks 207
phishing detection 148
  features 150
  with example 149-151
poisoning attacks 283
polynomial curve fitting 247
possession-based authentication 164
practical applications
  advanced statistical techniques,
    for enhanced security 254
  diagnostic tools, for correlation
    and causation 253
  responsible AI, implementing
    in cybersecurity 255
  techniques, for managing bias
    and variance 254
pre-trained AI model
  workflow 85, 86
**Principal Component Analysis
    (PCA)** 49, 289

privacy violation 280
probabilistic reasoning 45
production network 206
programmable logic controllers (PLCs) 204
prompting
  URL 223
Proof Key for Code Exchange (PKCE) 173
PyOD outlier detection library
  reference link 316
Python code
  data cleansing, with pandas 240
  data validation 241
  missing values, handling 241, 242
  writing, for good data quality 240
Python Imaging Library (PIL) 99

Q

quarantine 290

R

radial basis function (RBF) 129
random forest 62, 63
Random Forest (RF) 151
ransomware 191
ransomware attacks 207
ransomware detection, for ICS 216
  components 216
realistic learning environment 274
  Arms race problem 274, 275
  process, learning with data flow 276, 277
real-world automated cybersecurity tools
  Cisco Umbrella 26
  CrowdStrike Falcon 25
  Demisto 26
  Nessus 26

Snort 25
Splunk Enterprise Security 25
receiver operating characteristic (ROC) curve 260
rectifier linear unit (ReLU) 67
recurrent neural networks (RNNs) 55, 84, 131, 211
regularization methods 254
regulatory compliance 38
regulatory landscape 37
 bias 38
 data privacy 37
 new regulations 37
 transparency 38
reinforcement learning 50
reinforcement learning with human feedback (RLHF) 273
remote terminal units (RTUs) 204
replaced token detection (RTD) 86
resource constraints 12
 addressing 13, 14
 budget constraints 12
 infrastructure constraints 12
 manpower constraints 12
 technology constraints 13
 time constraints 13
Responsible AI 295, 312
retrieval augmented generation 223
return-oriented programming (ROP) 107
Role-Based Access Control (RBAC) 168-170
 strengths 169
 weakness 169

S

SandDroid 108
scatter plots and correlation coefficients 253

Secure Multi-Party Computation (sMPC) 153
secure sharing (SS) 153
Secure Sockets Layer (SSL) 10
Secure Water Treatment (SWat) 214
Security-Enhanced Linux (SELinux) 175
 and access control 175, 176
 database security enhancement 180
 enforcing mode 175
 permissive mode 175
 user authentication, in multi-user environments 178
 web server isolation 176
 writing, in Python to control Ubuntu files 184, 185
security information and event management (SIEM) 20, 21, 135
 alerting and reporting 20
 correlation and analysis 20
 data collection 20
 data normalization 20
Security Operation Center (SOC) 226
security orchestration and automation response (SOAR) 21, 23
 capabilities 23
security threats, LLMs
 data poisoning, training 227
 excessive agency 228
 insecure output handling 227
 insecure plugin design 227
 model denial of service 227
 model theft 228
 overreliance 228
 prompt injection 227
 sensitive information disclosure 227
 supply chain vulnerabilities 227
security violation 279

semi-supervised learning 49
semi-supervised learning methods 77
  assumptions 77
  label propagation 77, 78
sequential data 52
service-level agreement (SLA) 263
SIEM platforms examples
  IBM QRadar 21
  LogRhythm NextGen SIEM platform 21
  Splunk Enterprise Security 21
  Wazuh 21
SIEM systems 10
single-modal models 53
Snort 25
Somewhat Homomorphic Encryption (SHE) 153
spam detection 149
sparse interaction 68
Splunk Enterprise Security 21, 25
SQL injection 209
  databases 209
  web applications 210
Stanford Natural Language Inference (SNLI) dataset
  reference link 236
statistical foundation 244
statistical learning theory 46, 47
stochastic gradient descent (SGD) 150
strategic threat 190
supervised learning 48
supervised learning methods 59, 60
  convolutional neural networks 68-72
  deep learning 66-68
  logistic regression 61
  neural networks 64, 66
  random forest 62, 63
  SVMs 63, 64

supervision network 206
supervisory control and data acquisition (SCADA) 204
support vector classification (SVC) 127
support vector machines (SVMs) 47, 63, 64, 126
support vectors 64
symbolic processing 45

T

tactical threat 190
tactics, techniques, and procedures (TTPs) 190
targeted model 278
TensorFlow
  URL 96
Term Frequency/Inverse Document Frequency (TF/IDF) 138
textual data 53
textual input (prompts) 223
threat intelligence 190
  AI, working with 191, 192
  operational 190
  reports 33
  strategic 190
  tactical 190
  topic modeling 192
  types 190
time series data 53
time series features 139
  approaches 139
TI platforms (TIPs) 11
tokens 166
  biometrics 166
  MFA 167
  strengths 166
  weakness 166

tools and technologies, against threats 20
  EDR 22
  IDPSs 21, 22
  SIEM 20, 21
  SOAR 23
topic modeling 192, 311
topics 192
tqdm library 109
traditional malware detection techniques
  advantages 120, 121
  overcoming 120
traditional signature-based defenses
  disadvantages 136
traditional techniques, for vulnerability discovery
  dynamic analysis 224
  static analysis 224
  symbolic execution 224
transformer-based model learning 86
translation invariance 68
Transmission Control Protocol (TCP) 91
Transport Layer Security (TLS) 10
true negative (TN) 259
true positive rate (TPR) 260
true positive (TP) 259
trusted execution environment (TEE) 152
t-SNE 75-77
Turing test 44

U

UEBA anomaly detection 139-144
  use cases 145
unsupervised learning 49
unsupervised learning methods 72, 73
  K-means 73, 74
  t-SNE 75, 76, 77
UNSW-NB15 dataset 122

user authentication 161, 163, 171
  and access control 162, 163
  biometric authentication 164
  knowledge-based authentication 163
  multi-factor authentication 164
  passwords 165
  possession-based authentication 164
  technologies 165
  tokens 166
user authentication and access control 161
  exemplifying 171
  face recognition 186
  OAuth2.0, using in mobile application authentication 182, 183
  practicing, with Python 182
  SELinux, writing to control Ubuntu files 184, 185
user authentication, in multi-user environments
  MCS or MLS, using 179
  monitoring and auditing 179
  SELinux, configuration and enablement 178
  SELinux policies, configuring for user data protection 179
  SELinux user contexts, assigning 178
User Datagram Protocol (UDP) 91
User Entity and Behavioral Analysis (UEBA) 135
  AI, leveraging 137
  feature extraction 138, 139
  features 137
  features, categories 137, 138

V

vanishing gradients 67
variance 244, 247
  in polynomial curve fitting 247

managing 251
practical example, with polynomial curve fitting 248, 250
Variational Autoencoders (VAEs) 52
vertical federated anomaly detection 156
steps 156, 157
virtual environment
creating 89
developing 89
virtual machine introspection (VMI) 108
Virtual Machine (VM) 106
VirusTotal
reference link 107
visual network traffic analysis 89
background 90
example 99-102
libraries 97, 98
model training and testing 93-99
pre-processing 91, 92
tools 91, 92

W

Water Distribution (WADI) 214
Wazuh 21
web server isolation, achieving 176
SELinux Booleans, configuring 177
SELinux contexts, using 177
SELinux, enabling 176
SELinux, installation 176
SELinux policies, maintaining 178
SELinux policies, updating 178
testing and debugging 177, 178
white-box attacks 281

X

X (Twitter)
OAuth2.0, registering on 171

Z

zero-day and n-day vulnerabilities 207
zero-day exploits 210
hardware devices 210
software applications 210
zero-knowledge (ZK) attacks 282
zero-shot prompting 223
Zipfile library 110

‹packt›

packtpub.com

Subscribe to our online digital library for full access to over 7,000 books and videos, as well as industry leading tools to help you plan your personal development and advance your career. For more information, please visit our website.

Why subscribe?

- Spend less time learning and more time coding with practical eBooks and Videos from over 4,000 industry professionals
- Improve your learning with Skill Plans built especially for you
- Get a free eBook or video every month
- Fully searchable for easy access to vital information
- Copy and paste, print, and bookmark content

Did you know that Packt offers eBook versions of every book published, with PDF and ePub files available? You can upgrade to the eBook version at packtpub.com and as a print book customer, you are entitled to a discount on the eBook copy. Get in touch with us at customercare@packtpub.com for more details.

At www.packtpub.com, you can also read a collection of free technical articles, sign up for a range of free newsletters, and receive exclusive discounts and offers on Packt books and eBooks.

Other Books You May Enjoy

If you enjoyed this book, you may be interested in these other books by Packt:

Modern Generative AI with ChatGPT and OpenAI Models

Valentina Alto

ISBN: 978-1-80512-333-0

- Understand generative AI concepts from basic to intermediate level
- Focus on the GPT architecture for generative AI models
- Maximize ChatGPT's value with an effective prompt design
- Explore applications and use cases of ChatGPT
- Use OpenAI models and features via API calls
- Build and deploy generative AI systems with Python
- Leverage Azure infrastructure for enterprise-level use cases
- Ensure responsible AI and ethics in generative AI systems

Unlocking Data with Generative AI and RAG

Keith Bourne

ISBN: 978-1-83588-790-5

- Understand RAG principles and their significance in generative AI
- Integrate LLMs with internal data for enhanced operations
- Master vectorization, vector databases, and vector search techniques
- Develop skills in prompt engineering specific to RAG and design for precise AI responses
- Familiarize yourself with AI agents' roles in facilitating sophisticated RAG applications
- Overcome scalability, data quality, and integration issues
- Discover strategies for optimizing data retrieval and AI interpretability

Packt is searching for authors like you

If you're interested in becoming an author for Packt, please visit `authors.packtpub.com` and apply today. We have worked with thousands of developers and tech professionals, just like you, to help them share their insight with the global tech community. You can make a general application, apply for a specific hot topic that we are recruiting an author for, or submit your own idea.

Share Your Thoughts

Now you've finished *Artificial Intelligence for Cybersecurity*, we'd love to hear your thoughts! Scan the QR code below to go straight to the Amazon review page for this book and share your feedback or leave a review on the site that you purchased it from.

`https://packt.link/r/180512496X`

Your review is important to us and the tech community and will help us make sure we're delivering excellent quality content.

Download a free PDF copy of this book

Thanks for purchasing this book!

Do you like to read on the go but are unable to carry your print books everywhere?

Is your eBook purchase not compatible with the device of your choice?

Don't worry, now with every Packt book you get a DRM-free PDF version of that book at no cost.

Read anywhere, any place, on any device. Search, copy, and paste code from your favorite technical books directly into your application.

The perks don't stop there, you can get exclusive access to discounts, newsletters, and great free content in your inbox daily

Follow these simple steps to get the benefits:

1. Scan the QR code or visit the link below

    https://packt.link/free-ebook/9781805124962

2. Submit your proof of purchase
3. That's it! We'll send your free PDF and other benefits to your email directly

Made in United States
North Haven, CT
29 May 2025